W0042723

THE PAPUAS OF WAROPEN

1. The bridal couple on their festive tour through the village

KONINKLIJK INSTITUUT
VOOR TAAL-, LAND- EN VOLKENKUNDE
TRANSLATION SERIES 2

&

Prof. Dr. G. J. HELD

THE PAPUAS
OF
WAROPEN

PUBLICATION COMMISSIONED AND FINANCED BY
THE NETHERLANDS INSTITUTE FOR
INTERNATIONAL CULTURAL RELATIONS

SPRINGER-SCIENCE+BUSINESS MEDIA, B.V. 1957

ISBN 978-94-017-5654-9   ISBN 978-94-017-5928-1 (eBook)
DOI 10.1007/978-94-017-5928-1

Additional material to this book can be downloaded from http://extras.springer.com

EDITORIAL PREFACE

In 1947 Held's *Papoea's van Waropen* was published in Holland, from a manuscript that had been completed in 1942, and had since then come safely through the years of war and enemy occupation in Indonesia, mainly thanks to the perseverance of the author's wife, Mrs. G. J. Held.

Anthropologists, and other interested readers as well, soon agreed that this monograph, the result of long and patient field-work by an expert observer, deserved a wider circle of readers than could be reached by a publication in Dutch.

The Committee and the Editorial Board of the *Koninklijk Instituut voor Taal-, Land- en Volkenkunde* therefore have great pleasure in being able to offer an English translation of the Waropen study by this gifted anthropologist, a few years after his untimely death in 1955.

This publication was made possible by a greatly appreciated grant from the Netherlands Institute for International Cultural Relations, and by the voluntary editorial assistance rendered by Mrs. Held.

E. M. UHLENBECK

PREFACE

The study of the Waropen tribe, in the course of which the ethno-graphical data for this book were collected, was undertaken under the instructions of the Netherlands Bible Society at Amsterdam. The Direc-tor of the Department of Education and Public Worship permitted me to continue my work on the collected material also after my transfer to the Government Service. For this liberty I beg to express my gratitude.

Moreover, I would like to thank all those without whose co-operation it would have been impossible to complete this work: the officials of the Civil Service, in particular the Assistant Resident of Manokwari, the officers of the Government Navy, and the missionaries of the Utrecht Missionary Society, with whom I was permitted to collaborate during a few years. Dr. A. N. J. Thomassen à Thuessink van der Hoop helped me in making the map, for which I beg to express my sincere thanks.

Some of my readers may be interested to know why it was the Waropen tribe which was chosen as the object of research, although the field of the Utrecht Missionary Society, where the Netherlands Bible Society had stationed me, also provided opportunity for research elsewhere on Northern New Guinea. However, there was little reason to hazard an investigation of far distant or unknown tribes of the interior. One should not forget that travel in the interior is not so much dangerous as extre-mely costly. The repeated transfer of our small expedition, consisting of my wife and myself and a few servants, from our residence at Manokwari to the Waropen area entailed little expense, because we always found the Government Navy ready to undertake the voyage to the muddy shores of Waropen with us.

In other regions of Northern New Guinea where travel would have been easier and also more attractive on account of the beauties of nature, as for instance in the island world of Numfoor-Biak, it soon became ap-parent that the Christians, only recently converted from paganism, had difficulty in understanding our interest in their ritual and their myths as being of importance to the mission. Therefore we believed it better to start our investigations among the only coastal tribe in Northern New Guinea which at that time still remained pagan, i.e. the Waropen tribe.

On our first visits of reconnaissance, the gloomy tidal forest with its many inconveniences did intimidate us somewhat, but the kindness and the interest shown by the Waropen people themselves seemed promising. They made this first promise come true. The memories of many disappointments and inconveniences, of sickness and setbacks, have faded, but the memory of the joys of field-work has remained. That is the memory of the children going out boating in the calm of the evening, of the busy noise of the feasts, of the uproar during weddings, of the intimateness of all the goings-on in the village, with its ramshackle houses, its rickety prahus and its gay people. It is not the idyllic picture of the "happy native" we want to evoke. This idyll does not exist in actual fact. It is only the musings of somebody who has found many good friends he has had to leave again.

The enumeration of all their names would serve no useful purpose; the names of many informants I have mentioned in a collection of Waropen texts with translations. Some of them will be introduced in the following pages. I will only make an exception for my helpers at Nubuai: Adori of Ruma Pedei and Dusi of Ruma Sawaki, and at Napan: Maarten of Ruma Warami. The first-mentioned in particular assisted me with unflagging zeal; he helped me when the investigation had lost its vigour and when those periods of discouragement and aversion occurred which other ethnographers have experienced likewise.

My Numfoor clerk Cornelis Rumsayor has also always loyally assisted me. Those who believe that the Papuan world contains little talent should consider that most of the figures in this study have been drawn by him.

My wife, who also in other respects has rendered material assistance in this publication, always accompanied me. Due to this not only the stay in the village became much easier, but also the contact with the population became much freer and this in its turn considerably facilitated my research. Great satisfaction was obtained from the simple medical assistance to the inhabitants of the village under the direction of my wife.

It would be superfluous to provide a detailed description of the manner in which I undertook this investigation. Every reader will discover without difficulty its lacks and shortcomings. The description of a human civilisation in all its aspects is nearly always a task which exceeds the capacities of one single person, in particular when this person has to learn first from all the failures which he was unable to avoid during his first investigations in an area as difficult as New Guinea.

I have tried not only to give a cross-section of the culture of the Waropen such as it functioned at the moment of observation, but also to obtain some perspective in the description of this culture by looking for historical tendencies which help to determine cultural development. For if one assumes that there exists a functional connection within a culture at every moment of its course in history, it follows that this course is also determined in a certain direction. A culture does not merely consist of a number of cultural elements which are functionally connected at a certain moment. The shifts within this culture, the changing in the grouping of the cultural elements themselves, are not haphazard, but they are determined by certain causes and they conform to certain tendencies. I have tried to discover the results of some of these causes and the direction of some of these tendencies.

It will become clear from my explanations that I have aimed at the method of studying cultures such as this was developed by Professor J. P. B. de Josselin de Jong in his lectures at Leiden University. Due to enemy action the contact between him and his students here in Indonesia has been interrupted. We are still privileged to speak and to think freely about human civilisation. Freedom of science has been gained on our behalf by the very best amongst us. May it be given to us to satisfy the demands which the possession of this treasure of civilisation lays upon us.

Finally I would like to express again my warmest thanks to the Netherlands Bible Society which left me complete liberty in my work. I hope that the practical experiences of the Mission and of the Civil Service will demonstrate the value of the work done under the instructions of the Society. I may furthermore refer the interested reader to a Grammar and a Vocabulary of the Waropen language which the Royal Batavia Society of Arts and Sciences has been good enough to publish in 1942 as the first and second part of volume LXXVII of its Transactions. The publication of a collection of Waropen texts with translation is under preparation.

The foregoing was written in the spring of 1942, when the shadow of Japanese violence fell threateningly over Indonesia. Under increasingly difficult circumstances the publishers Nix at Bandung tried to see the work through the press. Our efforts were not crowned by success.

Then the long years of internment followed. Many of us met their death. A great deal of work was lost. That the present work was not likewise lost I owe in the first place to the perseverance of my wife. The manuscript belonged to the forty five pounds of earthly possessions she was able to retain amidst the unending misery of the Japanese camps for women. Mrs. E. Roberts *née* Visser guarded another part of the manuscript. Mrs. T. Holstvoogd *née* Hoorn was found ready again and again to co-operate in hoodwinking the inquisitiveness of the Japanese. The risks these women took will be appreciated by all our fellow-prisoners who know what the Japanese thought of the written and type-written papers they were able to confiscate.

When ultimately the manuscript still threatened to get lost in the flames which partly destroyed the unfortunate town of Bandung in the last months of the year 1945, it was Mr. G. K. Khouw who was able to save it with great trouble. Without his co-operation and that of my colleague Tjan Tjoe Siem and his family, not only this volume, but also the rest of my books and papers would have disappeared. My gratitude is due to them all.

After all the trouble so many people had taken for the sake of this report of my activities on New Guinea, I wished all the more, after the past five years, to make a more detailed study of some points locally. Under the circumstances, however, I considered it best to present the work as it was written five years ago.

I may be permitted to stress one remark. On p. 22 I wrote that for practical reasons I had restricted myself to a description of the culture of the non-christianised Waropen. I had hoped to find the opportunity at some later time to start the description of the problems which occur within a civilization which has opened itself to the influences of modern times. The contemporary history of 1940 is for many Papuas of the present the end of their pre-history.

I repeat again that my description of the Waropen culture of 1940 is not intended as a kind of social archaeology. To me personally the Waropen culture of those days is not that truly "primitive" culture, the sole object of study for the ethnologist whose main interest is "primitive peoples". My object was simply to give a description of a less complicated form of human civilisation, of course different, but far from dissimilar from any other human culture in the world. I hope that my description may be useful to those workers who — as I hope — will study the Papuans of the present before long.

In closing I may be allowed to express my gratitude to the firm of Nix at Bandung for their labours, so untimely interrupted. I hope that their publishers' business may soon flourish again. I was greatly pleased to find the firm of E. J. Brill ready to undertake the publication which they completed in record time in spite of many difficulties. A special word of thanks is due to the maker of the plates, who had to exert his utmost ingenuity on the photographic material which had been hidden under ground for a number of years.

And so I beg to dedicate this report to those who first commissioned me, i.e. to the Netherlands Bible Society at Amsterdam, who also during the last stages of this long road have kindly assisted me by word and deed.

CONTENTS

INTRODUCTION

CULTURAL AND GEOGRAPHICAL ENVIRONMENT

In Geelvink Bay we can distinguish three centres for the radiation of culture. First we have the culture of the Numfoor-Biak group which occupies the strategic points of the bay by taking in the extreme north-western tip near Manokwari (Mansinam and the Bay of Doreh) and the islands of Numfoor and Biak and further to the South the island Rumberpon and part of Roon.[1]

Behind this first line there lies on the Southwest coast of Geelvink Bay in the second place the centre of the Wandamen-Windessi group. In the third place we have on the East coast the centre of Waropen.

The influence of the Numfoor-Biak group makes itself felt towards the West past Manokwari on the North coast of the Vogelkop peninsula up into the Sorong area. Along the West coast of Geelvink Bay and in a southerly direction we find a few settlements of different tribes who have descended from the high mountains in the interior of the Vogelkop peninsula to the coast. Next comes Wandamen Bay as the second cultural focus, including Windessi and another part of the island of Roon, with an influence extending beyond the isthmus far into the Bintu region, on the Gulf of Maccluer.

Proceeding eastward from Wandamen Bay we find first in the southernmost part of Geelvink Bay again feeble groups of tribes from the interior and then, near the Haarlem and Moor islands which possess a language of their own, settlements of the Waropen tribe. This region of the Haarlem and Moor islands, together with the three Waropen settlements Napan, Weinami and Makimi, lies as a small cultural oasis on the long and lonely East coast of Geelvink Bay. It is only when having gone further in a northeasterly direction, and after having passed

---

[1] In the Dutch edition the author, explaining his spelling-system, says that for geographical names generally the spelling of the *Atlas van Tropisch Nederland* (1938) is followed; for the Waropen names, however, the author used the system adopted by him in his Waropen grammar and vocabulary. In this English translation we only deviate from the spelling in the original work in so far as everywhere the Dutch rendering of the *u* sound by *oe* has been relinquished. Double *o* in New Guinea words represents an *ŏ*. For Malay words the English spelling-system is used. Ed.

the small island of Nau, that one finds a more concentrated population and a succession of several rather large villages. These are the settlements of the Waropen.

The ring is closed by a population from the interior which lives along several rivers in the coastal area near the island of Kurudu. In the centre of the ring we have described, starting with Biak via Numfoor and past Manokwari to Wandamen and from there past the Moor islands via the island of Nau eastwards, and then via the island of Kurudu again to the Padaido islands, there lies the large island of Japen.

From the available data it is difficult to determine the position of the inhabitants of Japen. Undoubtedly Japen has undergone the influence of the three above-mentioned foci of radiation, hence in northern Japen from Numfoor-Biak, in southwestern Japen from Wandamen-Windessi, whilst the rather densely populated South-coast with Serui and the Ambai islands as its centre maintains a busy contact with the inhabitants of the Waropen coast, so that both populations influence each other. Still, one cannot say that from a cultural point of view the inhabitants of Japen are split up into the three above mentioned groups alone, because in this respect the Japen languages differ too much. Rev. F. Slump distinguishes on Japen at least ten different languages or dialects, including a so-called "language of the interior". Beside the more easily distinguishable cultural provinces of Numfoor-Biak, Wandamen-Windessi and Waropen, we should therefore for the time being leave room for a Japen group, although apart from its geography it cannot yet be clearly delineated.

In the foregoing I already mentioned the so-called tribes from the interior. This term refers to a number of tribes who do not speak Austronesian languages but the so-called Papuan languages, a confusing term used to indicate all non-Austronesian languages on New Guinea. The terms "languages and tribes of the interior" are slightly confusing because the speakers of the so-called Papuan languages also live in large tracts along the coast of Geelvink Bay, in so far as these have not been occupied by the three above-mentioned groups. For the rest, the speakers of the Papuan languages have occupied the interior and, at least in Geelvink Bay, they are greatly inferior in seamanship to the speakers of the Austronesian languages, whom in the following we shall indicate by the term "coastal population".

In daily life the people from the interior are easily distinguished from the coastal population, because the former feel themselves and are treated by the others as uncivilised mountaineers, as backwoodsmen

Sometimes there even exists a direct relation of suzereinty between the coastal population and the men from the interior. The Numfoor use the term for "man from the interior" also in the sense of "non-Numfoor" or "bumpkin" and so they occasionally indicate also the Waropen by means of the term *faksi,* just like the Waropen use the Waropen word *ghoa* not only for the people of the interior who do not speak an Austronesian language, but also for members of another tribe like e.g. the Windessi. Waropen mythology would have us believe that the Waropen tribe consists half of men from the interior and for the other half of men from the coast, but we are not dealing here with data that are historically reliable. The mythology only assimilates the contrast between coastal population and men from the interior in its view of the universe, a view which we will discuss later. Still, this does not mean to deny that in the Waropen tribe one might observe also traits from the interior.

As regards their culture, this non-maritime group of agriculturalists, which in the esteem of the Numfoor, the Windessi and the Waropen consists of uncivilised mountaineers, is easily recognisable. Somatological differences between the people of the interior and the coastal population provide little support, as long as the specialists are not agreed on the existence of these differences. The non-specialist often believes that in daily life he can easily distinguish between an interior type and a coastal type.

Anyhow, as regards culture it is certain that the people of the interior undergo more influence from the coastal population than vice versa. In their intercourse the men from the interior mostly learn the Austronesian coastal language, but the coastal population only rarely acquires the Papuan language of the interior. In general it is more the case that the coast radiates its culture into the interior than the reverse.

However, our knowledge of the interior is still very limited. The regions back of the Waropen coast in the eastern part of Geelvink Bay were visited in 1923 by the assistant-administrator E. Latumeten, again in 1934 by Detiger and Bierdrager, and in 1938 by J. van Eechoud.[2] These expeditions produced of course no extensive ethnographical material, as little as did the military posts around the Anggi lakes, concerning which we owe some data to Lt. V. J. E. van Arcken [3] and to the army surgeon K. F. Hordijk.[4] For the time being we are therefore unable to

---

[2] See the journal *Nieuw-Guinea* IV, p. 24.

[3] See the compilation *Nieuw-Guinee,* edited by W. C. Klein, I, pp. 93, 94; III, p. 1075, and the literature, indicated there.

[4] See the journal *Nieuw-Guinea* II, p. 241; III, p. 370.

determine the relation between the tribes of the coast and those of the interior.

The Waropen people themselves divide mankind into two kinds, the first of which are the *nunggu wano,* real men. These are the clan-members and in more extended sense also the fellow-villagers and fellow-tribesmen, and in the most extended sense the members of the other coastal tribes. *Nunggu wano* is also the translation of the word Papua, a word which has a somewhat disagreeable ring in the ears of the coastal population. Beside *nunggu wano* they distinguish *nunggu wetero* (*wetero* means "at random", "without more ado", "separate"), subdivided in *Ghoa,* people from the interior, and *Ande,* people of non-Papuan origin, i.e. Ambonese, Chinese, Europeans, etc. Sometimes they also speak about *Ande boma,* small *Ande,* and *Ande bawa,* big *Ande;* this indicates differences in rank and position. An *Ande boma* is e.g. an itinerant Chinese peddlar. All these terms are therefore, just like the word *Ghoa* with the additional meaning of "bumpkin", at the same time expressions indicating esteem. In the beginning they were therefore unwilling to explain the real meaning of the word *wano,* saying "*Wano* means 'creek' and so the *nunggu wano* are the Papuas who live on those creeks". Opposite to the *nunggu wano,* the real men, there are therefore the *nunggu wetero,* the separate, alien men, about equivalent to the "Barbarians". In daily intercourse one normally uses an indication of locality with the addition of *-ukigha,* people from; so e.g. *Mansinam-ukigha,* the people of Mansinam, or *Makim-ukigha,* the people of Makimi.

The Numfoor-Biak group closes, as it were, Geelvink Bay to the North with the result that all Indonesian influences from the West were captured there, in particular from the sultanates of Tidore and Ternate. When going East, the prahu traffic usually does not pass the mouth of the Mamberamo, so that contact with the area East of the Mamberamo was scarce. The interior deterred the coastal population because of its high mountains (the Vogelkop peninsula), or because of its vast marshes (the East coast of Geelvink Bay). Only with the hinterland of Napan contact seems to have been slightly closer. In this way Geelvink Bay constitutes for the inhabitants of its coasts a rather closed culture-area with an entrance towards the Northwest.

For the Waropen the region extending in the direction of Kurudu island forms the hinterland. In Geelvink Bay there is a rather strong radiation of culture from the population of Numfoor-Biak, who in their turn are again oriented towards the West, from where they receive cultural and political influences. The influence from Biak arrives in Waropen Kai mainly via the Padaido islands and the North coast of Japen, past the island of Kurudu. This influence may be traced historically; it appears i.a. in the spread of the doctrine concerning a future saviour, to be preceded by a *konoor*, a doctrine which has penetrated in the Waropen area via Kurudu. Also in other respects their mythology clearly shows Biak influence, like e.g. in the well-known myth of Manarmaki. Nor have the Waropen themselves failed to observe this connection, because according to their mythology they owe their customary law to the Numfoor where the original population of the mother-village on the Woisimi river had taken refuge when it had been nearly exterminated by the snake Roponggai.

The Wandamen-Windessi group likewise exerted influence on the Waropen, although this was perhaps not as strong as that of the Numfoor. Geographically the distance between Wandamen Bay and the Haarlem and Moor Islands is rather far. I do not know the course followed by the prahus from Wandamen Bay to Japen, but those of the Haarlem and Moor Islands followed the East coast of Geelvink Bay to cross over to Japen only near the island of Nau or near Cape Paradoi, often first going via the advancing Ambai Islands. Prahu traffic of people from Windessi and Roon with the Haarlem Islands certainly always existed. J. L. van Hasselt mentions a voyage by prahu from Manokwari to Roon and from there probably via the Auri Islands to Mios Num, West of Japen.[5] Probably he men of Wandamen often followed this route when travelling to Japen, or otherwise via Oransbari, Numfoor and Mios Num.

The contact which existed with Wandamen Bay via the Haarlem and Moor Islands was quite intensive, because curiously enough at the far end of this bay the village is situated which the Waropen consider to be their village of origin. This village, the present Ambumi, was formerly situated on the river Woisimi, where on most maps one finds the indication of a village Aropen or a region called Waropen.

From the point of view of culture the coastal groups in Geelvink Bay which I have mentioned show so many points of agreement, that Geelvink Bay — as far as the area occupied by these coastal groups is concerned — might be called one culture-area, in which each coastal group may be distinguished as a separate culture-province. Whether the

---

[5] J. L. van Hasselt, *Gedenkboek van een vijf-en-twintigjarig zendelingsleven op Nieuw-Guinea, 1862—1887* (Memories of 25 years as a missionary on New Guinea), p. 124.

island of Japen could be called a fourth culture-province, remains an open question.

These three culture-provinces mutually show many points of agreement; the kinship-groups are the *ruma* which among the Waropen are still strictly unilateral; different classes are in the process of coming up; religion is no longer the religion of the clan with its all-embracing initiation ceremonies; co-operation of clans linked in couples which we will find among the Waropen, we also found among the Numfoor where the ritual character is still more evident than among the Waropen. In the same way, clothing, housing, arms, decoration and ship-building show striking points of resemblance. But there are also occasional differences, and there Waropen Kai, in agreement with its more isolated geographical position, is more old-fashioned than Napan. Waropen Kai maintains a marriage system which would fit in with a circulating clan-system, whilst the other two coastal groups know the exogamy of blood-relations which, however, also still continues to be linked with group-exogamy. Furthermore, the Numfoor have a division of the clan into four *er*, which indicates a quadripartition of the clans as clearly as the division on Tobelo into four *hoana* which Dr. Rassers in his lectures at the Oegstgeest Missionary School quite rightly used to connect with the ancient clans. Shamanism, which occurs in the Numfoor-Biak area and which is of interest due to its connection with the clan and with the ancestors of the clan, is less conspicuous in the Waropen area. The ancestral images and the neck-supports, frequently found elsewhere in Geelvink Bay, are again much less common in the Waropen Kai area. In Geelvink Bay we therefore find not only unity, but also diversity among the coastal groups. Geelvink Bay is therefore an interesting field for study, because here, where comprehensive research of larger areas is much more difficult than for instance in Australia,[6] there is still some room for diachronic and comparative research, beside synchronic investigations.

The question whether in a broader sense we should call Waropen culture Melanesian or Indonesian is hard to answer. Professor J. C. van Eerde believed it possible to mark a borderline between Melanesian and Indonesian culture areas in Netherlands New Guinea at the Mamberamo

---

[6] "Survey work is of great value in Australia where the general principles of the social organisation, totemism, magic and economic life and of the languages are similar all over the continent, but this is not so in the islands to the north-east of Australia where Papuan, Melanesian and Polynesian languages, forms of social order and religion are all found". A. P. Elkin, *Oceania* VIII, p. 313.

river.[7] Although the Mamberamo river does constitute a border, even more for the Waropen than for the Biak people, it is not so simple to determine the delimitation. It is at any rate not true, as Professor van Eerde believed, that the spread of the sacred flute provides a useful criterion, because this object also occurs among the tribes of the interior on the East coast of Geelvink Bay.

Professor J. P. B. de Josselin de Jong, besides remarking, that "the characterisation of a type of culture and the delimitation of a culture-area or province are hazardous undertakings",[8] considers as "the structural nucleus of numerous Ancient Indonesian forms of civilisation" [9] the selfsame clan-system into which the Waropen marriage system also fits. Because this system — which we shall call in short the circulating system, in agreement with Professor de Josselin de Jong's terminology — was also observed by him in Melanesia, we shall have to consider the restriction to the Indonesian cultural-area as resulting from the frame of the inaugural address where it is to be found.

In general, religion e.g. in the area around Huon Gulf, as we know it from the descriptions of the missionaries of the Neuendettelsauer Mission [10] who have done so much for the study of languages and culture, seems to agree more closely with the clan-ritual, or it seems elsewhere in Melanesia to develop rather more in the direction of the organisation of societies than is the case in Netherlands Eastern New Guinea. But in most cases the data concerning Netherlands New Guinea are too scarce to make a more reliable survey possible.[11]

---

[7] J. C. van Eerde, "Indonesische en Melanesische beschavingsgebieden op Nieuw-Guinee", in *Tijdschrift van het Koninklijk Nederlandsch Aardrijkskundig Genootschap* ("Indonesian and Melanesian culture areas on New Guinea", in Journal of the Royal Netherlands Geographic Society), XXXVIII, p. 823 ff.

[8] J. P. B. de Josselin de Jong, *De Maleische Archipel als ethnologisch studieveld* (The Malay Archipelago considered as a field for ethnographical study), p. 4.

[9] *Ibid.*, p. 6.

[10] In R. Neuhauss, *Deutsch Neu-Guinea*, III.

[11] There was a time when an underestimation of the Netherlands exploration was coupled with an overestimation of the achievements of our Eastern neighbours. The growing interest for Netherlands New Guinea, under the sure guidance of the Study-circle for New Guinea of the Institute for the Moluccas and of the New Guinea Committee, i.a. by publishing the journal *Nieuw-Guinea* and the compilation *Nieuw Guinee* in three volumes, edited by W. C. Klein, have made it clear that our exploration of New Guinea has been effective all the same and that the achievements obtained in our part of this area in many respects will stand the touch of international criticism. It is, however, beyond doubt that we have lagged behind in the domain of ethno-

The language likewise provides little to go by. Wandamen-Windessi and Numfoor-Biak greatly resemble each other. Waropen with its two dialects, Napan and Waropen Kai, differs considerably from these two languages, and especially from the language of Biak, even in its phonological structure. While Numfoor possesses all kinds of combinations of two or even three consonants in different positions, the most complicated sound in Waropen is the oral-nasal (*mb, nd, ngg*), whilst other combinations of consonants are unknown in Waropen. A number of characteristics established by Dr. N. Adriani for the group of South Halmahera languages, among which he includes Numfoor, do not apply to Waropen.[12] In several respects, e.g. as regards the simplicity of its morphological structure, Waropen resembles the Melanesian languages. It also lacks the affixes, well-known in the Indonesian languages. However, little is known either of the other South Halmahera languages, or of the Melanesian languages, so that it is impossible to determine with any certainty whether Waropen should be included among the Indonesian or among the Melanesian languages.

This means that neither cultures nor languages and races in Netherlands New Guinea have been studied sufficiently to provide a definite answer to the still always thorny question: Melanesian or Indonesian.

After this summary indication of the cultural-provinces we should now first give a closer definition of the Waropen area geographically. Beside the village of Ambumi in Wandamen Bay, already mentioned in passing, a hundred miles further East one finds on the mainland opposite the Haarlem and Moor Islands the three villages of Makimi, Napan and Weinami, and again sixty miles farther North the villages of Waren, Sanggé, Paradoi, Mambui, Nubuai, Woinui, Riséi-Saiati, Wonti and Sasora. East of Wonti there existed up to 1939 a village called Demba, established as a police-post, but deserted since then.[13] So the villages on the Barapasi river do not belong to the Waropen area.

The name Waropen is also in use in the administrative division. There Upper and Lower Waropen indicate two districts on the East coast of

---

graphical exploration — not only in New Guinea as a matter of fact. See A. P. Elkin, "Anthropological research in Australia and the Western Pacific, 1927-1937", in *Oceania* VIII, p. 306 ff.

[12] This division has been adopted by Dr. S. J. Esser on the new language map in the *Atlas van Tropisch Nederland*, published by the Royal Netherlands Geographic Society in co-operation with the Topographical Survey in 1938, sheet no. 9b.

[13] Sasora village was not made a direct object of my research because only a later communication by Rev. ten Haaft proved that also this village is Waropen.

Geelvink Bay in the sub-division Serui, between the Wapoga and the Mamberamo, which therefore also include elements of the population which belong to the tribes of the interior, like e.g. the dwellers on the Barapasi and the Kerema rivers. The administrative district Waropen is therefore on the one hand larger than the tribal area of the Waropen coastal population, and on the other hand it is smaller, because the three villages Napan, Weinami and Makimi, as well as Ambumi on Wandamen Bay, administratively belong to the sub-division Manokwari.

In the remainder of this study the term Waropen will apply to the coastal population living in the villages enumerated above. In case a distinction has to be made between the area to the East and that to the West of the Wapoga river, I shall call the eastern part Waropen Kai and the western part Napan. The name Waropen Kai is also in use among the Waropen themselves when wishing to indicate the stretch of coast East of Waren. What we call Napan, the Waropen usually call Makimi, but Napan is better known and may be found on the maps, also because an administrative assistant is stationed there. That the village on the Woisimi in Wandamen Bay was Waropen seems to have been heard already by P. van der Crab in 1871, although he calls many villages Waropen which in actual fact are Wandamen-Windessi.[14] This village on the Woisimi which maintains many relations with the villages of Makimi, Weinami and Napan is therefore an important link between the culture of the Wandamen and that of the Waropen.

When we examine the region in which the Waropen live, it appears that they have settled exclusively in the areas of the mangroves or of the tidal forests. This is all the more striking, because the other coastal tribes seem to have settled by preference on what Professor H. J. Lam has called the sand-beach. Concerning the flora of both these areas professor Lam writes: "sand-beach (with a porous and highly permeable soil, with much salt and lime) .... Here the common beach-associations occur, most of the members of which are spread by the marine currents and which therefore occupy a wide area; the association Pes-caprae and the Barringtonia association, with occasional casuarina trees (Casuarina equisetifolia) .... mangrove or tidal forest (bako-bako, Rhizophoraceae, etc. and the Nipah association), in low-lying, muddy coasts and estuaries under the influence of the tides (the soil hardly accessible to the air,

---

14 P. J. B. C. Robidé van der Aa, *Reizen naar Nederlandsch-Nieuw-Guinea ondernomen in de jaren 1871, '72, '75, '76, door P. van der Crab en anderen* (Journeys to Neth. N.-G. undertaken in the years 1871 etc. by P. v. d. C. et al.), p. 97.

saturated with salt or brackish water with a great deal of lime). For the
spreading of the varieties of this combination of plants the same applies
as before".15

It is difficult to say whether the Waropen have been pressed back
into the tidal forests because in the many creeks and channels they found
a safe refuge against their attackers, or whether they looked there for
an easy base for sallies against the neighbouring tribes, at the same
time obtaining free access to the fresh-water marshes rich in sago and
situated behind the tidal forest. However that may be, it remains a fact
that everywhere they avoid the sandy beaches and only settle there, still
reluctantly, under pressure from the Administration. This is the case
e.g. in the Napan and in the Ambumi areas. Ambumi village was formerly situated on the Woisimi river and even now it moves gradually
towards the sea. Napan village formerly lay on the channel between
Nusariwe and the mainland. The dense growths of sago served as bases
for victualling for the crews of the prahus, enabling them to roam for
long periods outside their own village and rendering the whole stretch
of the coast unsafe.

Concerning agriculture K. van der Veer writes: "The Waropen coast
opposite Japen is rather densely populated, but the soil of the tidal
forest is highly infertile whilst fresh water is only obtainable at six miles
away from the coast. Toddy prepared from the nipah-palm here often
replaces drinking water. Red peppers (Capsicum frutescens L.), considered indispensable as a condiment for the fishdiet, are grown in small
'hanging gardens' laid out on poles over the water. Sago is collected
outside the zone of brackish water, but real agriculture is out of the
question".16

This means that the tidal forests make real agriculture impossible
and restrict the food-supply exclusively to fishing and the processing
of sago. This forced the people to a quite lively intercourse with other
tribes which produced all kinds of goods unobtainable to the Waropen.
The custom that brothers have to help their married sisters when
processing the sago is the cause that people of one village continually
are in the other, people from Waropen Kai in Napan, or on Japen, as
far away as Ansus. For the women, plaiting developed as a home-
industry which again led to export. For the supply of the highly valued
shell-bracelets, of bush-knives and of many kinds of storegoods, the
Waropen are dependent on other tribes.

---

15 In W. C. Klein, *Nieuw Guinee* I, p. 189.
16 See W. C. Klein, *Nieuw Guinee* II, p. 495.

The red earth which is used for baking pots does not occur in the Waropen area, so that this craft has remained unknown here and stone-ware (sago-molds, cooking-pots) have to be imported, mostly from Japen. Formerly there was rather frequent contact with the tribes from the interior because according to my informants in those days it was from them that the much prized tobacco was obtained and later also birds of paradise. Since tobacco has become obtainable through the store and since there is no longer a trade in birds of paradise, it is now in particular plaited mats which are obtained from the Barapasi and occasionally garden-produce from the people of the interior. Drums were obtained especially from the Moor Islands where still to-day the population excells in wood-working. Perhaps the tidal forests do not produce the kinds of trees suitable for the making of drums.

The nipah-palm grows so abundantly in the mangrove region that the population can dispose without any trouble over unlimited quanti-ties of palm-wine made from this palm. For this reason drunkenness is the order of the day, or rather of the night. Every evening a few small prahus go out to fetch the bamboo tubes with the fermented juice in order to distribute these to those many persons who in actual fact are habitual drunks. I cannot judge the influence of this drink on the physique, but on a superficial view the deleterious influence does not seem to be very great. However, with so frequent an abuse of drink, a coarsening of manners is unavoidable and the struggle started by the mission against this popular abuse is fully justified. It is true that at the same time good drinking-water is scarce, but that it would only be pos-sible to obtain it at six miles distant from the settlements must be based on an exaggerated piece of information.

As long as we stayed in Waropen, our drinking-water also was brought in by prahu. For Nubuai the source lies some distance behind the village of Mambui, at a rough estimate not more than three or four miles away. But it always remains a handicap that all drinking-water has to be transported in bamboo tubes, whilst one has moreover to wait until the falling tide is practically at its lowest, otherwise one gets brackish water. Even so it has a muddy taste and a cloudy colour. Moreover, in the old days it was so easy to lay an ambush for the unsuspecting women and children who came to get water. For this reason a Waropen often quenches his thirst with toddy and occasionally with ordinary river-water.

For the traveller the tidal forest is not very attractive. The sand-beach and the variations in colour changing from the blue sea via a white fringe of foam into yellow sand against a background of green foliage make a pleasing impression and seem to invite the sailor to land. The uniform coast of the tidal forests, however, which disappears to the left and to the right towards the horizon like a narrow green band, washed by opaque muddy water which is sucked eddying and whirling by the current of the tides, then towards the sea and then again towards the land, seems inhospitable and closed. The low muddy shore already makes it impossible for ships drawing as little as six feet to enter the wide channels on which the villages are situated, unless the master is thoroughly acquainted with the narrow fairway which the constantly changing currents have scoured in the treacherous mudbanks. With slightly unfavourable weather a heavy sea runs on the mudbanks, which makes it impossible to enter and which paralyses the prahu traffic. And once past the surf, the silent, rapidly flowing water, ringed in by a labyrinth of rhizophores, does not tempt to further penetration. In the narrower creeks and channels the water moves without a ripple, covered by streaks of greyish-brown dirt like trails of a worn carpet. Here and there dead trees which have lost their bark and have been burnt white by the fierce sun thrust out their fantastically formed branches over the water, shuddering mysteriously under the pressure of the tidal current.

At low tide the forest runs dry. Then a leathery, stinking layer of mud remains, from which the turbid water is sucked away with a chuckling noise, as audible in the deep silence as the rustling, scratching and shuffling of the millions of crabs and crustaceans which the women collect to serve as food. Along the wider channels the inhabitants have cut down a fringe of the mangrove forest and on the opened mudbank the houses are placed on poles at a height that they just remain free of water considering the difference between the tides of six to nine feet. Underneath the houses the empty shells of consumed crustaceans in the course of the years form shell-mounds; in the long run these constitute quite a hard floor. However, even on this hard floor one should walk carefully, because the rotting stumps of ancient poles riddled by the teredo can cause serious wounds to the unprotected foot. Behind the house the tough mud starts again, mud into which one may sink to the ankles or even over one's knees. In several places the inhabitants are still able to point out shell-mounds indicating the location of former villages.

In the beginning I believed that I might credit a rather low intensity of the malaria to the tidal forest, because we personally always remained unaffected in the Waropen area and hardly ever saw any mosquitoes. Another reason was that at the daily distribution of medicine which was well attended, there seemed to appear a much smaller number of people suffering from fever than we were wont to see elsewhere. However, Dr. H. de Rook reports that malaria is also of frequent occurrence on the Waropen coast [17] so that we may at most assume that the Waropen have become more or less immune to this disease. But there still remains from a hygienic point of view the inestimable advantage that twice a day the village is thoroughly cleaned by the tides.

The tidal forest has a charm of its own. The water with its meandering creeks lies there like a fantastic mirror and throws back the sunrays which play a glittering game of light and shadow with the dark-green leaves. In the wide channels on which the villages are situated floats the image of clouds in constantly changing colours, especially in the morning and in the evening. Then the water is red with various shades of colour, just like the sky is red. The houses from which the smoke spirals upward become black silhouettes on frail spindly legs, so that they actually resemble a centipede, as myth has it.

The tidal forest is a strong line between sky and water. Across the smooth water dark prahus move, carrying women returning from the sago-forest, or young people playfully chasing each other, with laughter and the tinkling of white bracelets on the dark arms of the girls. The tidal forest has many colours and much of the cleanliness of a quiet Sunday. The Waropen also sees this beauty, although he does not express it in lyrical effusions, but in a contented humming of the well-known mythical songs. Once it happened to me that someone remarked without being asked that he admired the colours of the setting sun.

Also at night the water has a life of its own. Then the fishermen sail through the darkness with smoking torches, an image that has also been taken up in the myth of the torch-earring. The silence of the night resounds with the plopping of fishes who again and again jump out of the water, and the rustling of the stream past the poles of the house. The rickety houses creak. To the Waropen, night is the time when ghosts and suangi roam about. On moonlit nights the light flings white ribbons across the water. The girls sit in the open inner room of the house, singing marriage songs. The boys call out jokes to which the girls reply with giggles or with indignation.

And when the evenings are dark, the water may start to phosphoresce. A child is sitting in front of the house and throws sticks into the water.

---

[17] See W. C. Klein, *Nieuw Guinee* III, p. 837.

Where the stick comes down, silver circles arise. A man comes rowing past the house and at every quiet beat of his oars ripples of shiny metal roll around his prahu.

Then the river may run in spate when the water returns from the interior carrying whole bundles of reed and trees, pressing them against the houses so that they lean over. Both in mythology and in daily conversation these floods form a subject that is discussed with gusto. The people are afraid of a catastrophic flood which once will bring about the end of the world.

In this way the water has become a living element, peopled with mythical monsters and vague images of the imagination. The water brings the fish and the lobsters, it carries the prahus, but it also hides all kinds of secrets. On the other hand, the sago-forest is farther away and because of the many mosquitoes it is unpleasant to stay there. For this reason people remain there no longer than necessary, and even so one prefers to have the women work at preparing the sago. The division of labour, the way of life and certainly also the thinking of the Waropen are influenced not a little by the environment in which they live.

Climatological differences do not exercise an inordinate influence on the culture of the Waropen. There is very little agriculture so that it matters little whether it rains much or not. When it rains one only leaves the house from sheer necessity. And because the rainfall is high, there are many days when "heaven obstructs man" (*doragha iora nuo*). Rain played an important part in the slave-hunts in olden times, because the attacking party preferred to attack under the cover of a heavy downpour, which not only rendered the attackers longer invisible, but also made the attacked stay indoors by their small fires, whilst their prahus were filled with water so that it became more difficult to escape.

Moreover, there is not the clear-cut difference between the period of the eastern monsoon and that of the western, as elsewhere in Indonesia. For the prahu traffic between Waropen Kai and the Napan area, the beginning of the eastern monsoon period is the most suitable time. During both monsoons, periods occur when the prahu traffic is completely interrupted by strong winds. When crossing, e.g. from Waropen Kai to Japen or from Napan to the Moor Islands, the prahus always avail themselves of the sea- and land-winds, also taking the position of the moon and the height of the water into account.[18]

---

[18] For the climate see Dr. C. Braak in W. C. Klein, *Nieuw Guinee* I, p. 164, esp. pp. 169 ff.

HISTORICAL INFORMATION

The Waropen possess few objects which inform us about their historical connections. The well-known Papua axe is found here, characterised by the oval or lenticular cross-section.[19] However, this type of stone axe has already become rare in the Waropen area, although formerly it must have been of a much more general occurrence, being also used as a battle-axe. This we still observe in the models made of *gabagaba* in use during the Serakokoi-ritual.[20]

Probably since long the Waropen have obtained their bush-knives and the heads for their arrows and harpoons from the people of Biak who have applied themselves to metal-working and who still wander around Geelvink Bay as itinerant smiths. Stones are moreover not found in any quantity in the tidal forest, and moreover the many sea-shells provide an easily obtainable material for scratchers and cutting instruments. Bone is occasionally used for knife-handles, sometimes for spoons, whilst the bones of the bat are made into sewing-needles. Knives, spears and arrows made of bamboo are likewise still constantly used. Metals are liked as wearing apparel, all kinds of valueless trinkets from the store as much as the highly valued flat, silver wrist-rings, which the men of other tribes also love to wear. Brass foot- and finger-rings are also used. Furthermore, mention should be made of various types of beads and ancient porcelain (*rewanggu*), all objects that are highly valued as more or less ceremonial means of exchange throughout Geelvink Bay.

It is impossible to determine the routes along which these objects reached the Waropen coast, but in all probability the Numfoor and the Wandamen have acted as intermediaries in their diffusion. Occasionally there was also direct contact between the Waropen and the world of the Moluccan islands. So for instance there are flags of which it is said that they are from Tidore, at least as far as their model is concerned. There is a myth which tells us that the famous captain Amos — well-known to the first missionaries — was guided to the Waropen by the brother of the divine trickster, but all in all this traffic cannot have been very intensive (Texts 49).[21] The enthusiasm to go and visit the dange-

---

[19] See Dr. A. N. J. Thomassen à Thuessink van der Hoop's contribution to F. W. Stapel, *Geschiedenis van Nederlandsch Indië* (History of the Netherlands Indies) I, p. 50.

[20] See below, p. 192.

[21] I refer here to an unpublished collection of Waropen texts. [Now published (with Dutch translation) in Verhandelingen van het Koninklijk Instituut voor Taal-, Land- en Volkenkunde, The Hague, no. XX (1956), Ed.]

rous and notorious Waropen in their tidal forests cannot have been very great.

On the sacred island of Nusariwe in the immediate vicinity of Napan there are a few caves or caverns which according to the population served as dwellings to the people during the mythical primeval period. These caves lie at some little distance from the coast, but an investigation provided nothing particular. No traces of drawings or human habitation were found. One of these caves, however, contained remains of a human skeleton, a skull and some other parts. The population avoids these places and only a few have ever been there. According to my informants the cave containing the skeleton had been discovered only relatively recently by a woman. The bones are lying simply on the ground and perhaps they are not so very old, so that possibly they are the remains of some unfortunate person who was surprised by death in this rather lonely spot.

Historical data concerning the Papuas are scarce and in fact they tell more about the ideas of the travellers who visited them than about the Papuas themselves.[22] The first mention of the name Aropen is in connection with the voyage undertaken at the orders of the Government in 1705 by Jacob Weyland with the *Geelvink,* the *Kraanvogel* and the *Nova Guinea.* Having passed between Biak and Japen, Weyland reached the northeast corner of Japen on April 28th and the most easterly point of his journey on May 17th.[23] Close to this point a settlement is reported, situated in the "Kaaij" area, from which two natives were taken, who proved to live in enmity with the four inhabitants of Japen who also were on board.[24] These Papuas were handed over in Batavia; three of them went to Holland and of these two returned in

---

[22] Most of the data I owe to Colonel A. Haga, *Nederlandsch Nieuw Guinea en de Papoesche eilanden, Historische bijdrage* ± *1500—1883* (N. N. G. and the Papuan islands, an historical contribution etc.), in two volumes, further referred to as Haga, *Ned. Nieuw Guinea,* and to A. Wichmann, Entdeckungsgeschichte von Neu-Guinea (History of the discovery of New Guinea), in *Nova Guinea,* I and II, further referred to as Wichmann, *Nova Guinea.*

[23] The object of this voyage was the eastern tip of Geelvink Bay. Haga supposes that Weyland sailed up to the Mamberamo, but the Waropen Kai region as it is known at present does not extend beyond Kurudu, so that we should have to assume either that Weyland did not reach the Mamberamo, or that the Waropen area formerly extended much further to the North than it does now. Another possibility is that the men from Waropen were on a voyage, because they were "fished up". Haga, *Ned. Nieuw Guinea,* I, p. 166.

[24] Haga, *Ned. Nieuw Guinea,* I, p. 167.

2. The tidal forest

3. Village street at Weinami

4. Maarten of Ruma Warani

5. The albino child is afraid of the camera

1710, obtaining civil rights on Banda. The story does not tell us whether these people came originally from Japen of from the Waropen.[25]

Feuds among the people from Japen and the Waropen existed until quite recently. Weyland also mentions a settlement "Erropang". Haga says that from this word "one easily recognises Aropen. It was reached on May 30th; its inhabitants 'did not dare to withstand our people', but fled away in prahus".[26] Until the end of June the party continued to sail through Geelvink Bay.

The opinion on the inhabitants of Geelvink Bay was not very favourable. They were called "an unapproachable and untractable, purely wild and malevolent people".[27] Haga says: "What the report has to say for the rest about the population is of no great importance. Nearly everywhere people or traces of habitation were found, whilst the population seems to have been densest near the eastern tip of Geelvink Bay. Both sexes went practically stark naked 'and their shame only covered with some tree-leaves'. The arms were bow and arrow, assegais and an unwieldy type of bush-knife or chopper of iron. The wooden houses were built high above the ground".[28]

Five years later we find their name again in a memorandum on the state of the administration by Claasz, describing the sphere of influence of Tidore and mentioning i.a. a coastal settlement called Roppon.[29] Several regions on the West coast of Geelvink Bay are easily recognisable, "Arraffak" and "Wandamin" can hardly be anything but Arfak and Wandamen. To our regret Haga is not able either to identify the other names and so we do not know whether Ropen (which Haga likewise connects with Aropen as he calls it) refers to the village on the Woisimi or to the region near Napan. In that case we might conclude from this notice that the Waropen village on the Woisimi already possessed a certain renown in this area as early as 1710.

After this first contact the Waropen tribe sinks back into oblivion for many years. Some expeditions passed it, however, e.g. an English expedition in 1840.[30] Japen, for instance, was visited in 1850 and in

---

[25] *Op. cit.* I, p. 175.
[26] *Op. cit.* I, p. 167.
[27] *Op. cit.* I, p. 165.
[28] *Op. cit.* I, p. 174. The statement by Dr. A. N. J. Thomassen à Thuessink van der Hoop in F. W. Stapel, *Geschiedenis van Nederlandsch Indië* I, p. 49, that "the Papuas of New Guinea did not know any metal tools at all before the arrival of the Europeans" is evidently not quite correct in this generalised form.
[29] Haga, *Ned. Nieuw Guinea*, I, p. 193.
[30] *Op. cit.* II, p. 74.

1869. But the Waropen are only mentioned again in 1871, when Van der Crab travelled i.a. to Ansus on Japen where there appeared to exist differences with the Waropen. From there he went to Napan, evidently not the Napan we know, but a village with the same name on the Jauer peninsula, some distance to the Southwest.[31] The voyage proceeded via Roon to Wasior in Wandamen Bay, and of this locality Van der Crab remarks that it is Waropen. Although, as Haga remarks, this information is incorrect, one can hardly do anything else but deduce from it that the village on the Woisimi enjoyed a certain notoriety. Knowledge of the name Waropen is not to be explained in any other way.

In 1873 the naturalist A. B. Meyer made a voyage along the East coast of Geelvink Bay, "called Kai by Meyer and Aropen by others".[32]

Next, in 1879, we hear again of enmity between the men of Japen and the Waropen, on the occasion of a voyage by the district-controller J. van Oldenborg, to whom it was reported that "men of Aropen had attempted to steal the coat of arms during a night attack, but had not been successful".[33]

Only in 1881 the Waropen area was really visited for the first time, when we leave the possibility of a visit by Weyland and his men in 1705 outside of our consideration.[34] The visit was intended for the villages near Napan; it was made by the above-mentioned district-controller Van Oldenborg and Lieutenant-Commander M. A. Medenbach.[35] From the village of Ansus on Japen they visited the East coast of Geelvink Bay and saw the villages of Wajunami, Painan and Making. "When on April 5th this coast was sighted, it showed itself as a low-lying marshland, covered by a dense growth of rhizophores and containing a great number of inlets of which it was believed that these could only be mouths of rivers.

"In the afternoon some small prahus were observed and the party succeeded in coming into friendly contact with their crews. This was certainly due to the interpreter taken on board at Dorei, who understood the Wandamen language spoken by these Papuas. They appeared to belong to the Aropen tribe, living in a village called Waju Nami and trading with the Wandamen, whom they perfectly resembled in

---

[31] *Op. cit.* II, p. 213 ff.
[32] *Op. cit.* II, p. 258. Wichmann, *Nova Guinea* II, 1st part, p. 167 expresses quite an unfavourable opinion on Meyer's reliability.
[33] Haga, *Ned. Nieuw Guinea*, II, p. 365.
[34] *Op. cit.* I, p. 167.
[35] *Op. cit.* II, p. 393.

their outward appearance. These Wandamen seemed to exercise some sort of supremacy here, at least, they had recently appointed one of their men as *korano,* but at the time of the visit of the *Batavia* this chief was on the island of Moor.

"The houses, mostly built over the water, did not provide anything remarkable. The natives assured us to be living in peace with their neighbours and also that during these last years robbery by the people of Jappen had become less frequent.

"On a slightly elevated place on the shore, at the mouth of a river, we thereupon put up the Netherlands coat of arms, to the great joy of the inhabitants who helped us with diligence. The commander of the *Batavia* fixed the position at 135° 48' East and 3° South.

"A Dutch flag was placed near the coat of arms and some gifts were distributed among the population, whereupon the voyage was continued on April 6th".

The remark that the Waropen understood the Wandamen language proves that already at that time there was quite some contact between these two regions. This is confirmed again by the mention of the appointment as *korano,* although this must not lead us to any conclusion about the supremacy of the Wandamen.[36]

Six years later the same place was visited again, by A. G. Ellis and F. S. A. de Clercq. They found Makimi (12 houses) and Wainami (11 houses).[37] "According to the information collected by De Clercq they constituted at the same time the southern border of the Waropèn region. As other settlements going from the South to the North there were enumerated to him: Makimi, Wainami, Warèn, Sanggei, Woju, Pareirori, Nuwoái (Wichmann notes that this place is better known as Kai), Risei, Wóti and Wapori. On the 7th the *Java* anchored between the islet of Náu and the village of Warèn. When the investigation had been continued up to the 11th, the island Aberé (Kurudu) was visited".

In the names of the villages enumerated here we easily recognise the now extant villages, so that we may safely assume that during the last sixty years there have not been many changes in the distribution of the villages of the district Waropen Kai. In the name "Wapori" we may perhaps retrace the name Kaipuri, literally Hindermost Kai, which is often used to indicate the region around Kurudu. Perhaps it even refers to the village of Sasora. The villages in Wandamen Bay are, however, not mentioned.

---

[36] See below, p. 85.
[37] Wichmann, *Nova Guinea* II, part 2, p. 429.

The Waropen Kai district was visited for the first time in 1888, by F. S. A. de Clercq. At first sight it may seem strange that the nearby island of Japen, from where the mountains on the mainland are visible, was visited so much earlier than the populous Waropen region, but we should not forget that the villages on the sandy beach are visible at sea from afar, whilst on the uniform mangrove-coast the presence of even large villages is only betrayed by a few fishing prahus. De Clercq visited Nubuai village which he indicates as Kai, presumably after the most important clan in accordance with the custom of those days. Here he counted fifty houses.[38]

The first contacts which the Woisimi villages had with the Government were not of a pleasant nature, for already in 1889 they received their share in the punishment meted out to the inhabitants of Wandamen Bay. It is not clear whether they had deserved this punishment, or whether their villages were burnt down because the leaders of the punitive expedition did not know that they had to do with again another tribe of Papuas.[39]

In general, however, the Waropen could not be called peaceable, because L. A. Oosterzee, as far as I know the first district-controller of the permanent Government settlement in Northern New Guinea, notes in 1899 that the natives on the Mamberamo estuary suffer from the attacks by people of the Waropèn area.[40] In 1903 A. Wichmann and J. W. van Nouhuys again visited Napan.[41]

Contacts with the Waropen tribe became gradually more frequent. If in those days also some time had been taken to bring order in the archives, the memoranda on the state of the administration, etc., we would have been able in all probability to report several more visits. The missionaries who had temporarily established themselves in Jauer, on Roon and on Mioswaar as early as 1867, i.e. several decades before there was any effective influence of the Administration, must certainly have had some contact with the Waropen. People at Ambumi told me, that they had asked the missionary A. van Balen for a guru, but that he had referred the stationing of a guru to his successor, D. B. Starren-

---

[38] *Op. cit.* II, part 2, p. 463.

[39] "Een hongi-tocht op Nieuw-Guinea een halve eeuw geleden" (A hongi-raid on New Guinea half a century ago), an anonymous contribution to the journal *Nieuw-Guinea* IV, p. 152.

[40] Wichmann, *Nova Guinea* II, part 2, p. 704.

[41] A. Wichmann, "Bericht über eine im Jahre 1903 ausgeführte Reise nach Neu Guinea" (Report on a voyage to N.G. undertaken in 1903) in *Nova Guinea* IV, p. 138.

burg. In 1906 therefore the Waropen villages in Wandamen Bay went over to Christianity; since then they had more contact with the Administration which appointed village-chiefs and moved the whole village from the Woisimi (Texts 181). I cannot say with any certainty at what time the villages of Napan, Weinami and Makimi went over to Christianity, but probably this also happened during the period of the missionary Starrenburg. The latter was kind enough to give me a short word-list of Waropen, probably the first attempt of noting down this language.[42]

The area of Waropen Kai was only completely visited during the military exploration. "Captain Ten Klooster and the naval Lieutenant Doorman entered and charted all rivers and creeks (of the coastal region between Wapoga and Apauwer) so that we now possess a complete picture of the Mamberamo-delta and also of the notorious Waropen coast (where of course we again came to grips with the pirates)".[43]

"On the East coast of Geelvink Bay, and especially in Waropen, several larger villages are to be found, like Waren (30 large houses), Kai-Nubuai, Wonti (at least 165 adults) and Sasora. The villages on the Barapasi river together count more than 200 adults. On the whole, however, the region is thinly populated, like that on the South coast of Geelvink Bay".[44]

It appears from this notice that the Waropen were known to be aggressive. The Waropen tales also mention more than once the punitive expeditions of the vessel the *Pionier*, or as they call it, *Foineri* (Texts 181, 185, 186). In a diary of 1913 which I happened to obtain, I read that the village Nubuai was burnt down by a punitive expedition in the year 1913. The certificates of appointment which the Waropen carefully preserve show that the Administration appointed the first village-chiefs in 1918. About eight years later the administrative post Demba was established; here an Administrative Assistant was stationed, supported by a few Regional Assistants and a group of armed police. In connection with this establishment, Resident W. A. Hovenkamp writes: "The opposition in the Waroppen expected by some has in no

---

[42] Something may have been noted down from the Waropen caught in 1705; cf. p. 16.

[43] *Verslag van de Militaire exploratie van Nederlandsch-Nieuw-Guinee* (Report on the military exploration of Neth. New Guinea) *1907—1915*, p. 62.

[44] P. 240 of the *Verslag* (Report) quoted in note 43.

way come about". In 1939 the post at Demba was abolished again, to be transferred to Waren, which up to that time had been the station of an Administrative Sub-Assistant.

Until a few years ago the Waropen Kai district was still considered so dangerous that passports were compulsory when travelling. As soon as this compulsion had been removed, Christianisation was started, beginning at Waren. It became definite in 1938 with the occupation of Nubuai and the surrounding villages, so that by now the whole of the Waropen tribe has gone over to Christianity.

This transition to Christianity is the limit to which our ethnographical investigations in the present work extend. They will therefore refer practically without exception to pre-Christian conditions. We have imposed this restriction upon ourselves because by discussing the present-day situation we would unavoidably have to touch on contemporary history and so we would be compelled to speak about missionary work in general on a much broader scale. This restriction is therefore purely methodical in origin. It does not result from the opinion that it would be impossible to write an ethnography of christianised tribes, or that the study of the sociological problems connected with this transition would be less important than of those pertaining in pre-Christian conditions. For an understanding of modern conditions it seemed, however, necessary to me to obtain as clear as possible a picture of the more ancient pagan culture, because of its uninterrupted historical connection with the present Christian culture.

NUMBER AND OUTWARD APPEARANCE OF THE WAROPEN

It is only since very recent times that we have somewhat reliable figures at our disposal concerning the population-density of the Waropen. In 1705 Weyland found that the population of Geelvink Bay seemed to be densest in the eastern corner, i.e. in the Waropen Kai district. Also at present we find in this region rather large villages, although we should not assume as certain that this was also formerly the case. It is difficult to rely on first impressions, because, as we saw on p. 21 the officers of the military exploration party were of the opinion that the region in general was thinly populated; perhaps they were referring to the area East of Sasora.

Concerning the population we possess the figures of the census of 1930 and of those of the end of 1937 which we reproduce on the next pages, those for 1930 on the left, those for 1937 on the right of each column.[45]

I would not venture to draw any conclusions from these figures. According to our records the total population would have decreased by 500 persons during a period of seven years. This need not indicate an absolute decrease of the number, because it is quite possible that many persons moved elsewhere. The strong decrease of Demba village may well be the result of the artificial construction of this post which was constituted (if I am well informed) by directing a clan from Wonti to this place. In 1930 the ratio of the sexes within the Waropen tribe was men : women = 3240 : 3438; in 1937 this was 3066 : 3107.

Dr. P. M. L. Tammes of the Coconut Experimental Station at Menado was kind enough to inform me that the shift to be observed in the ratio of the figures for the sexes lies within the limits for mistakes according to the calculus of probabilities, so that it does not warrant any special conclusions regarding the factors which might have caused such a shift. I would not dare to say anything either about conclusions to be drawn for the future regarding the striking difference in the number of girls, for also in this case the criterion adopted for the division between adult and non-adult of course plays a role. The same applies to the ratio existing between the total number of adults and the total number of non-adults among the Waropen, a ratio which in and by itself does not seem to be unfavourable consi-

---

[45] The figures of the Administration agree rather well with the results of our count for Nubuai village which was completely noted down genealogically. Because in this village practically everybody marries rather young I have considered all unmarried persons as non-adults. By way of comparison I have placed the Administration figures for 1937 to the left of my figures for 1938.

Clan	Men		Women		Boys		Girls		Total	
Kai	48	44	56	61	32	60	35	48	171	213
Apeinawo	79	80	92	94	90	71	56	58	317	303
Nuwuri	44	43	51	43	38	36	23	22	156	144
Pedei	65	91	80	92	53	65	38	54	236	302
Sawaki	180	184	230	187	172	160	148	140	730	671
	416	442	509	477	385	392	300	322	1610	1633

The difference lies mainly in the number of men. It is possible that during my count some men were reported by different houses so that they came to be counted twice. It is also possible that the population in its reports to the Administrative Assistant was somewhat sparing as regards the number of adult men, c.q. taxable men.

VILLAGE	CLAN	MEN		WO
		1930	1937	1930
SASORA				
DEMBA	Fafa	*44		*51
WONTI	Ghairo Wanda Kai Bunggu	115	58	112
		242	244	255
RISEI	Womorisi Wainarisi	134	128	132
SAIATI	Imbiri Daimboa		81 35	
WOINUI		108	116	125
NUBUAI	Kai Apeinawo Nuwuri Pedei Sawaki	24	21 48 79 44 65 180	24
		426	416	501
MAMBUI	Ghopari Ghama		57 65	
PARADOI	Tao Sirami Watofa Satia Sawai	156	122 35 19 39 28 28	149
SANGGEI		147	149	139
WAREN	Papirandei Watofa Saimua Wairoi Imbiri	36	43 48 14 27 26 46	53
		154	161	174
TOTAL WAROPEN KAI		1542	1458	1664
NAPAN		70	70	53
WEINAMI		45	43	30
MAKIMI		41	56	39
AMBUMI		33	29	52
TOTAL NAPAN		189	198	174
GRAND TOTAL		1731	1656	1838

* The figures for Sasora village, which are not completely available, have not been included in the grand total.

MEN	BOYS		GIRLS		Total per village and clan	
1937	1930	1937	1930	1937	1930	1937
	*20		*20		*135	
69	116	55	110	46	453	228
257	253	230	260	218	1010	949
156	106	107	96	80	468	471
64		52		61		258
40		31		34		140
104	110	83	124	95	467	398
25	16	9	11	.9	75	64
56		32		35		171
92		90		56		317
51		38		23		156
80		53		38		236
230		172		148		730
509	377	385	398	300	1702	1610
68		38		27		190
80		53		38		236
148	103	91	133	65	541	426
38		31		19		123
20		17		18		74
42		39		34		154
36		29		23		116
25		16		14		83
161	112	132	136	108	534	550
59	29	32	38	24	156	158
60		38		49		195
19		12		11		56
37		21		19		104
30		23		18		97
68		45		38		197
214	150	139	170	135	648	649
1702	1372	1263	1476	1080	6054	5503
74	40	50	43	54	206	248
36	38	38	18	20	131	137
45	31	30	33	33	144	164
37	28	29	30	26	143	121
192	137	147	124	133	624	670
1894	1509	1410	1600	1213	6678	6173

dering conditions in New Guinea.[46] For that matter, one does not often find a surplus of women in New Guinea.

It is to be hoped that before long we may have more figures on population trends in New Guinea so as to allow specialists to draw their conclusions. For the time being the available data are too few and in discussions on this subject sentiments are too much irritated when the one side talks about a decrease of the population and the other about an increase, or at least about a retardation of a decrease, which would have been unobservable also before the effect of Western influences on the Papuas.

No anthropological observations were made so that only superficial statements on the outward appearance of the Waropen are possible. Albinism is far from rare. Several individuals have a blond skin and reddish-blond straight hair.[47] In general the colour of the skin varies between black-brown and brownish yellow. Abnormally small or abnormally tall persons are rare; most adult individuals are of average size.

The Encyclopedia of the Netherlands Indies says of the Waropen in volume III, page 321: "In general the men are strongly built with regular features, some with marks of burning or tattooing; all have the internasal septum perforated; they wear their hair so short that it will just carry a forked comb. They do not wear a beard or moustaches, pulling out the hairs by means of a double shell...... The women are anything but well-shaped; they are of small stature and soon take on a withered appearance. Their breasts are strikingly small; many women have these as well as the whole of their back tattooed. Most women are marked by scars burnt on their backs and shoulders". To this not very flattering description the author of the article adds: "The Papuas of Waropen stand far below their neighbours on the Southwest coast of the bay; there is no civilisation here worth mentioning".

It is not clear which Waropen the author saw, because the Waropen, just like the other pagan inhabitants of Geelvink Bay, used to wear their hair dressed high, whilst the women, except of course the aged, are conspicuous because of the well-known mop-type of hairdress.

---

[46] On this point cf. *Volkstelling 1930* (Census of 1930 in the Neth. Indies, V). In view of our present knowledge of the population of New Guinea it is difficult to differentiate between the tribes or between the coastal population and the inhabitants of the interior when considering the census-figures. In the grand totals this causes a levelling of possible differences in the constitution of the various elements of the population. This is particularly regrettable when differences in race enter into the matter.

[47] See the boy on photo 21.

In general the women are not particularly small either and as far as Papua women go, rather good-looking. Among Europeans the Waropen coast is often referred to as the Bali of New Guinea. But we should not forget that when the article quoted above was written, there had been only little contact with the Waropen and that during long years they had a bad name, because of which they appear in many reports as pirates, drunkards and cruel people.

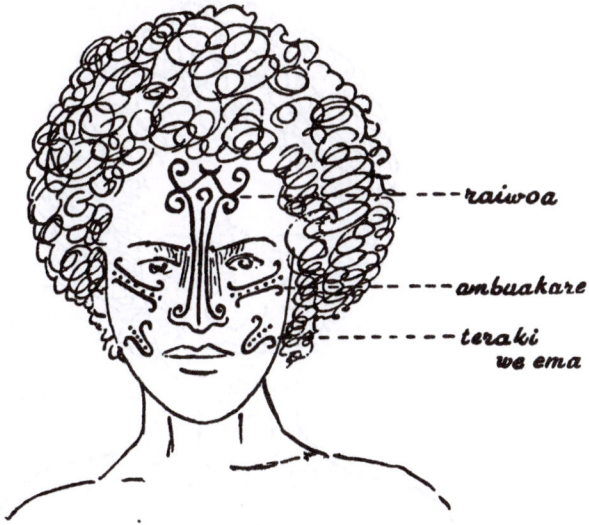

In order to complete the information provided by the Encyclopedia we may add that older people do not pluck out the whole of their moustaches by means of a double shell (*mundai*), but that occasionally they leave a small tuft at the corners of the mouth. Women remove the bodyhair, which usually is not very dense, from their armpits. Growing boys often grind the cutting edge of their upper teeth smooth with a piece of pumice-stone. Both sexes perforate the nasal septum as well as the ears; women, occasionally pierce the shell of the ear at several places along the edge.

During the years of puberty the girls undergo extensive tattooing (*onda*, also: painting on prahus, letter, writing), both on the chest and the legs and on the arms and the face. The pattern is first indicated on the skin with blacking, whereupon it is pricked in by means of two fishbones, tied close together to a couple of pieces of wood, which are softly tapped with another piece of wood. Then the small wounds are rubbed again with blacking so that they become slightly inflamed, with the result that the motif is indelibly fixed in the skin. The whole process is rather painful and for that reason it is executed in stages. My informants were unable to give me a certain meaning for these tattoo-marks, like e.g. an indication of the tribe or of status, neither was I able to find a myth concerning their origin.[48]

The Waropen woman, just like her occidental and oriental sisters, subjects herself to the dictates of fashion, albeit with as many sighs as they. I was assured that tattoo-marks looked so well, especially on

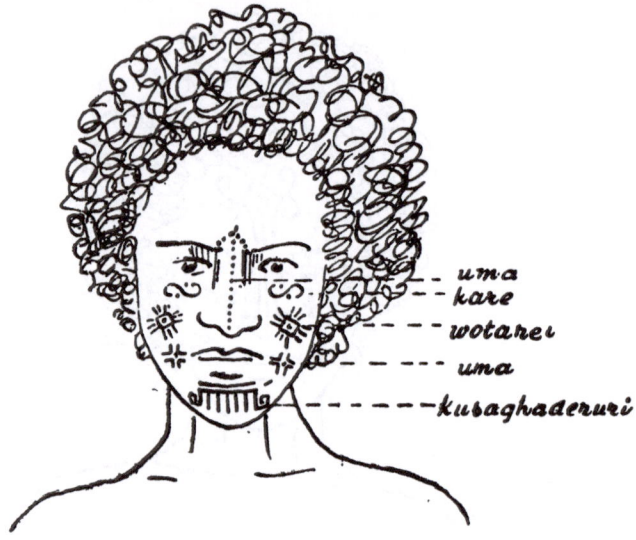

women who after childbirth have to stay indoors for a long period, due to which their skin stands out more whitely against the marks.

In the extensive tattoo-marks on the bodies of girls (*raiwonda*, bead-design) which in contrast to the face-marks are identical for all women, the Waropen recognise the design which should also be apparent on the bark of the foot of a *bora*-tree which becomes visible at low tide; this tree is also used as post in a house (*masa*) or for making tops (*mbumbu*). "One tattoos according to the foot of the drying *bora*-tree" (*kikonda ebaba fono uruwa bora*). I was, however, unable to discover any tattooing design in the bark of this tree. A very faint indication of a religious meaning of tattooing might be deduced from the aversion of the Christians against this practice.[49]

---

[48] According to J. L. van Hasselt the tattoo-marks among the Numfoor often represented pictures of objects used by deceased persons, see the *TBG* XXXI, p. 592.

[49] See however also p. 294, note 226.

The tattooing of the men is less gaudy than that of the women. In modern times some of them wear their name in large scrawls on their chest or on their arms. Others have a pattern of two small crossed flags with a crown in the centre, or a cross like a medal, or a female figure. Occasionally they have the face tattooed. The men's tattoo-marks seem to serve no other purpose than that of decoration. Like our seamen used to do, young men often have themselves tattooed when on a voyage, e.g. when working on a plantation or somewhere else outside their normal home.

There are no real specialists for tattooing; it does not demand any special supernatural precautions or initiation. Anybody who feels like it may apply the marks, so that I also observed young girls tattooing each other. Of course, one person is more skilful than another and so there are women who have a certain fame and who therefore are often called upon for this purpose.

One old woman told me that formerly tattooing was the work of slaves. Although nobody was able to confirm this information, there does seem to exist some aversion against applying the marks to strangers which may be deduced from the rather high payment demanded. I was assured that women skilled in tattooing sometimes let themselves be paid a whole sago-palm for applying the complete marks.

From time to time the skin is rubbed with some coconut oil and coquettish girls will sometimes dye their nails red by tying crushed leaves of a certain plant, ra (balsam?), to them. The hair needs special attention because otherwise it gets completely tangled, as is the case with old women who do not take care of their appearance and who have not cut their hair short. The mythical snake is also represented as a creature with the hair in long tangles.

The young women embellish themselves by constantly combing the hair carefully with a bamboo comb and by washing it from time to time with lime-juice and rubbing it with coconut oil. Picking vermin is moreover the daily work of all members of the sex and the sole attention which a married couple may show to each other in public. When the vermin becomes too troublesome, mud is made to dry in the hair, which is then washed out again. Baldheadedness does not often occur among men; popular belief ascribes it to sexual overindulgence, so that allusions to baldness are strongly resented and may easily result in violent quarrels.

In order to be able to parade a beautiful head of fluffy hair, the young women often tie it up in tufts, enveloped in a piece of nipah-leaf. These tufts, or as the Waropen call them characteristically "hair cudgels", are often inserted in engraved bamboo tubes (*amanikarudo*) and even in a lamp-chimney, as was done by one very coquettish girl. These dancing tufts — usually two, one in front and one behind, and sometimes four, in all four directions — enveloped as they are with the lightyellow leaf, make a frivolous impression, and the young men with whom the girls stand in a "joking relation" often tease them by pulling loose these coverings. Christian women do not wear the hair like this, but tie it in a knot drawn back as smoothly as possible by means of coconut oil, just like the Ambonese nyora (the wives of the gurus).

In order to be especially beautiful one scatters small white pigeon-feathers on the wide mop of hair, like brides use to do. In the finely frizzed hair in which minute drops of oil glitter, the white feathers look as nice as the red Hibiscus flowers which other young women stick in their hair.

Although the young men do not dress their hair as conspicuously as the women, they devote no less care to it. Only rarely do they wear it in the wide mop which was perhaps the fashion formerly. Most young men have the hair cut short around the head, with a forelock combed high which a few of the younger people already divide in a parting in the Western style. A single one tries to imitate the smooth rich locks of the Ambonese or Javanese dandy by means of oil or pommade. The younger men wear the hair so long that they can put a bamboo comb (*sura*) in it, which is adorned by a rosette as large as a fist and made of all kinds of coloured paper which the Ambonese like to use for making paper chains and decorations (*kertasi bungga*, from Malay: kertas bunga, flowered paper). When on top of this rosette again the highly prized feathers of the bird of paradise have been fastened, a prahu manned by young men with their waving colourful combs makes a gay impression.

The Waropen Christians who like to differ from the non-Christians in outward appearance, a phenomenon to be observed among many peoples in Indonesia, no longer wear the high locks and the comb. Schoolgoing children usually have the hair cut short completely.

The bodily cleanliness of the Waropen is not unsatisfactory. Only the Christians wash themselves purposely with some regularity, but also the others by necessity so often come into contact with water, that dirtiness is usually found only among people who have to stay

long indoors, hence among the very old who also avoid the fierce sun.
If breaking the isolation after the mourning period is called "throwing
away the dirt" we will have to understand this not as a pure metaphor.
The Christians also soon learn to wash their clothes and the line with
fluttering washing already belongs to the picture of the Christian
villàge. However, also elsewhere there are women who possess enough
domestic sense to start washing the clothes of the family. Many young
people take great pleasure in washing themselves in a heavy downpour
of rain.

In spite of all this one should not have any exaggerated ideas about
hygienic conditions. Lice are present in great numbers, both on the
people and in the houses. These houses where the sick and the healthy,
children, dogs and pigs live together, are never thoroughly cleaned.
Moreover, spitting and blowing the nose are allowed freely, even very
freely. Fortunately the strong movement of the tides results in a radical
removal of the village dirt and therefore in this respect the situation,
taking all in all, is not quite unfavourable.

In the old days — and old men still maintain the custom — clothing
was limited to a strip of dried banana-bark (*umame*), drawn between
the legs and held between a piece of string tied around the hips. In
old-fashioned circles it actually is not good form for the nobles to pay
much attention to their clothing. Older people look down with a
certain contemptuous mockery especially on the less gentle young
men who take great trouble to obtain all kinds of wearing apparel
from the Chinese stores, even white sun-helmets, although according
to ancient custom the head-covering is really the privilege of those of
gentle birth. The younger pagan Waropen wear a sarong, held up by
a leather belt (*bana puri*, from Malay: band perut, belly-band), or a
hip-band made of beads, and mostly a tricot singlet. Among Christians the sarong has been replaced by European trousers. The flamboyant and fantastic costumes prove that the people have not yet
become quite accustomed to wearing clothes. However, in those villages
where Christianity has been established for some time, tastes become
better balanced. Schoolchildren usually wear dark or striped shorts
and a jacket. The Waropen women have not yet acquired any great
skill in sewing and moreover it is not very easy to keep the cheap
stuff in good repair, so that clothed people often look ragged.

Nearby all men wear on their wrists the well-known armlets of
akar-bahar (*fandana*), a few also wear the highly prized bands made
of some kind of silver (*saraka*). On the upper arm and on the wrist

one may often observe finely plaited armbands (*aisa*) which are perhaps also a sign of light mourning. Under the *saparo* and the *sarako*
one often wears simple plaited bands (*wangga*). Older men usually
also wear some kind of amulet (*aiwo*) on a piece of string, around
their neck, the younger often have in its stead the key of the highly
popular clothes-chest (*burua*).

It seems that in former times the women were also dressed in leaves,
at least we read in the report of 1705 quoted above that both sexes
were "only covered with some tree-leaves". Very occasionally the
younger women cover themselves with leaves, but this only happens
when for the rest they are withdrawn from sight by water or mud.
The normal costume for women consists of a loin-cloth which usually
hangs down in front half-way down to the knees, like an apron. Most
women, especially the older ones, wear blue cotton (*rari*); girls will
also wear red cotton, the colour worn on festive occasions. Younger girls
wear a girdle (*ghono*) consisting of strands of red and yellow strings,
collected behind on both sides in a three-cornered piece of bead-work
with a fringe. Young women moreover usually wear another belt of
beads (*korurawo*) slightly lower than the girdle just above the posterior.

The arms are mostly richly adorned with bands of shell (*saparo*)
or of celluloid (*ponisi*). Also on their legs many girls, just like the
young men, wear bands of copper (*rewano*) or of celluloid. The soft
tinkling of armbands in the silence of the evening betrays the approach of a young girl who comes rowing in. Around the throat she
usually still wears a thin string of fine, light-red beads, and on the
fingers rings of tortoise-shell (*enikambo*), of iron or of the fruit of
the sago-palm (*rewakambo*). She also wears decorations in her ears
that have been pierced — usually on the occasion of an initiation-
ceremony — often a glass hanger (*dimbo*) which the Waropen are
able to make themselves out of ordinary bottle-glass, sometimes all
kinds of other things which one can get hold of. The girls fond of
show were always after our wooden clothes-pins and metal paperclips.
Carved nuts of the motoa (*kagharo*) which are still kept, I have not
seen in use as ear-pendants, although they were formerly used for this
purpose.

Older women take less trouble with their appearance. Mostly they
wear the apron pulled tight between the legs and do not wear other
ornaments but a few strings of ancient beads which possess a certain

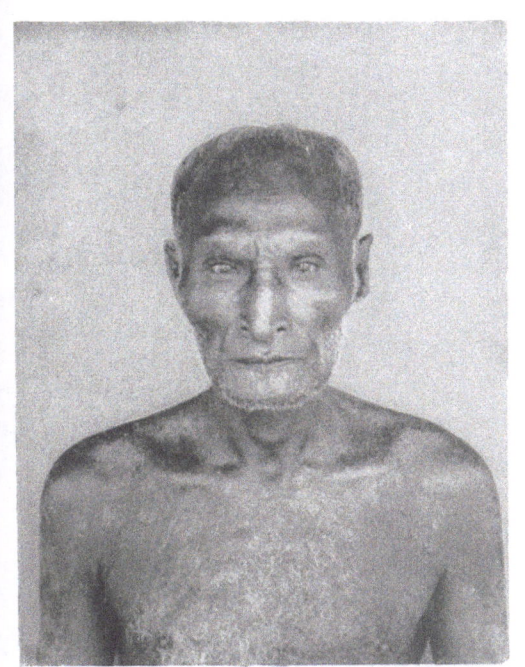
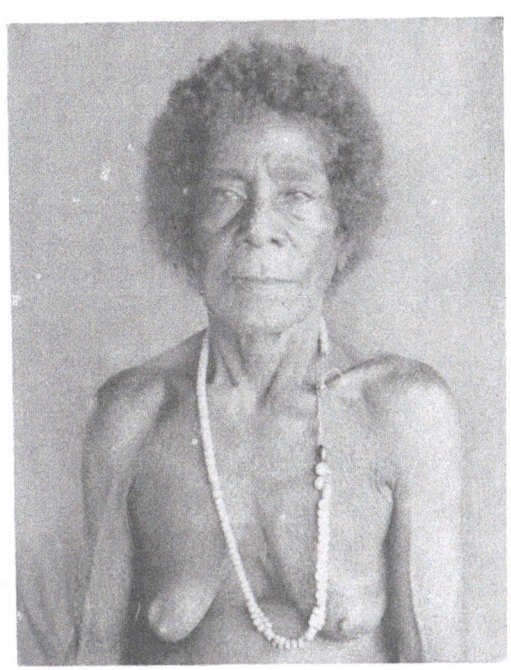

6, 7. Dedui of Ruma Erari and his wife

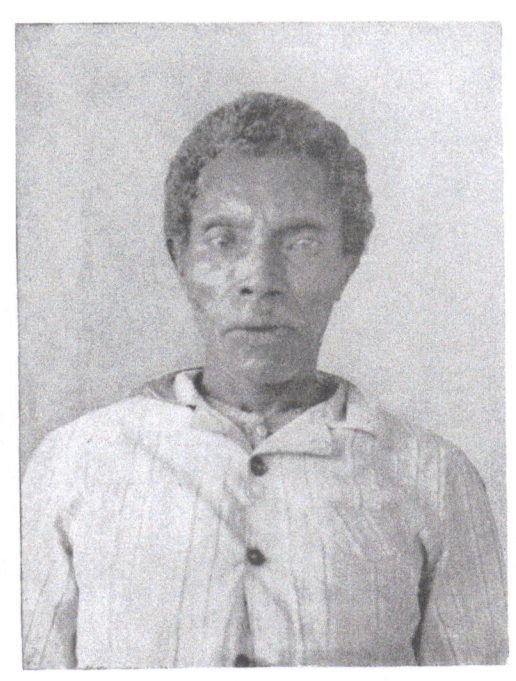
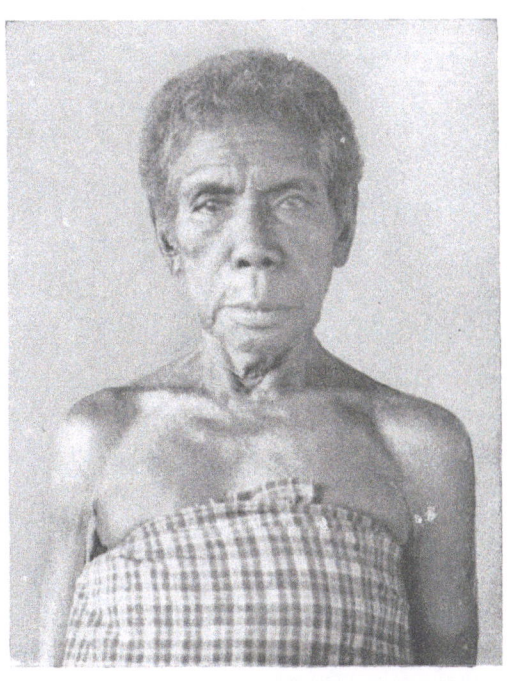

8. Josef of Ruma Waratanoi

9. Hanna of Ruma Waratanoi

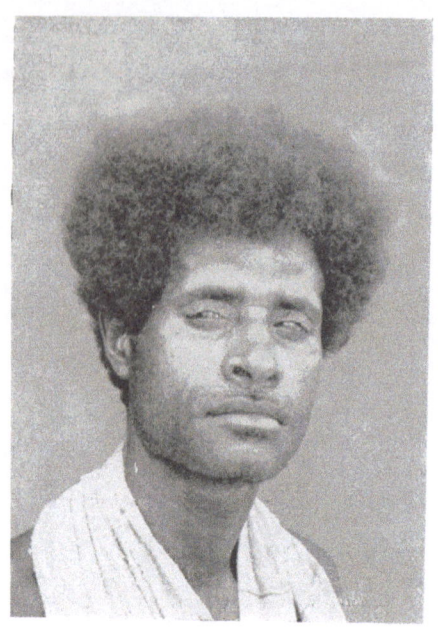

10. Man from Nubuai

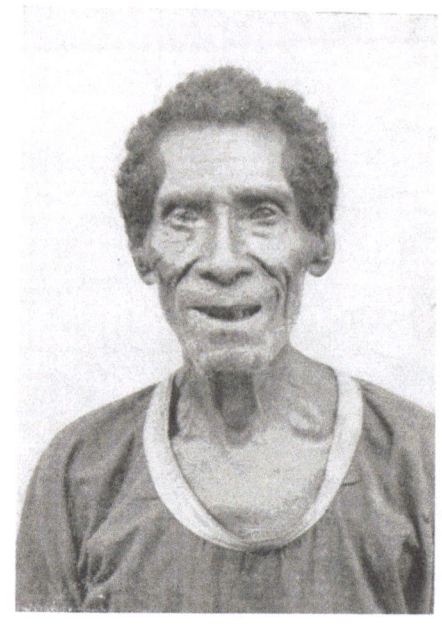

11. The oldest inhabitant of the
Waropen area

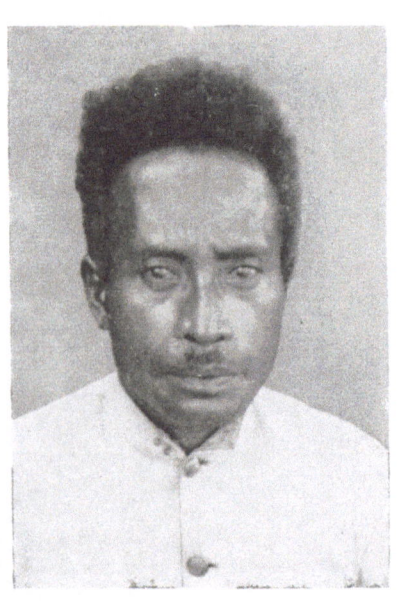

12. Muda of Ruma Gharami

13. Children's types

barter-value.[50] They also often possess a sarong to protect them against the cold. The younger children, both boys and girls, go completely naked. When about ten years old the boys are given clothes to wear, whilst the girls get a three cornered piece of coconut-shell (*ghaipiarei*), sometimes decorated with some bead-work and a few motifs.

The girls like to have a little box or a mirror at hand. The man one rarely sees without their carrying-bag (*rowu*), in which they have the ingredients for chewing areca-nut, a small mirror, a shell to pluck the beard, a spoon, and several more of this type of personal possessions. For tobacco or areca-nut or money they mostly use an engraved small bamboo tube or perhaps a separate little bag, finely plaited (*rerewino, finado, manggowusa*).

GENERAL IMPRESSION
OF THE BEHAVIOUR OF THE WAROPEN

As an individual psychological investigation falls outside the scope of our research, we shall have to content ourselves with enumerating some impressions which we give for what they are without any pretensions. It is difficult to give a description of the nature of the Waropen in general terms. In spite of himself the observer compares his own reactions to those of the persons observed, persons living in a completely different culture, following a different code and applying norms which differ from those he is accustomed to apply himself.

It is my personal impression that most Waropen are quickly moved superficially, but that nevertheless their emotions do not go very deep. When there is a direct cause, the reaction follows sometimes quite

---

[50] There are different beads which are given a certain barter-value.

kerefeisamumu,	flat, white,	value:	1 string	±	3 tumang of sago.
rawo nibe,	yellow, smooth,		1 —	±	1	—	— — —
ranggaghai,	yellow, ribbed,		1 piece	±	1	—	— — —
somadai,	white,		1 —	±	4	—	— — —
kitotomi,	speckled,		1 —	±	5	—	— — —
woai garo,	genuine old beads of various kinds;						
	often more than 10 tumang each.						

The value of the big *saparo,* shell-bands which can pass around the elbow, was put at 4 to 6 tumang, that of the common, smaller specimens at 2 to 3 tumang. However, they were also offered for sale to us for some tobacco. See also W. K. H. Feuilletau de Bruyn, Schouten- en Padaido-eilanden, (Schouten and Padaido Islands), *Mededeelingen Encyclopaedisch Bureau* (Reports of the Enc. Bureau), XXI, Batavia 1920, p. 172.

suddenly. In this way the women who remained behind when their husbands or brothers were leaving to go and work as contract-coolies, took their departure rather laconically up to the moment that the steamer left. At that moment there arose lamentations and desperate calls as if this were a separation for always.

In sudden anger people will sometimes maltreat each other rather seriously, but whenever there is a certain lag between stimulus and reaction e.g. when during a quarrel one first has to row towards one's opponent or one has to go and look for him, the fury shown has something artificial, causing those affected to take the dangerous threats with weapons, the cutting into the poles of the house and the yelling of curses rather laconically. During raids, the fury of the participants had to be roused by toddy before the action started.

The sight of a Waropen quarrel makes the observer fear the worst at every moment, but only rarely people let themselves be carried away completely by their emotions. Once I saw a man who was involved in a difference concerning the ownership of some sago-trees with an older woman, take up his oar with the evident intention of splitting the woman's head. I had become uneasy and asked the woman's son who stood looking at the quarrel from close by whether it was not time to separate the parties, but he replied laconically that the people quarreling stood in the relation of mother-in-law and son-in-law so that it was out of the question that the man would really hurt her.

During the frequent conjugal quarrels it is especially the women who make a terrible noise. A (valueless) piece of pottery is flung into smithereens in utmost fury; other pieces of pottery fly out of the house (to be picked up out of the soft mud or the water without much trouble afterwards); the walls of the conjugal room are kicked open; the leaf-thatch is torn from the roof — at first sight irreparable conflicts. Still, I once saw a woman who had let herself go as far as she could, viz. the destruction of the conjugal room, quieting down for a moment at the first word of admonition from another female inmate of the house, the roof over whose room she had started to demolish also by mistake. When she had determined the correct measure of the part of the roof which belonged to her, she continued immediately to destroy the roof over her own room, howling with fury.

Manners demand a greater calm of the men than of the women. A man will try to hide his emotions as much as possible. For the inventions of occidental technical science like wireless and airplanes the man only shows a shortlived interest. This does not mean to say that his interest has not been aroused, because it often appears later that a lively interest does exist.

The women are less concerned at making an outward show of quiet and dignity than the men. In public, women among each other are much livelier than men, and during quarrels the shrill voices of the women sound above those of the men. If during a feast there arises a difference about the division of the goods, the women usually start vociferously voicing their displeasure. One rather often sees the woman in extreme anger assuming the posture of the challenger in a fight, when with some weapon in hand and with shrill drawn-out curses one "dances in the way" of one's opponents (owaura), as the Waropen call it.

It is especially the older married women who act in this way. Younger women will quarrel less, but sooner burst into tears; in doing so they will also go further than a man. A young woman whose husband had destroyed her bamboo tube during a conjugal quarrel (I should add that this water-tube had belonged to her deceased father) continued to wail the lament for the dead during a whole night, mourning in turn her father and the broken bamboo tube.

Very often certain ways of behaviour do not find their explanation in the individual predisposition, but in social codes. In general the Waropen has a violent aversion to running blood. His whole treatment of wounds is limited to plugging the wound with leaves, shells, paper, etc. If it is sometimes necessary to open an abcess, this often arouses an emotion wholly unrelated to the stimulus. It is not the pain which is particularly feared, because in themselves the Waropen are also less susceptible to bodily pain than Europeans. Against our admonition to kill animals for slaughter with one quick blow the Waropen place their opinion that this would be cruelty towards the animals. A boy of about eight years old grew angry at a cat which had eaten his fish. Rapidly he decided to hang the animal, without one word of protest of the by-standers. Our suggestion that domestic animals which one cannot or will not keep should be killed quickly obviously aroused aversion. A quick dead is considered just as gruesome as an open bleeding wound.

It would be incorrect in my opinion to call the Waropen cruel in view of information of this kind. This behaviour is determined in the main by their ideas concerning illness and death. Other behaviour like-wise should rather be explained from the completely different culture. Gratitude in our sense of the word will not be found in this potlatch-society. The gift is accepted without a word of thanks. It would be highly improper to show how much one enjoys the present or how much one needs it. This does not mean to say that therefore there would be a complete lack of appreciation. If the giver states frankly what he wishes to receive in return, nobody considers this unseemly. Small friendly presents like some tobacco or a few areca-nuts a man of position may ask without damage to his prestige, whilst of people of lower status this would be considered as begging. By means of a present which is excessively large according to Papuan standards, or which is not offered in a dignified fashion, one insults the recipient. For the investigator the giving of presents is a delicate point which he has to watch carefully.

The not very strong emotionality is accompanied by a rather great activity. The Waropen only rarely sit dreamily gazing into space, but they are usually busy with one thing or another. The men, when at home, always have something to tinker on a prahu or their fishing gear; when not engaged in their household, the women usually have some wicker-work near at hand. In contrast to this there is the fact that most people will quickly take up some piece of work, but often lack the tenacious perseverance to finish work once begun. Fishing as done by the Waropen demands a great deal of immediate activity, but no care which returns day by day such as is needed for a slightly developed agriculture. The men prefer to leave the women to do monotonous work which demands perseverance, like beating sago. The men's task is limited to preparing the tree for beating, but then they let the work lie. When building houses, the gathering of the materials takes endless time. Then the house is suddenly erected by as large a body as possible, but then again the finishing touches are long in coming. Numerous houses in the village never actually reach the stage of completion. Also during our investigations we had to count with this quickly stimulated but rather soon exhausted activity, by constantly changing the subject under discussion.

A curious phenomenon, which may perhaps fit in with this generalising picture of a superficial emotionality and activity, is a certain lapse in thinking which causes fluctuations in the attention. It repeatedly occurred during an investigation that for no evident reason all interest for the subject had suddenly disappeared. The informants had sudden lapses which temporarily rendered them completely incapable of actions which they normally were able to perform. A simple action performed several times without mistakes seemed suddenly completely unknown. If one forcefully insists, the confusion only grows, but usually a short rest is sufficient to allow one to continue in a normal fashion.

A phenomenon which may be connected with the strongly implicit nature of different cultural traits (see p. 40 below) is the lack of skill of the Waropen. Their tools for processing sago, like beaters, spoons, bowls, etc., are often carved, as are the decorative pieces for the prahu, the *ruwa*, the neck-supports and the small bamboo tubes for tobacco etc., but the motifs are executed in a slovenly fashion and they are endlessly repeated. The women have a primitive technique for basket-work and formerly they produced all kinds of bead-work. It is in this bead-work that they often achieved pleasant colour-combinations, but in general neither the men nor the women possess many technical abilities and even in this limited field they show little skill.

Their power of observation is trained in a certain direction through daily use. At a great distance they recognise rowers by their style and prahus and ships by their form and rigging. They observe the presence of fish by even the slightest movement of the water and often they knew, so to say by intuition, the kind and the size of the fish. To the contrary they have trouble in reading a photograph if it is not very large. Such pictures are studied from all sides until they really see what is on them.

In daily intercourse the Waropen are frank without immediately becoming presumptuous. Only children will occasionally go beyond the proper limits and they do not obey easily. They are, however, rarely malicious and although children everywhere stand in front, they do not trouble anybody with mischief. Children simply pay no heed to the adults and if need be they do not hesitate to strike back when their elders give them a punch or a blow on the back. Adults do not like their children to cry and they take little trouble at active, conscious education. Social pressure is evidently so strong that practically every person still becomes sufficiently trained for community life by means of the daily work. No power in the world is able, for instance, to compel children to take medicine if they do not want to. The extreme is to threaten them with a piece of a fantastically formed fungus, or with the "white man", or with the police, who have to assume the function of bogeyman among the Waropen.

However, at a rather early age the children already know how to behave. The girls have anyhow less time to play outdoors than the boys, because they have to take care of their smaller brothers and sisters and to help the women with their work; boys over ten or twelve years start to join the adults. Boys over this age are in a certain sense counted among the adult and the married men. Only men of middle age or older constitute again a more or less clearly separate age-set above them.

For the girls things are slightly different. They can remain real children for a much shorter period; soon they have to start helping the adults in their work. Nevertheless, they do not so closely join the older women who have already had children, but form a separate stratum of younger girls. The young women pay great attention to their appearance. They lead the way at the dances, and only later, when they too have had children, they gradually pass over to the group of older women whose main activities consist in care of the gardens, processing the sago and supervising the ceremonial barter at feasts. When the Waropen speaks about the *waribo,* he means all males who are not either children or old men, hence including the married men. When he speaks, however, about the *wiama,* he especially means the young women, in a greatly extended sense also the younger girls and the young married women. A rough estimate in age-limits would be therefore that *waribo* refers to men between 15 and 35 years, whilst *wiama* refers to the women between 10 and 25. From this it follows that women are considered old sooner than men.

All these groups follow a social code of their own, which exercises the least pressure on the young children. The *waribo* should be ready to make distant journeys and to follow the raiding party. They perform the actual labour in the village and it is this group which seeks the closest contact with modern times. The older men are the makers of plans; they often take the initiative for undertaking long journeys, building houses, etc. It is to them that the younger men listen when the rules of life are discussed. In daily life this gerontocracy is not very noticeable and sometimes the younger people express themselves with great contempt about the older ones who do not wish to move with the times. Nevertheless, they consider it sufficient reason for explaining the motives of their actions to say: "This or that has to be like this, because otherwise the older men become angry at us, the younger men". The older men are usually more dignified in their bearing and do not come forward so readily.

The *wiama,* i.e. the young women, are noisy and inclined to giggling in their behaviour. However, they do not venture to press forward so soon as the *waribo,* and when so they are more easily forgiven. The older women, on the contrary, are always busy with their children or with the household. These differences in behaviour were typically shown during our investigations. The children and young women were practically of no value as informants. The young men showed the greatest interest in the investigation and functioned as daily helpers. Still, they had to obtain most of their information from the older men, who are usually willing to make pronouncements on social institutions, but prefer not to go to any great trouble in doing so. The older women are the best informants on mythology and family affairs. However, they are only rarely to be persuaded to join in the kind of instructive conversation one may have with the old men; they return to their work as soon as they have given the desired information.

It is therefore impossible to establish a uniform rule for the behaviour of all Waropen. What is acceptable for one person, is disapproved in another. When men make water, even in the vicinity of women, nobody will consider it as objectionable, but nobody can break wind or defecate in public. Women perform all natural functions in hiding. Everybody may spit or cough or sneeze or yawn as much as he wants to, as long as he does not disturb anybody. Hiccups are considered slightly objectionable, whilst vomiting is rightout ridiculous. To eat with smacking noises is considered as uneducated. But for the rest an old man, for instance, need not be as careful about etiquette as a younger person. An old man may ask somebody for some tobacco, because he is already so old, and a young girl may do so jokingly, because she is still so young, but in actual fact such things are considered as not quite fitting. Just like among some tribes of Indians, the Waropen often indicate something by pursing their lips.[51]

Also in mutual intercourse the Waropen maintain these restrictions. People will still accept a remark by an old influential man. A remark made by a young girl is not taken quite seriously, whilst children hardly count at all, but for the rest people are careful and polite with each other in their intercourse. One avoids plain speaking or exposing somebody to ridicule. Is quite permissible to make fun of a person who is blind or who stutters or who has some physical defect, as long as the person in question is not somebody to whom for some reason consideration is due. For the rest, the people show enough patience with deformed or crippled persons, even though they freely laugh at them.

---

[51] See R. H. Lowie, *The history of ethnological theory*, p. 262.

VILLAGE LIFE

If one were to expect to find among the Waropen, and among the inhabitants of Geelvink Bay in general, a society which would be fascinating in any case because of outward originality and colour, one would be sorely disappointed. Judged by its outward form, the civilisation of Geelvink Bay is definitely poor and monotonous, and it produces an impression which is quite different from that of e.g. the Marind-Anim in Southern New Guinea, described by Wirz.

As we saw already, the Waropen present very little of a romantic nature in their outer appearance. On their bodies they have neither fanciful finery nor complicated painted designs. No fantastic feasts, no impressive temples. As the villages lie there on the silent water, with their leaning houses built on poles, and a fleet of rickety prahus, one notices little of busy life and excitement.

Not only outward, but also the inner picture of Waropen society gives this same impression. The higher stratum hardly differs from the common man, so that up to the present the existence of classes went unnoticed. In religious ritual the supernatural objects are so inconspicuous, that it may take quite some time until they are discovered. The initiation-demon who among so many other people has been made into a striking figure, is found here as an unsightly roll of sago which is only noted among the other provisions at a feast when the informants have pointed it out.[52] Various lesser ceremonies are performed with hardly any outward show, whilst in the danced myths the mythical figures hardly ever appear. The people dance and sing evidently without completely understanding the myth they chant. Masks are unknown.[53] The Waropen village sanctuary, the *dama*, which we shall discuss below, did not even possess the carved poles which support the Numfoor *rum sram*. The ancestor-images, well-known among the Numfoor and the Biak, only occur sporadically here.

For these reasons it is quite possible, that many things in Waropen culture have escaped observation. The custom exists, for instance, of hanging a piece of gaba-gaba in the trees close to the creeks, when the sago palms have reached the stage that they can be processed. Once it was

---

[52] I do not know in how far the impersonation of the initiation-demon during the festivities at Manokwari in 1938 was due to alien inspiration; cf. "Slangenfiguren in de Geelvink-Baai" (Snake-figures in Geelvink Bay), in *Cultureel Indië*, II, pp. 138 ff.

[53] However, I did see a wooden mask *(sakaiba)* worn during marriage festivities in the vicinity of Manokwari.

observed that these pieces of gaba-gaba had been carved in the shape of an ancestral image, from which I would conclude that this custom is definitely religious in origin. When one considers, moreover, that the Waropen show little religious emotion, it must be clear that village life is quite humdrum and ordinary.

Still, in Waropen culture one finds implicit the same characteristics which we also know from other cultures. The initiation ritual which is explicit among other peoples and therefore more conspicuous, is in fact also to be found here, showing many of the general characteristics of these rituals. Just because the typical characteristics of many Melanesian cultures, like the place of the clan-sanctuary, the ceremonial co-operation among the clans, etc., were so strongly implied in daily life, so little had become known of the Papuan cultures in Northern New Guinea, in spite of the devoted labours of several missionaries.

In the following chapters it is especially the cultural aspects discovered during my research which will be discussed. The concatenation of descriptions of feasts and ceremonies and institutions risks obscuring the implicit character of these occurrences. Still, I believe that this "implicitness" should be considered as an important trait of Waropen culture. Against the background of ordinary daily life these feasts etc. do not stand out very strikingly, just like an informal family celebration contrasts much less with normal family life than an official party, although in both cases it may concern a marriage, the social consequences of which are equally large.

Daily life in the village flows along in an even, humdrum way. The work which the people do provides little variation. The fishermen perform their task with very simple tackle nearly always quite close to the shore and they run hardly any risks. The women's work in the sago-forests demands more perseverance than insight. Practically the only element of real danger which might provide some stimulus, viz. the permanent threat of a raid, has likewise practically disappeared in modern times; in this way the mutual quarrels and the voyages — where, for that matter, the Waropen show themselves far less intrepid than the Biak — are the principal sensations they have. It is curious that in spite of this monotony the Waropen are still a much gayer and more cheerful people than the Biak. In the villages there is frequent and hearty laughter. The people in general are not melancholy, and evidently the emotions of village life are quite sufficient for the Waropen. Most of them in the long run feel ill at ease outside their village and only leave for the plantations in large groups together. There is little of a spirit of adventure among the Waropen, particularly among those of the Kai area. They prefer to be snugly and calmly among their own relations, in their own village.

The climate does not make a clear division between the seasons. Their work brings little variation. Every morning, towards sunrise, the damped down fires are stirred and one villager after another comes shuffling along, shivering. Some continue to sleep for a while, because the Waropen are past-masters in the art of sleeping; they sleep as well in a room on a mat as on the outriggers of their prahu over the water, and equally well in the night as during day-time.

The first to come into action are the old women. They collect the tools for working the sago, they open the slats in the floor and let themselves down into the prahus which lie underneath the houses during the night. Rinsing the mouth with some river-water is the highest demand as regards the morning toilette. People will eat if there are still some leftovers of yesterday's sago-mash or a sago-cake with some fish; if there is no fish at all, people prefer to wait before they eat, so that they often have their first meal towards noon.

The men only start working when the women have gone, unless their presence is required in the sago-area. Their fishing is controlled by the movement of the tides. Traps can only be placed and the nets laid out when the tide is about to come in, because they want to catch the small fish that are carried back to the sea by the ebbing tide. Their seros and nets are generally not higher than five feet and because the difference in the tides is at least seven feet, it follows that the time available for this type of fishing is only a few hours out of every twenty-four, apart from those few days every month that one could place two seros within twenty-four hours if needs must be. Neither is it possible to fish with line, spear or poison with every tide, and it is only rarely that people go far away from the village for this purpose. And of course, the women can only go and gather molluscs in the tidal forest at low tide.

As a rule we may therefore assume that everybody who has work to do outside of his house will start working in the early hours of the morning. A number of people of high rank do not bother greatly about working. About ten o'clock several people come home again, e.g. the men who have been fishing close by, or the women who had only little to do in the sago-grove, or who had only gone out to collect fire-wood or material for plaiting, or something similar. Women with babies, the very old and occasionally the unmarried young women, do less work outside. They always have some work at hand, weaving mats, preparing fish or sago, or simply doing nothing.

There are men who spend long days in making prahus and who therefore might pass as specialists in this field. The division of labour has not progressed so far that these men have been freed completely for their special work, but nearly every day their houses resound with the familiar knocking of their adzes. Then there are men who particularly occupy themselves with carving and who may also be considered somewhat as specialists; many others pass their time with making fishing-tackle, making nets, etc.

Towards noon, activities slacken noticeably. Most people take their meal at this time and then go and sleep for a few hours. Only several hours after mid-day, activities increase again. The fishing prahus leave again, others go into the nipah-forest to collect the tubes with palm-wine, hung there earlier in the morning. Girls and children collect fire-wood and water and towards dusk one sees most of them going home, both the group which went out early for a full day's work (i.e. especially the women and a few men who worked in the sago-forest) and the second group who only had to do an afternoon's work.

Then follows the pleasantest hour of the village-day. The housewives stir their fires. From every roof there flutters a light-blue plume of smoke. Amorous young folk go boating on the water in front of the houses amidst shouts and laughter. From different sides voices call the family members home. People sit down to their evening meal, practically the sole fixed daily meal they know. During meals people are vaguely grouped according to families, but often the men sit together in small clusters; this is even the rule when there are guests. Then toddy is handed round and people chew areca-nut and tobacco. From all houses there arises a hubbub of voices; here and there a dog whimpers. A single man beats his drum for a while, but when darkness has fallen completely, silence follows soon. Occasionally the glare of a bamboo torch prepared with resin flares up in the pitch-dark village —, people who still move around in their house or who have to pay a visit —, but for the remainder the village is now dark and silent, until the next chilly dawn with its activity of leaving prahus.

Often the monotony of this kind of life is interrupted by a feast, more often of course in the larger villages than in the small ones. Then the thumping of the drum and the singing of the male and female dancers resounds for days on end. There are innumerable occasions when one may beat the drum, quite apart from important moments in the life of the family, such as deaths and burials or birth and sickness. There is not the slightest reason "to do things quietly". On the contrary, the chance to do something special for a day or for several days is grasped with both hands. As I explained above, one may participate in everything without let or hindrance; fasting or spiritual exercises are unnecessary, tiring dances with many taboos for the officiants are unknown. The participants may take part as much as they like and with as great an exertion as they please.

And then there are the voyages. The Waropen likes to travel often and he does it with pleasure, and nearly always we find people from elsewhere in the villages with whom one may pass a night chatting. The Waropen is, however, far from being an intrepid explorer who seeks danger for danger's sake. A quiet sea, a calm wind and a fair voyage to a relative in a village not too far away, that is the ideal kind of travelling in his eyes. The custom that brothers have to go and help their sisters to process the sago when they have married into other villages, causes a great deal of travelling about. And even so, people do not hurry and do not have any important objects. In order to get some areca or to bring some sago or to assist at a small feast, a prahu often sails out cheerfully, to return later even more cheerfully with joyous songs, happy to re-enter the well-known waters on the journey home.

In this way daily life is quiet and free from shattering emotions, of an even tenor, like the creeks in the tidal forest. No epidemic, no contact with alien peoples has been able thoroughly to disturb this peace. What will be the reaction in the long run to the increasingly intensive contact with the West remains a question which the future will have to answer.

CHAPTER ONE

SOCIAL ORGANISATION

BIPARTITE CLASSIFICATION

Between water and land, between sea and forest, the whole of Waropen life runs its course. The sea opens the way towards the outer world, the land runs into the unknown mountain regions, where the inland tribes and the magicians dwell, a region of which most people only have hazy ideas; there Mount Ghamusupedei is situated, believed to be the dwellingplace of various mythical beings. For the Waropen, the contrast between sea and land is of the greatest importance, not because he takes pleasure in contemplating two-fold divisions in nature, but because it is on the sea that the men go fishing, and on the land that the women work on the sago. Land and sea complete each other, like a meal is only complete when sago and fish are combined. As is often the case in a division of labour between the sexes, it is also here that the most monotonous work, viz. processing the sago, is given to the women, whilst the men devote themselves to fishing. This division is strictly maintained; the man may help in cutting down the sago-palm and splitting it, the woman may assist in fishing with the sero, but that is all. Beating the sago is as little a man's work as fishing with the spear is a woman's task.

Sea and land not only determine the division of labour between the sexes, but also the indication of places. The Waropen do not use the four quarters, but they only consider the image of the creeks in the tidal forest where they build their villages and where the women always work upstream, i.e. in the forest, and the men downstream, i.e. on the sea. The land is *re* and the sea is *rau*. They will rarely omit to indicate whether a movement is conceived as being towards the land or in the direction of the sea. The man who from a point further downstream calls a friend to come to him, will call: *"Awo ma ghero!"* (Come down!). Instead of the contrast land — sea, people likewise often use the terms above, *sa*, and below, *ghero*. In these cases the upper course of the rivers, *ghaidouri*, is always seen in opposition to their lower course, *ghaidoghairo*. People live *ghairorisa*, on the downstream side, or *uririsa,* on the upstream side.

This contrast is maintained also outside the village; in that case the coastal area extending eastward is considered as the highlands or the interior, whilst that stretching southward is considered as lowland. The region towards the East is called *Kaipuri,* where we find the word *furi,* behind. When in Waropen Kai the term *Risiwako* (from *risi,* side and *wako,* below) is used, this usually indicates the Napan area. And when the Napan people speak of *Risifono,* they mean the direction of Waropen Kai, whilst *Risiwairo* indicates the direction of Wandamen Bay, especially the village on the Woisimi. The uplands are also occasionally called *Risifono.* Hence, if somebody from Nubuai says: *rawo fono,* I row towards the back, he means that he is going in the direction of Kurudu. The world known to the Waropen practically does not extend any further than Kurudu.

Fono is likewise the back of the house which faces the river; behind the house is the forest, in front the water flows. Often the river describes such curves that the front of the house faces East, but even so the quarters do not count, because the only thing of importance is that the water is considered as the front and the land as the back. The eastern bank of river is not distinguished either from the west bank as the low side and the high side; one only speaks of the upper and the lower course, or of land and sea.[54]

This antithesis between land and sea, between above and below, is likewise the basis for a mythical bipartition. The forest, the higher side, is associated with the women; the sea, the lower-side, however, with the men. It is said that upstream and quite close to the last houses of the village a female sea-monster is to be found, and a male one close to the first houses downstream. I could not understand clearly whether the people suppose that this is one single being, male downstream and female upstream, because the information on this point was rather vague.

The Waropen villages are usually not situated immediately on the mouth of the rivers, but some distance upstream. Now the burial-grounds

---

[54] In his manuscript Numfoor dictionary F. J. F. van Hasselt writes that the Numfoor distinguish "*swandirwuri,* litt. the head of the beach, or *warwuri;* — this term is applied to the whole of the eastern (presumably 'western' is meant) part of the main coast of Geelvink Bay" —, and "*swandiweri,* litt. the leg of the beach, everything situated West of Doreh Bay up to Sorong". [See *Noemfoorsch Woordenboek* (Numfoor Dictionary) by J. L. and F. J. F. van Hasselt, 1947, p. 229, svv. swandĕrwuri and swandĕrweri, Ed.]

of the dead are mostly found below the village, although this rule also has its exceptions. Close to the river-mouth they imagine the dwelling-place of a mythical being which assumes the form of a sea-eagle (*manduko*) at sunrise, to change into a mythical sea-creature (*rina*) at sunset, a creature occasionally thought of as an ordinary fish, but usually as a winged ray. The winged ray which moves its fins up and down like wings when swimming, is somewhat reminiscent of a bird. The Malays also call this animal "bird-ray". So, before arriving at the village we find near the burial-ground for the dead, where vague ideas likewise locate the village of the dead (*inggoinu*), a winged double creature. And behind the village, in one of the creeks of the sago-area, there lives another double creature, viz. the bisexual snake, which we will discuss more fully when treating of the initiation ritual.

The geographical image of the village is therefore continued into the mythical world without transition. Geographically the village extends from the river-mouth on the coast to the creeks in the sago-swamp. From a mythical standpoint this area is occupied for the lower part by the birdfish and for the upper part by the bisexual snake. Below there are the burial-grounds for the dead, above one finds the houses of the living. In complete conformity with this idea, mythology explains that originally the women lived in the forest and the men in a men's house on the sea-shore. In Waropen the term *nu* (probably related etymologically to the Malay word banua, world; cf. also Numfoor *manu*, like in Manokwari, lit. old village) therefore not only serves to indicate the built-up area and the sago-grounds, but the whole of the area in-dicated here.

This parallelism between mythical and geographical opposites which will be discussed more fully below had to be mentioned here, because otherwise the arrangement of the clans would not become clear, as the clans are not only grouped together geographically, but also mythically. Without some general idea of the mythical grouping based on the geographical arrangement it is impossible to describe the relations in rank between the different clans.

DIVISION AND ORDER OF THE CLANS

The Waropen villages of the Kai district are divided into two, four or five *da*. A *da* is a local, non-exogamous group formed by a number of *ruma*, houses, which, however, are exogamous. Although the *da* is not concerned with the marriage regulations and although for this reason the strict definition of a clan, which demands traditional one-sidedness, does not apply, I shall continue to translate *da* by clan, also because ethnology does not possess any better term; in so doing I follow the example given i.a. by Professor de Josselin de Jong.[55] For usually a clan organisation constitutes not only the basis for marriage regulations, but also for social co-operation, mostly likewise of a ritual nature. And although there may not exist any relations between the *da* and marriage regulations, its connection with the ritual co-operation between groups is quite clear.

This is shown in the first place by the distinction "above" and "below" which also applies to the *da* and because of which the *da* at the lower end of the village occupy a more prominent position than those at the upper end. Nubuai village is divided into two parts by a line which runs from the Mambuighaido to the Ghaifoni; this causes the clans Apeinawo and Kai to have their position downstream and those of Nuwuri, Pedei and Sawaki to have theirs upstream.[56] The Kai clan, whose members mostly keep slightly apart from the other Nubuai people, consider themselves as the most important. It was their ancestor Gharopendi who descended from heaven to cut open the Nubuai river from the sea up to the dividing-line on the Ghaifoni. His daughter married the clan-ancestor of Pedei. According to tradition the clans Nuwuri, Pedei and Sawaki opened the Nubuai river starting from the interior and so in fact they are originally men from the interior. The clan-ancestor of Sawaki, Maseiori, was actually a slave of Gharopendi. Members of the three inferior clans occasionally admit that properly speaking they are men from the interior, but this does not prevent them as inhabitants of the coast from looking down on true people from the interior, like those who live in Saifoni village, a strongly waropenised village of people from the interior.

It seems that the superiority of the Kai — after whom formerly the whole village was often called — in ancient times also assumed other forms. In this respect several informants assured me that this clan had possessed the monopoly of making and using the *faiano*, fishing net, and the *ea*, sero, which others could only use with the permission of the Kai, or against payment, so that the members of the Kai could assert extensive "seignorial" rights on the most important method of fishing. The members of the oldest house of the Kai also claim a particular type of ritual lament which will be discussed below (p. 189).

---

[55] J. P. B. de Josselin de Jong, *Studies in Indonesian Culture*, I, p. 5.
[56] See the sketch of Nubuai village on p. 49.

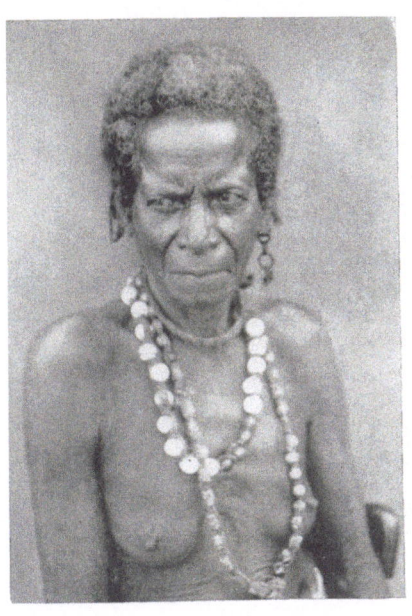

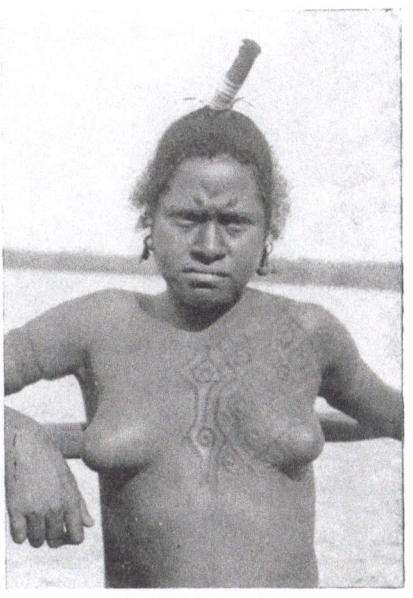

14. *Mosaba* with valuable beads   15. Tattooing half completed

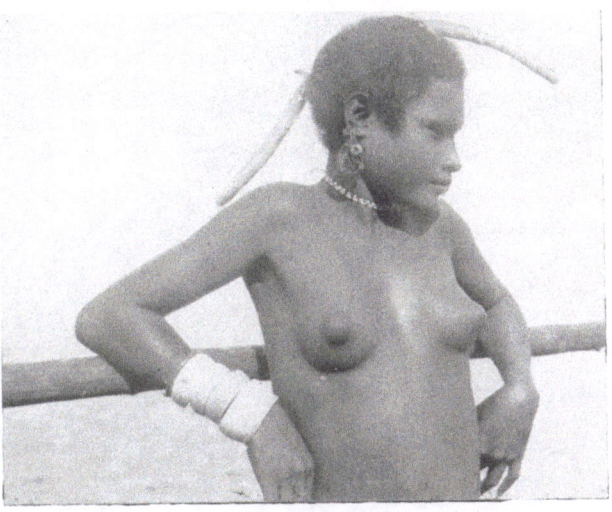

16. Hairdress

17. Old-fashioned front part of a house

18. The Apeinawo clan (Nubuai)

The ritual co-operation of the *da* becomes clearly evident from the curious grouping in pairs of these parts of the village. In Nubuai village Apeinawo – Sawaki constitute one couple and Pedei – Nuwuri the other. Members of such a couple are related like the "head" and the "tail" of the triton, *buriworai* and *buriferai*. Apeinawo is therefore related to Sawaki as *buriworai*, the head of the triton, to *buriferai*, the tail of the triton. The "tail" is the conical tip which is blown, whilst the "head" is the mouth of the shell.

Here follows the schematic indication of the grouping of the clans in a few villages; the arrow connecting the clans points from the "tail" to the "head".

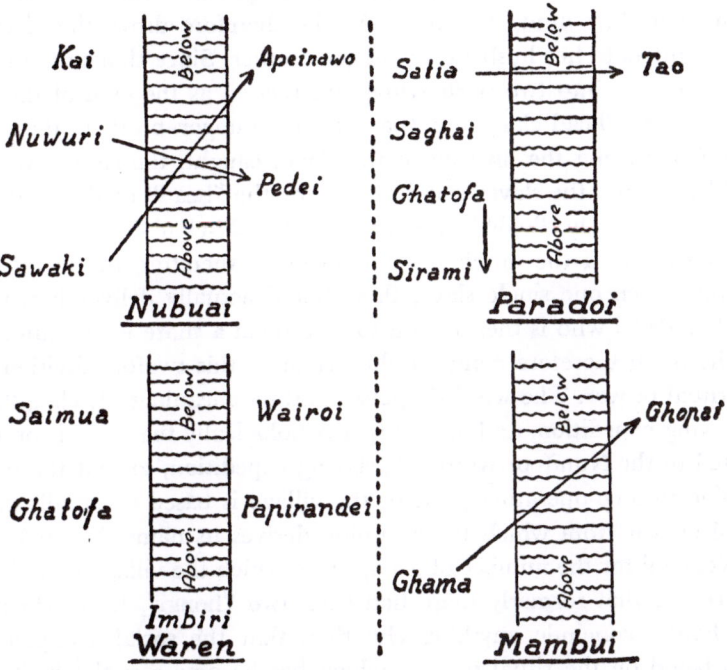

The relation between the "head" and the "tail" is one of rivalry, the one clan always glorying in always being braver and stronger than the other. As late as 1936 there was a bloody fight between Sawaki and Apeinawo due to trouble over women. In their songs the members of the clan boast of the greatness of their clan. The Apeinawo have made a long journey to the Mamberamo river and have left a sign there intended to taunt the Sawaki to go and fetch it. The Sawaki people in their turn have left such a sign at Manokwari where they went once when the chief of their clan was in prison there. This rivalry causes a certain amount of irritation which the informants do not like to stimulate by providing information. My first data on the grouping into "head" and "tail" I obtained from an old man inside his own house. For this reason I was only able to note the arrangement in a few villages, although it was always admitted that the grouping as such did exist.

The myth concerning the origin of this arrangement (which is the same in Mambui and Nubuai; Texts 176, 177) relates that the present component parts formerly formed one whole, but that the wife of the oldest member of the senior branch became tired of the many feasts she had to arrange for her husband and his companions when they returned from their slave-hunts (cf. p. 209). She therefore drove the victorious slave-hunters to her husband's younger brother. Since that time the clan has been split into two parts which are related as the clan of the elder brother, the "head-clan, and that of the younger brother, the "tail"-clan. Concerning the division of the slaves caught, the ruling was then established that the slaves caught either by the "head" or the "tail"-clan should be equally divided and that the "head"-clan should obtain one half plus one, if the number was uneven. If therefore the members of Nuwuri catch one single slave, they should actually deliver him to the head of Pedei who is then bound to give them a share in the ransom.

The myth therefore connects the origin of this twofold division with the ritual of war. The word *da* possesses three meanings: 1. clan, 2. raid or slaving expedition, and 3. enemy. Etymologically the word is probably related to the Numfoor word *rak*, slaving expedition, so that the present division into co-operating parts of the village is based essentially on the ritual of war from which the grouping derives its name. When I asked in Weinami for the number of groups into which the village was divided, I was told that formerly there had been two "hongi", from which one can hardly conclude anything else than that the social grouping had been based on the ritual of war, which has become a potlatch-ritual in the wider sense. The organisation of the *da* is not that of a number of war-groups inside the village, but a division of the whole of the village population. The linking of the clans mentioned in the myth starts from unity and rivalry, especially in economic-ritual relations.

In the village history of later years this rivalry on an economic basis is no longer so conspicuous, because the Government was forced to bring this matter under control. In the old days this relationship between the clans was not confined to the ritual of the raids, but it extended to the whole field of ritual, as is still shown by the position of the head of the clan. In order to understand the general nature of the grouping of the clans we have to go outside the area of the Waropen in order to study the organisation of the Numfoor.

In Doreh Bay four Numfoor clans are arranged in couples as *rwuri*, head, and *purari*, tail, just as among the Waropen. Now in this Numfoor group the component clans possess one single clan-sanctuary, the well-known *rum sram*. The relation between head and tail in this case does not refer to the triton-shell, but to this sanctuary. The "head"-clan belongs to the "head" of the *rum sram* and the "tail"-clan to its "tail". In this case there is not only co-operation between the "head" and the "tail" (when building the clan-sanctuary), but also between the "head" of one couple with the "head" of the other, or between the "tail" of the one with that of the other. If e.g. repairs have to be made at the "tail"-end of the clan-sanctuary of couple A, the "tail"-clan of couple B has to assist likewise. In the same way the "heads" of both couples work together. Now, in the Numfoor ritual we also observe "head" and "tail" in a certain part of the initiation-ritual, where the snake of initiation is represented by a length of rattan. The "head"- and the "tail"-clans place themselves at the head and the tail of this snake, represented by a piece of rattan, and start a tug of war; in this case the outcome of the contest has the nature of a prognostic.[57]

In this ritual, "head" and "tail" function in exactly the same way as the component clans, who elsewhere are also occasionally connected by a fixed relation of connubium. In the initiation ritual the "head"-clan and the "tail"-clan are related like the head and the tail of the initiation-demon, or like the "head" and the "tail" of the clan-sanctuary, like two rivals in the same indissoluble relationship which we know since long from the image of the rivalising phratries.

In my opinion the curious grouping of the Waropen clans in the head and the tail of the triton-shell follows the more obvious example of the Numfoor. In a mythical sense the triton equals the young men's house and this again equals the initiation-demon. The rivalry of the clans is not only concentrated on the ritual of war, but on the whole ritual of initiation in a wider sense. Therefore the myth tells us succinctly that head and tail of the devouring crocodile change after its death into two clans (Texts 28). In actual fact this is exactly the same

---

[57] See "Slangenfiguren in de Geelvink-baai" (Snake-figures in Geelvink Bay) in *Cultureel Indië*, II, p. 138.

as the explanation found in other myths, where the mother who has been killed is divided among the different *ruma* (house, branch of a family), whereupon society comes forth from her (Texts 95).

As a local group the *da* still exists only in the Kai district and, as may be seen on the map of Nubuai, even there the *da* is about to lose its nature as a local group, because the houses of one clan are also being established within the area of the others. In Weinami the *da* no longer exists as a local group, but the people there still knew that formerly there had been two *da*, i.e. only two groups who co-operated in waging war. These groups were Warami, with the houses Saiori, Sakuatorei, Kawiai and Romboifai, and Waratanoi, with the houses Wengge, Samberi, Marei and Samisanoi. Waratanoi, said to have been founded by the well-known mythical figure Waiseri or Kurisera, was the "head" and Warami was the "tail". According to my informants, Weinami and Napan formerly formed one double village, like Risei — Saiati still do at present. Makimi village was said to have consisted of one single house, Mandatanoi.

However, it seems that in this area the *da* already had lost its character of a local group as far as people's memory went, because according to my informants even in earlier days there had already been houses which had not joined a *da*, e.g. a house called Aiomi. At Weinami the word *da* was simply translated as "hongi" (raid) and people denied that the *da* should ever have been a part of the village. The only conclusion to be drawn from this is, I believe, that in this village the organisation of the *da* was only that of a war-league and no longer a division of the whole village. Also in the present village of Ambumi the word *da* is only known in the sense of "slaving-expedition"; a *da* is formed by a number of prahus, combined for this purpose. The ancient village on the Woisimi was moreover not divided into a number of *da*, but only into a number of *ruma*.

The present-day village of Ambumi consists of three parts, Jerinusi, Ambumi and Karami, each under a chief of their own, but this organisation is artificial, being due to a combination of parts of the ancient village on the Woisimi which had been settled on the islets of Jerinusi and Karami due to the influence of the Administration; a fourth part of the ancient village is constituted by the village of Wamori. In these four fragments of the ancient village the old *ruma* have been absorbed in such a way that the people were unable to indicate which *ruma* belonged to each of these four villages or parts of villages. The division into these four villages — which again form four local groups — is such an arbitrary disruption of the ancient traditional genealogical division into *ruma* that every individual is more or less uncertain, whether he must consider himself as a member of the Government-village to which he belongs, or as a member of his old *ruma* which has been scattered over the different villages.

However, it is certain that a bipartite division also existed in the ancient village on the Woisimi. What else could be the meaning of the myth which tells us that after the deluge two brothers in a prahu were left, who became the ancestors of the village? To remove any doubt, as it were, then the marriage is mentioned of "a girl from the back of the prahu" (Texts 22, 23). Also among the Numfoor the prahu is the starting-point of the social organisation. Some vague historical memory of such a twofold division has also remained, as the people of the present village of Jerinusi — mainly formed by the ancient *ruma* Marani, Samberi and Wokuai — were still able to tell me that anciently Marani was "below" (*wairoi*), whilst Samberi and Wokuai were "above" (*uri*). Although these indications are not correct geographically as far as the present-day village of Ambumi is concerned, a certain relation in rank seems to be connected with them. Samberi and Marani are said to have been closely connected in former days.

In the Kai district the clan therefore still continues to function as the base for a rivalising bipartite division; in Weinami this only lives on in historical tradition, whilst in Ambumi it has even disappeared from the latter. To my mind this series reflects a sociological shift. In the Kai district the grouping into clans has proceeded farthest, in Wandamen Bay we find the simplest form and in Weinami an intermediate stage. Perhaps the grouping at Kai is at the same time the most old-fashioned and, from an ethnological and historical point of view, the oldest, whilst that in Wandamen Bay is the youngest. In Wandamen Bay influences of other tribes as well as those of modern times have made themselves felt more strongly than at Weinami, where in their turn they were stronger than in the Kai district.

An important point which demonstrates again and all the more clearly that the *da* actually is a clan is the consciousness of having descended from one and the same clan-ancestor who sometimes is an animal; in this way a tortoise is the ancestor of Pedei and of Nuwuri at Nubuai (Texts 11, 12). In daily life the feeling of relationship between the members of the clan is no longer felt so consciously, but it is still so strong that in the end the members of the clan consider themselves as relatives due to their common descent from one ancestor.

When discussing the social classes we will return to this point (see p. 67).

Finally, it should be stressed that the number of *da* per village is two, four or five. If the grouping of the clans is connected with an ancient division into phratries, one may in fact expect two or four clans, linked in couples. A marriage organisation may likewise quite well coincide with a grouping in ritual clans; in that case we would therefore be dealing with exogamous clans. When such a system of exogamous, ritual clans again assimilates a classificatory system, we arrive at the exogamous, totemistic clans which are to be found in some localities in Australia.

It is quite as conceivable that the Waropen *da* once was of importance for the marriage organisation, as that it will assume this importance again at some future date.

In a circulatory marriage system an uneven number of clans is also not unusual, because such a system always needs three groups in order to be able to function. One would therefore be justified in considering the fifth clan in the Waropen organisation as an intermediate clan, a clan situated between the others and maintaining relations equally with all. Such a position is actually occupied by both the fifth clan at Paradoi, the Sawai, and by the fifth clan at Nubuai, the Kai. The separate place is often explained by the population by saying that this clan was added later, from outside, although the mythology relates in the case of the Kai that it was this very intermediate clan which was the first clan of the village, its ancestor having descended from heaven (Texts 11, 12).

RUMA, FAMILY-BRANCH

As stated above, the *nu* or village is divided into two, four or five *da*; each *da* in its turn is divided again into a number of *ruma,* but this number is not regular, like that of the *da*, but arbitrary. A large *da* has many *ruma*, a small *da* only a few. Genealogically speaking, a *ruma* is a group of paternal relatives where the relationship has to be viewed in the words of Rivers as "a relationship which can be demonstrated genealogically", but in this case not exclusively because of marriage — as stipulated by Rivers — for among the Waropen traditional relationship is also created by adoption.[58]

---

[58] W. H. R. Rivers, *The History of Melanesian Society,* I, p. 16. This small, unilateral community I indicate by means of the term "family-branch" and not by one of the words for clan, because the *ruma* are not arranged in a certain organisation.

The social organisation of the Papuas of Northern New Guinea is often described in terms derived from those of the Numfoor. As misunderstanding is often caused by applying the Numfoor terminology without more ado to the organisation of the other tribes, I will give here a brief description of the organisation of the Numfoor. The Numfoor tribe is divided into four main sections, called *er*, but these are of little importance as regards the actual grouping. In daily life the division into *keret* is decisive. The *keret* is often considered as the counterpart of the Waropen *da*, although in the *keret* genealogical relationship is more important and ritual co-operation is less so than in the *da*. Probably the *keret* was formerly at the same time a local group. In the names of the Numfoor *keret* we often find the element *rum*, house and in this way we have a *keret* Rumfabe, a *keret* Rumbruren, etc. These *keret* are again sub-divided into exogamous *sim*, rooms, which may therefore be compared to the Waropen *ruma*; such *sim* always possess composite names, like e.g. Rumfabe Wanambasiwiri. As will also be shown for the *da*, one family-branch is considered as being the oldest and most important of the clan. This family-branch may be indicated by adding Fakendawer after the name of the clan, from *faker*, bottom-layer, and *rawer*, front; e.g. Rumfabe Fakendawer.

Because the value of this Numfoor terminology has not been correctly understood, the word *keret* (variously spelled *kered*, *geled*, etc.) has been used to indicate both clans and family-branches or other groups. In reports concerning the Waropen the word *keret* indicates both *da* and *ruma*. Of course, the application of the terminology of one tribe to another should not be encouraged.

In this connection J. G. Detiger says of the Waropen coastal population: "However, each of these villages does not form one kèrèt but several who live together in separate wards". Because further on "kèrèt-chiefs" are mentioned, the word *keret* here evidently refers to the *da*.[59]

The Waropen word *ruma* has two meanings, viz. 1. genealogical group of relatives, and 2. living-quarters. In this way two people may live in the same *ruma* (quarters) and yet not belong to the same *ruma* (family-branch). With the first meaning all *ruma* which are considered as independent possess a name of their own. The division into genealogical

---

[59] Detiger reports likewise of the tribes of the interior that these are divided into "parts of tribes", these being "separate autonomous communities, formed by a few related families" who "form separate settlements". For these groups he again uses the Numfoor term *keret*. — See *Adatrechtbundels* (Bundles of Adat Law) XXXIX, p. 423.

groups often provides clearer indications than the location and the arrangement of the living-quarters. The Waropen care little for order and regularity when building their houses and even the inhabitants of the village have some difficulty in making out who lives in the confusing conglomerate of houses, wholly or partly finished and partly already dilapidated. However, as soon as one knows the father or the mother or another relative of the inhabitant, this relationship enables people to know exactly who is the person concerned. On the sketch-map of Nubuai (which does not have any pretensions at topographical accuracy) the reader will observe that e.g. the *ruma* Sawaki of the *da* of the same name consists of at least five separate living-quarters.[60]

A number of years ago the Administration abolished the so-called "rumah panjang" (Malay; lit. long house) in an attempt to create family settlements. This resulted of course in a thorough revision of the plan of the village and in a break-up of the genealogical group. The Waropen understood this measure to mean that it was prohibited to collect more than two brothers with their wives, children, parents, etc. in one house. However, as soon as possible they tried to bring the living-quarters of one genealogical group together again and during the period of my investigations several houses had already started to assume again the type of the old "rumah panjang". The chief of Sawaki even returned to the old-fashioned construction of the front of the house which also possesses a religious meaning. In other Waropen villages the Administration subjected the plan of the village to another revision, due to which the houses were placed in one line and became smaller in type. In Napan the houses were transferred to the mainland, where they continue to stand. This has also been done at Ambumi, but there the whole of the village is gradually shifting back into the water.

To the Waropen the *ruma* represents much more clearly a genealogical group than a dwelling. They do not see the village as a number of houses with names, but as a number of family-branches. They find it easier to say: "So-and-so belongs to this family-branch and he lives there and there", than to determine: "In these quarters such-and-such a family-branch is living". The same applies to the family-branches; the Waropen are rarely able to see them as a whole. They do not classify people by arguing e.g.: "Gharori belongs to the *ruma* Erari and therefore he lives at Apeinawo in the first house", but rather by saying: "Gharori

---

[60] Formerly the number of living-quarters seems to have been smaller, at least De Clercq in 1888 counted 50 houses in Kai, undoubtedly living-quarters.

is the son of Ekamai, and Ekamai belongs to the family-branch Erari which lives over there at Apeinawo".

Often a family-branch is better known by the name of some prominent old man of this branch than by the proper name of the branch. In this way it was clearer to most of my informants when I said: "He lives in Sirakoi's house" than when I said: "He lives at Maniwuri". Usually there are in every house one or more *manobawa*, great or old men, who may be considered as the chief of the family-branch. Such *ruma*-chiefs do not have a separate title. The Waropen express the leadership of the *ruma* by means of the word *onéa*, command, e.g. *Sirakoi ionéa rumani*, Sirakoi commands this house. Together with the *serabawa* these *manobawa* constitute a council of elders (cf. p. 75).

When I was noting down the location of the different *ruma*, my informants had to accompany me because they were unable to indicate which family-branches lived in the different houses without making personal enquiries.

The *ruma* are subject to a constant process of disintegration on the one hand and of combination on the other. It happens quite often that a member of the *ruma* falls out with the other members; when the quarrel runs high, part of the *ruma* detaches itself and establishes itself in a separate dwelling. In one case such an off-shoot is able to obtain recognition as a new, separate family-branch, in another the new group remains known as a separate part of the *ruma*. In this way three separate *ruma* Erari are to be found in the Apeinawo clan, having separated because of quarrels, but they continue to consider themselves as one.

In other cases the section which has seceded is gradually recognised as a separate family-branch. In this way Mamurani separated itself from Rumaniowi when one of the men of Rumaniowi had been killed by his brother-in-law (Texts 179, 180). Mainei left Tanatirewo after a quarrel about a slave. Aibini is the name of a neck-support, after which the members of this house called themselves after having separated from Maniwuri, because Aibini opposed a raid on people with whom he was related. Sasarari is a branch of Woisiri; they seceded after a quarrel about a torn loin-cloth, an occurrence of which the name *sasa*, to tear, and *rari*, loin-cloth, still reminds us.

These secessions are to a certain extent outweighed by combinations. It occurs that a family-branch which secedes, attracts also part of another *ruma*; both these detached parts together form again a new family-branch. On a smaller scale combination takes place by means of adoption and because of living-in. The lines who are only related because of living-in are considered among the youngest and hence the least important of the family-branch in question. A man may go and live in the *ruma* of his wife, e.g. because of a quarrel or simply for company's sake, or, as occasionally happens also, he and his wife go and live in a third *ruma*.

This process of disintegration and combination is not always limited to one's own clan. It sometimes happens that family-branches join the *ruma* of another *da*. In this way parts of Kai have placed themselves under the protection of Tanatirewo at Sawaki after serious quarrels. Mamurani has broken away from Maniwuri at Pedei and at present it is considered part of Apeinawo. Manai from Pedei was a family-branch of Tanatirewo. And it is not impossible that Refasi from Apeinawo formerly formed one single unit with Refasi from Sawaki, just like the two *ruma* Ruatakurei in both clans, although nowadays nobody knows anything more about a former unity. On the map of Nubuai one will find the house of Sapari of Apeinawo above the clan Pedei, and again houses of Pedei in Sawaki.

The constant disruption of the genealogical framework due to this process of combination and disintegration is counter-balanced by the tremendous weight of tradition which continuously repairs and consolidates the traditional genealogical relation. When a number of different family-branches has been living together for a long time, the tradition will grow that these family-branches constitute one single *ruma*, i.e. that they belong to one genealogical frame. In the case of some man or other who has been living for a long time in a strange *ruma*, the children of this *ruma* will in the long run consider him as a (classificatory) brother of their (classificatory) fathers and then he will also be adopted into the traditional genealogy of this *ruma*. The constant re-grouping of the exogamous groups of relatives within the *da* is facilitated by the feeling of relationship which after all unites all members of the *da* due to their descent from one mythical ancestor. However, transition from one *da* to another still remains rare.

So long as the notion remains alive that the members of one *ruma* are each other's relatives, so long the *ruma* will continue to exist as a group of relatives. The same is true for the clan; provided that the members of the *da* still feel that in the end they are descendants of one clan-ancestor and that therefore they are mutually related, so long the clan will continue to function, and so long the clan will remain capable of absorbing new elements by means of the construction of a traditional relationship.

Additional material from *The Papuas of Waropen,*
ISBN 978-94-017-5654-9, is available at http://extras.springer.com

NUWURI   PEDEI
{ Ienusieiwo   { Pedei   Saidui   { Aiomi
  Marini   Aibini   Manai   Rumaniowi
  Nuwuri   Dori   Ghanadi   Gharami }
        SAWAKI   APEINAWO
{ Tanatirewo   { Sasarari   { Erari   Korisanogha
  Maniwuri   Woisiri   Refasi   Mamurani
  Aibini   Refasi   Pandori   Sapari
  Ruatakurei   Windewani   Numboi
  Sawaki   { Duwuri   Ruatakurei   KAI
{ Maie   Imbiri   { Koghi   Manieghasi
  Maighane   Kaisandumagha { Ghaneioi   { Bindosano
  Soin demi   Kandenafagha { Kaiwai   Samanui
                   Keghi }   Roghi
                                                { Nuwoarai
                                                  Niki
                                                  Saroi

List of *ruma* at Nubuai.
The houses constituting one unit, formerly or
at present, have been connected by brackets.

It seems, however, that even in the Kai district the clan is beginning
to lose this capacity. The separate existence of houses from Pedei at
Sawaki and from Apeinawo at Pedei is still considered as an anomaly;
the house Maniwuri, which according to some originated from Pedei,
according to others belongs to Apeinawo since ancient times. But I
believe that in the long run also in Kai the consciousness of the traditional relationship due to descent from the same clan-ancestor is bound
to become increasingly blurred and due to this the clan will merely
continue to exist as a vague notion of relationship between the different
houses or as a sense of obligation for ritual co-operation in certain matters.
As at Napan and at Ambumi, the clan in the Kai district may continue
to lose the character of a local group, to change into an alliance between
certain genealogical groups for certain undertakings, e.g. for purposes
of war.[61]

---

[61] There is a preference for deriving the names of the *ruma* and the *da* from
capes, islands etc., known in mythology. If clans or family-branches possess
identical names, this does not necessarily indicate a former unity. The names of
the houses at Ambumi were: Warami, Saiori, Sakuatorei, Kawiai, Romboifai,
Waratanoi, Wengge, Samberi, Marei, Samisanoi, Aiomi. The houses of the
village on the Woisimi were: Siwerori, Kiri, Adueri, Wakumuni, Urusi,
Maniai, Runaki, Marani, Dimawi, Korei, Wopairi, Wokuai and Samberi.

THE CLAN, THE FAMILY-BRANCH AND
THE MARRIAGE SYSTEM

The concept of traditional relationship which determines the limits of the *ruma* and, after all, likewise those of the *da* is so elastic, that it is far from simple to define accurately the limits of the *ruma* and, to a lesser extent, those of the *da*. During my investigation I collected the genealogies of all family-branches at Nubuai. In general the population was quite ready to provide the necessary information, although I cannot say that the people gave themselves extra trouble to make the genealogies as exhaustive as possible. Usually it was the old women who were able to provide the best information concerning family relationships. The genealogies thus established are probably quite reliable because the total number of inhabitants as established from the genealogies differed but little from the total at which the Administration had arrived by a count.

After all that has been said above, it will not be considered astonishing that quite often certain family-branches were reported by two different *ruma*; the constant shifting of the family-branches causes a continuous ordering and re-ordering. Some family-branches wish to be considered as independent, without this declaration of independence being accepted by public opinion.

The investigation showed that although several *ruma* wished to be considered as independent, they still realised that in fact they belonged to another *ruma*. The total number of *ruma* which the villagers recognised in daily life was 47. Among these *ruma* there is a number which the people themselves do not distinguish positively; these have been connected by a bracket in the list on p. 59. When we combine these family-branches, we arrive at a total of 36. And if we continue to combine those *ruma* of which the historical unity is still remembered, we reach a total of not more than 25.

It is necessary to stress this point because according to Waropen theory the *ruma* is exogamous; all informants stated unhesitatingly that marriage within the *ruma* is prohibited. The marriage system in force among the Waropen makes marriage with the mother's brother's daughter obligatory; other possibilities, like e.g. marriage with the father's sister's daughter, are excluded.

This so-called circulatory marriage system has been often discussed and this is not the place to enter further into this matter. I only want to mention that this system requires at least three groups in order to be able to function, viz. a group from which ego receives wives, and a group which ego's group provides with wives. Within a clan-system exogamy only functions in correlation with some kind of arrangement of the clans. Because among the Waropen we not only find a linking of the clans in couples, but also exogamy of the family-branches and moreover a marriage system which presupposes an organisation in groups which provide wives and groups which receive them, an investigation of the different marriage relations is an evident task. For it would be quite possible that the actual marriage relations within the village resulted in a connubial relationship between the clans similar to that pertaining between the clans and the marriage classes among several Australian tribes.

For this reason I investigated all marriage relations in the genealogies of Nubuai, even those of people already deceased, in order to arrive at as large as possible a figure. In view of the tradition, the house Mamurani has been counted again among the clan Pedei and not among the Apeinawo, and similarly Manai from Pedei among the Sawaki. The results of this count have been combined in table I on p. 63. These figures confirm the correctness of the statement that the clan is not an exogamous group. On the contrary, the clan is quite endogamous, as is shown e.g. for the clan Kai, where 31 of the 66 women married into it came from the selfsame clan. In the second place it proves to be impossible to arrive at a certain arrangement of the clans from these marriage relationships; the small size of Nuwuri has of course an unfavourable influence on the figures.

Nor is it possible to construct a grouping of the *ruma* into those which provide women and those which receive them; in actual fact even the exogamy of the *ruma* is not maintained. When we keep to the 46 family-branches, there prove to be only 26 in which not a single endogamous marriage was reported. And when the *ruma* are combined, then only 9 of the 25 are strictly exogamous. Hence, there are reasons to assume that formerly the *da* had a smaller number of *ruma,* or in other words, that at that time the exogamous groups were larger. Traditional relationship must have been of greater importance in those days. In this way, the largest group which is still kept together because of a feeling of relationship, viz. the *da,* might also have been exogamous formerly. The limits drawn by the traditional exogamy of relatives of the *ruma* are becoming more and more narrow; actual blood-relationship of the smaller family-branch is coming to replace the traditional relationship of the larger family-branches. This is the result of the exclusiveness of the growing social classes.

INCOMPLETE UNILATERALITY

In the Waropen organisation we have therefore not found the orthodox situation of groups providing wives and groups receiving them.
In other words, one does not find that Manieghasi of Kai provides
wives to Erari at Apeinawo and that they receive wives from Ienusieiwo
of Nuwuri. On the contrary, in actual practice it often occurs that two
houses stand in mutual connubium.

At first sight this situation can hardly be brought to agree with the
demands of exogamy, nor with the m.b.d. marriage (from now on this
abbreviation will be used for mother's brother's daughter) or with an
arrangement of the clans. But one has to consider first of all that the
starting-point for the Waropen is not a traditional arrangement of the
clans or the families, but a traceable blood-relationship and in the last
resort the family. He does not say: "A man from the family-branch
A marries a woman from branch B", but always: "The daughter of
the brother marries the son of the latter's sister". Once this relationship is given, it is of less importance whether the exogamy of the
family-branches is violated or not. This is the orthodox type of marriage and every Waropen maintains that all marriages follow this
identical scheme. In the Napan region the decay of the connection
with the exogamy of the group has gone further still and the stress on
the family-relation is even stronger. Also in this area the m.b.d. marriage
is prescribed, but not between children of actual brothers and sisters.

In the case of a truly unilateral organisation, the children of
brothers and sisters implicitly belong to different family-branches.
If, therefore, the Waropen cling to the m.b.d. type of marriage and
are still unable to maintain a strict exogamy of the family-branch,
there can be only one explanation for this phenomenon, viz. that the
family-branch is not strictly unilateral.

Like I said above, the *ruma* is paternal. Tradition establishes the
relationship between the members of the *ruma*. If a stranger has
been living in a *ruma* for a long time, people in the long run do not
know any better than that he also belongs to this *ruma* and he is
given a place in the genealogy of the *ruma*. Now the Waropen marriage as a rule is patrilocal, i.e. the woman goes to live with the
husband. But it also often happens that for some reason or another
the man goes and lives with his wife's *ruma*. The reason for such
a matrilocal marriage is very often a quarrel, but sometimes simply the
circumstance that there is more room in the wife's *ruma*. Also in

		Kai	Apeinawo	Nuwuri	Pedei	Sawaki	Total	Other villages
Kai	T	31	4	4	8	19	66	21
	P	31	18	4	21	32	106	21
Apeinawo	T	18	31	3	35	24	111	42
	P	4	31	5	30	23	93	38
Nuwuri	T	4	5	9	9	14	41	27
	P	4	3	9	9	17	42	16
Pedei	T	21	30	9	47	55	162	48
	P	8	35	9	47	47	146	67
Sawaki	T	32	23	17	47	161	273	45
	P	19	24	14	55	161	280	68

I *Exchange of wives between the clans within Nubuai village.*
   T — wives taken from
   P — wives provided to

	Wonti	Risei	Saiati	Woinui	Mambui	Paradoi	Sanggei	Waren	Weinami	Napan	Makimi
T	8	11	7	19	64	15	7	18	1	6	2
P	5	17	7	6	34	22	10	20	3	21	6

	Kaipuri	Saponi	Moor Is.	Nau Is.	Barapassi	Ansus	Panduamin	Wooi	Manawi	Serui	Ambai	Remainder Japen
T		17				4						4
P	2	11	7	2	1	5	2	2	4	6	3	14

II *Exchange of wives between Nubuai and other villages.*
   T — wives taken from
   P — wives provided to

polygamous marriages the woman often remains in her own house.

Matrilocal marriages therefore occur in many genealogies. If a woman continues to live in her own house, her sons are the potential husbands of her brother's daughters who live in the same house. And once the fact has been gradually forgotten that in this case the relationship had been traced matrilineally, people only remember that a man found his wife in his own *ruma*, a fact which often aroused some astonishment and shame when I pointed out the demands of *ruma*-exogamy. In such cases the traditional relationship usually prevails again over the brother-sister relationship which made the m.b.d. marriage possible, for in the long run the descendants of such a matrilocal marriage are considered again simply as members of the *ruma* of their mother. In other words, in the patrilineal series one finds a woman, from time to time, but without a permanent invalidation of the unity of the *ruma* being the result. Gradually the sense of relationship within the *ruma* simply counteracts this departure from the patrilineal line, with the result that a *ruma* does not become systematically endogamous once a married woman with her children has continued to live there.

It is not only the rather frequent matrilocal marriages which disrupt the patrilinearity of the *ruma*; it also happens frequently that after her husband's death a woman will go to live again with her brothers, which leads to the same situation. Already after one single generation, when the widow had died, people often turned out not to know to which house her husband had actually belonged. Sometimes the people in another house then remembered that the man had belonged to them, from which they concluded that his children in fact belonged to their *ruma*.

The classificatory scope of the terms of relationship is the cause that the change-over from the patrilineal to the matrilineal tracing within the *ruma* and back again proceeds without a break. If there existed terms of relationship which exactly distinguished the different degrees, it would be possible to follow the unilateral tracing of relation more accurately, but Waropen terminology is very flexible, distinguishing mainly between generation, sex and marriage relation.

It has now become clear why the Waropen groups in spite of the type of marriage in force still do not assume a fixed relationship of givers and takers of wives. This is possible because the tracing of relationship in the *ruma* is incompletely unilateral, or, as professor

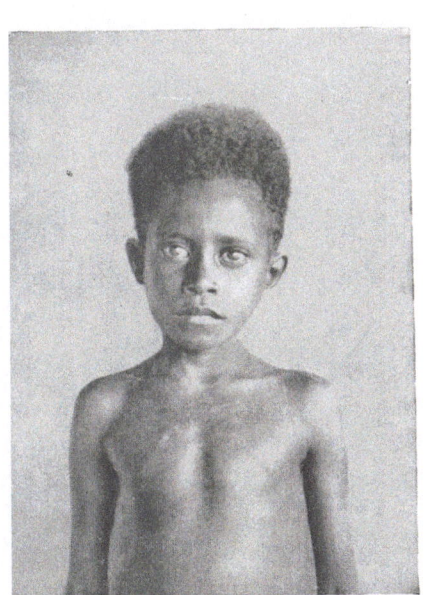 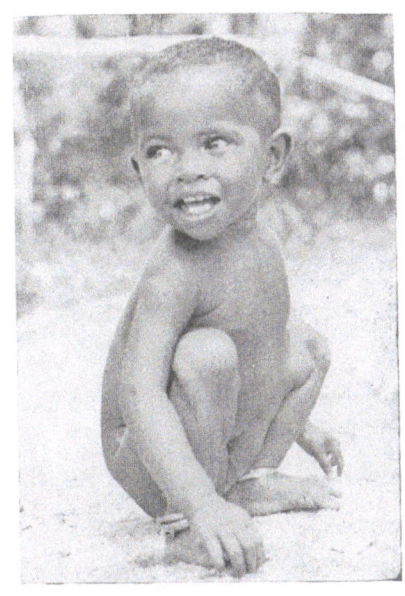

19, 20, 21. Children's types

22. Young men from Nubuai

23. Adori of Ruma Pedei

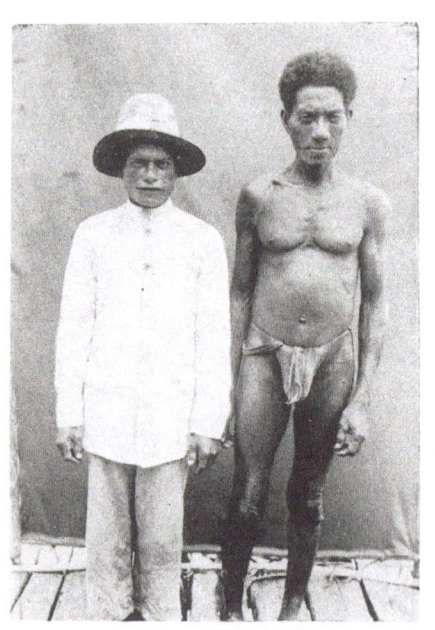

24. Gharori of Ruma Erari and
his eldest brother

25. Umesi of Ruma Sawaki (left)
Munamberi of Ruma Nuwuri (right)

Ter Haar has called it with a characteristic term: "alternating unilateral".[62]

If in the Waropen organisation there existed a connubial relation between the groups such as it has been observed among exogamous clans, the incomplete unilaterality might have quite well resulted from a complete double unilaterality, viz. a patrilineal tracing of relationship beside a matrilineal one.[63] If we assume for instance that formerly the family-branches were united in two *ruma*, matrilineal phratries (A, B), and that furthermore there existed four patrilineal *da* (1, 2, 3, 4), we might establish the following scheme for such a system:

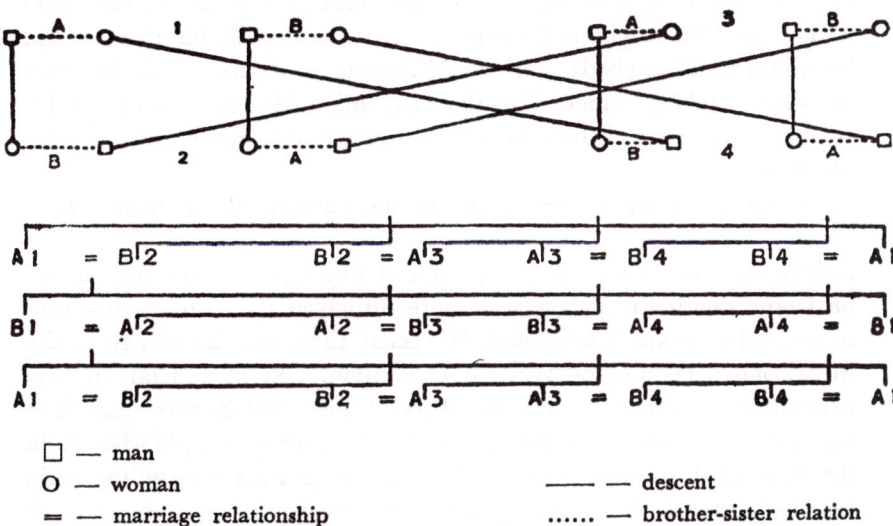

□ — man
O — woman   ——— — descent
= — marriage relationship   ...... — brother-sister relation

The system can of course function only with an even number of clans. We would then arrive at eight groups in all. Would the preference for the number eight in Waropen ritual be a reminiscence of this?

---

[62] See B. ter Haar, *Beginselen en stelsel van het adatrecht*, p. 148. [In the English translation of this work, Adat Law in Indonesia (New York, 1948), p. 157. Ed.].

[63] See on this point F. A. E. van Wouden, *Sociale structuurtypen in de Groote Oost* (Types of social structure in the eastern part of the Archipelago), 1935. From Thurnwald's data in his study on Banaro society (*Die Gemeinde der Banaro*) I believe it can be shown that double unilaterality also existed among the Banaro.

SOCIAL CLASSES

In Waropen society the individuals enjoy different degrees of
respect. Leaving the distinctions due to sexual differences outside the
discussion, we arrive at three criteria:
1. personal qualities,
2. advanced age,
3. descent.

To the personal qualities which enabled a man to acquire social
prestige in olden times belonged the quality indicated by the Waropen
word *kako*, rough, hard, cruel, i.e. the Waropen idea of martial virtue.
A man of this type is e.g. old Munoi from Sawaki at Nubuai, who
as late as 1936 frightened people of Apeinawo with his threats that
he would shoot anybody who would dare to row past Sawaki in order
to fetch drinking water. For the remainder Munoi, in spite of his
repute, is not a rough type at all according to our notions, but a quiet
old man.

Knowledge of the ancient manners and customs also enhances one's
position, especially when such knowledge goes together with better
insight, as is often the case. A person of this type was the old Sirakoi
on whose death the village was more dismayed than on the occasion
of any other death I witnessed. In recent times the knowledge of the
Malay language and of Western civilisation, and in general an easy
manner in dealing with the members of the Civil Service, also lead
to a certain enhanced prestige. It is often the younger people who attain
the best results in this respect, but in the general esteem they still
remain upstarts. It is also possible to acquire prestige because of
knowledge of magic potions and medicines. Riches were never indicated
as a basis for social prestige, probably because riches are considered
rather as the result of position than as its cause.

A factor which greatly contributes to social position is age, as long
as it has not degenerated into evident senility. This is probably not
only due to the greater experience which aged persons have collected,
but also to the importance attributed to age by itself. The mythical
beings are nearly always old. The old men are designated by the term
bawa, great, and when the Waropen say, in order to explain a certain
action: "We act like this because otherwise the great become angry at
us", one does not know whether by means of "great" they mention
the still living old people in one and the same breath with the mythical
ancestors to whom the present civilisation is due.

Hence, personal qualities and advanced age produce a heightening of social status, but only to a limited extent, viz. as long as one does not also belong to people of gentle birth due to one's descent. The prestige of the above-mentioned Sirakoi may have been quite considerable, but he did not satisfy the requirement of high descent and in this way he could at most be a *manobawa*, a great man, but never a *serabawa*, a great *sera*. *Sera* means head, noble, god; it is the same word as Numfoor *sren*, which also means: clean, pure and sacred.

The *sera* is the male descendant of the oldest family-branch of the clan, and theoretically the *serabawa*, the great *sera*, is the oldest male of the oldest line of this family-branch, but in actual practice the personal qualities and the age of the person concerned are taken into account. The idea of social class is therefore a further elaboration of the principle of seniority which is likewise already present in the structure of the total marriage groups. Finally, a difference in status is already present within the family, because the elder brother is more important than the younger.

Beside the *sera* we find as his inferiors in status the *waribo*. The word actually means "young man", but it is more suitable translated as "companion"; for the *waribo* are all fellow-clansmen who assist their clan-chief in the slave-hunt and e.g. when he is building a house. These men are called *waribo* because they are descendants of the younger family-branches of the clan and not because they would be younger in years than their *sera*.

It is not possible sharply to distinguish the *sera* as a social class from the *waribo;* the feeling of common descent is still so strong within the clan that people hesitate to indicate differences in social status. When the clan-chief of Apeinawo was once asked to indicate exactly which people in his clan were *sera* and which *waribo*, he replied that the whole of his clan consisted of *sera*, because they were descended from the same clan-ancestor. In actual practice the class of the *sera* includes the closest relatives of the clan-chief who are living in his *ruma*, the *seraruma*.

As far as outward behaviour is concerned one does not notice much of a difference in status between *sera* and *waribo*. The *sera* themselves often adhere to the statement I have also heard from other prominent old men in Geelvink Bay, viz. that truly high status need not be proved by outward show. Munoi, whom I mentioned before, is considered as a true *serabawa*, but he still wears a loin-cloth of banana-bark and the only indication of his dignity is that he always wears an old trilby, because in olden times it was a privilege of the prominent that they wore a head-covering, a privilege which is very often violated these days by the younger people with their pith-helmets and their trilbies. Other prominent old men likewise care little for their clothes. In the myth of Manarmakri (see p. 303) not a single outward sign indicates that the lame old man is a god. In the same way, the women of *sera* status, the *mosaba*, do not wear a more striking dress than the women of an inferior position; however, they do often wear strings of valuable old beads.

It is therefore possible to say that there does exist a difference between the class of the *sera* and that of the *waribo;* however, clan-solidarity and the feeling of clan-relationship are still so strong that the limits between these two classes are still more or less unstable.

A difference in status between members of the "head"-clan and those of the "tail"-clan is hardly noticeable. Each clan still feels itself so strongly as a unit, that my suggestion to recognise e.g. the clan-chief of Kai as being the first, was considered completely unacceptable. The members of the "head"-clan feel themselves superior, but this superiority is only a feeling of religious superiority and not of social superiority. In Nubuai the chief of the Sawaki clan is highly respected simply because his clan is by far the largest. At Ambumi and Mambui the chiefs of the "head"-clans enjoyed the greatest prestige. It may be said in general that membership of a "head"-clan does contribute something to the determination of social status, but it is not decisive.

Beside the *sera* and the *waribo*, the Waropen know a third type of people, viz. the *ghomino*, the slaves. However, in the Waropen area a separate class of slaves does not exist. This is easily understood, because fishermen differ from agricultural or pastoral people in not having many tasks they can leave to slaves. Slaves are rather to be found among the Numfoor, for instance, who also possess a more developed type of horticulture. Most of the *ghomino* who were caught were quickly returned to their relatives against payment of a ransom. My informants even assured me that in the Waropen area not a single slave was to be found. Presumably we will have to understand this in this way that the occasional slave whom his relatives failed to ransom was in time adopted into the genealogical framework. Concerning one of the chiefs of Nubuai I was repeatedly told that his father had been a slave; for this reason this man could never become a *serabawa,* although he happened to be highly respected because of his wide travels and his knowledge of Malay.

The possibility of a marriage between a free man and a slave-woman was absolutely denied. As long as a woman who has been captured has not yet been adopted into the relationship of a *ruma,* she is of course unable to bring along a dowry and so she is not a very good party. A slave who has been captured has better chances as he may acquire social prestige because of extraordinary qualities. The marriage between a free woman and a slave is considered to be reprehensible; among the Numfoor to such a union the rule was applied that *botor iwef pir bepon,* the (valueless) bottle destroys the (valuable) old plate. My informants believed furthermore that the child of a *waribo* and a *mosaba* could never be a real *sera,* but the child of a *sera* and an ordinary woman is again a *sera.* Descent through the father is therefore decisive.

Summarising the above we may therefore say that a distinction between two social classes, i.e. between that of the *sera* and that of the *waribo,* is in the process of development, a distinction which is still vague due to the influence of the feeling of clan-relationship, but which might become more clear-cut in the course of time. Hence, there are two conflicting tendencies in the constitution of groups among the Waropen. On the one hand one likes to see one's *ruma* as large as possible because a large *ruma* provides authority. On the other hand one likes to belong to a small and distinguished *ruma.* The traditional feeling of relationship and social exclusiveness are the two contrary factors which eventually decide the size of the *ruma.*

CHIEFS

Although the difference in status is real enough, it is not astonishing that earlier investigators were misled by the unstable character of the distinctions in status, and so their information is either incomplete or incorrect. J. G. Detiger, for instance, writes in *Adatrechtbundels* (Bundles of Adat Law) XXXIX, p. 425, under the heading "Native chiefs and social classes": "The foregoing clearly shows that no village-chiefs with a power rooted in native law are to be expected here. In the anarchic Papuan society of this district one will also look in vain for kèrèt-chiefs endowed with authority. Chiefs installed according to customary law as these are to be found in the Hollandia District East of Dèmta do not occur here. The official kèrèt-chiefs (called k o r a n o, m a j o o r, s e n g a d j i and often incorrectly k e p a l a k a m p o n g — i.e. village chief) owe their function to the Administration. Many of these have formerly been appointed by the traders, to be confirmed in their authority later by the Administration. Others again are returned gaolbirds who have acquired a certain amount of experience and a wider view due to their stay in the prison at Ternate or at Manokwari; so they rise slightly above their fellow-villagers in cleverness and knowledge of Malay and therefore they have been appointed in their present position by the Administration, as was formerly also done in Papua. A point already brought forward in the reports of my predecessors is that it is clear that such chiefs *ex officio* possess no authority over the population and that they are little more than organs for the transmission of the orders of the Administration, or interpreters. Orders given by the kèrèt-chief on his own are not obeyed, whilst instructions of the Administration given through the kèrèt-chief are only followed when the people know that non-compliance will entail punishment. This circumstance, combined with the low strength of the personnel available, render the Administration of this district very difficult.

"Papuan society in this area does not know classes. However, in some villages people were still able to point out those persons whose ancestors had been slaves, caught during a raiding-party. During quarrels it happens that these descendants of slaves are insultingly reminded of their origin. Still, they do not occupy a subordinate position in the village, and in their rights and duties they are fully equal to those originally born free.

"One custom deserves special notice; it is directly connected with the raiding-parties and with the disappearance of this a-social custom it is likewise bound to disappear. Once the decision to hold a raid has been taken, the participants turn to the so-called s è n g (formerly on Jappèn) or t è r r a.[64] (among the Baudji's in the interior of the Waroppen area), who provides them with victuals, prahus and other necessities. Against this we find the obligation to hand everything captured during the raid to the 'sèng'. On their return he presents a feast to the participants in the raid, at his personal expense. The 'sèng' who does not participate personally in the raid is usually one of the oldest and most prosperous men and as such he enjoys a certain prestige among his fellow-tribes-men. However, this position does in no way entitle the 'sèng' or 'terra' to exercise authority over the members of his tribe".

The *Encyclopedia of the Netherlands Indies* says concerning this subject in vol. III, p. 321: "Except when there is a 'manseren'... the

---

[64] In this term we recognise of course the Waropen word *sera*.

highest authorities in these regions are the family-heads; for the rest, the people here live in complete anarchy... The village is headed by a person who bears the title of 'korano' or 'majoor' or something similar, but he only has some say in his own family; the others disregard his orders, unless they consider it profitable. A few villages possess a head-man who is called 'manseren' in this part of the coast. He is a demi-god who treats sick persons with magic. Prisoners who have been caught are lodged in his dwelling. He need not work; his subordinates do so for him, like they also protect him against dangers, carrying him away if an official appears, threatening disaster. His power is practically absolute. Other village authorities are the 'kepala mambri', i.e. persons who have caught many prisoners or who have killed many people and who therefore belong to the heroes; they enjoy great prestige. Their houses, as well as those of the manseren, are to be recognised by the skull-trophies and by the blocks used for chaining slaves to be found there. There are also soothsayers who have to divine whether a raak (raid) will be successful or not. The manseren need not participate in the raak".

The informant (a Numfoor, judging by the Numfoor words *manseren,* nobleman, lord, and *mambri,* courageous) to whom we owe this notice evidently provided correct information, but it has been reproduced in a garbled fashion and so it is not clear.

We will not attempt here to give a closer definition of the concept "chief" on which one may make the existence of a class of chiefs depend. But if one were to follow A. Knabenhans and speak about a "Häuptling" (chieftain), beside an "Altenrat" (council of elders) and a "Volksversammlung" (tribal meeting) even among the Australians [65] where an indication of social classes is still completely out of the question, one is fully entitled to speak of a system of chieftains among the Waropen, where it is based on certain hereditary privileges and on descent.

Among the Waropen the chiefs derive their leadership not directly from special personal qualities or from their wealth — enabling them to buy a place in one of the hierarchically organised men's clubs — but from their descent of the senior family-branch of their clan, of which the *serabawa,* the great *sera,* or the *seratinggu,* the true *sera,* has to be the oldest representative. Whenever the oldest representative

---

[65] A. Knabenhans, *Die politische Organization bei den Australischen Eingeborenen.* (The political organisation of the Australian Aborigines), 1919.

of the senior branch does not yet possess the requisite age, a younger brother of his father acts as *serabawa*. At present, for instance, Umesi is the clan-chief and simultaneously administrative chief of Sawaki at Nubuai, although young Mui is the representative of the senior branch. Mui it already acquiring prestige and in spite of his youth he already has two wives. At Apeinawo the very aged Ekamai was the *serabawa*. Old Dedui, who represents one of the junior lines, now shares the highest authority with young Gharori, the representative of the senior branch; he is fast rising in standing and he has also been appointed by the Administration.

As Detiger correctly remarked, we are not concerned in this case with village-chiefs, but with clan-chiefs. The *serabawa* is considered to be the most direct descendant of the mythical ancestor of the clan and for this reason he enjoys religious respect, as already indicated by the name *sera*, nobleman, divinity. It is often believed that they are able to heal the sick. Ghamabai, younger brother of the chief of the Sawaki clan, is considered to be an expert in the preparation of charms and potions. Concerning the clan-chief of Nuwuri I was assured that more than once he had revived people who had been killed by the *sema*. In the myths supernatural powers are ascribed to the *sera;* one often finds: "If I am a true nobleman, I can perform such and such a miracle", e.g. the bending of heavy trees. Undoubtedly the notice in the Encyclopedia quoted above concerning the "manseren...... a demi-god who treats sick persons with magic" refers to this.

It is peculiar that the relationship between the *serabawa* and the ancestor of the clan is not strictly determined patrilineally. The present chief of the Nuwuri clan at Nubuai, simultaneously Administrative Chief of this *da,* actually belongs to the senior noble line of the Sawaki clan via his father's father Sabami. Sabami went to live with his wife at Nuwuri. In a strictly patrilineal sense it is not the present chief Munamberi who is the obvious *serabawa* of the Nuwuri clan, but another man, called Faidari, who is actually considered by some people to be the rightful chief of Nuwuri. Faidari, however, knows so little Malay that his chances are small. According to some people Munamberi ought to be *serabawa* of Nuwuri because his father's father would have been *serabawa* of Sawaki if he had not married into Nuwuri.

Occasionally it is quite difficult to follow the tracing of the genealogies. A number of years ago the chief of the Apeinawo clan was a man who through his father belonged to the *ruma* Saidui of the Pedei clan, and through his mother to the *ruma* Sapari of Apeinawo. It is far from clear how a man who neither through his father nor through his mother was related to the senior branch of the *seraruma* of Apeinawo could be *sera* of this clan. My informants believed that this relation must still have been established in some way or another, because otherwise the man could never have become *serabawa*.

The religious authority of the *serabawa* is shown more by a number of privileges conferred upon him than by certain external powers which we might expect in chiefs with administrative influence. Prerogatives of the *serabawa* are e.g. the following:

the right to receive the first slaves caught by the members of his clan or by members of the "tail"-clan;

the right to be presented with tobacco by newly married people;

the right to a fuller funerary ceremonial, the so-called Serakokoi;

for Kai: to a special dirge and a fuller marriage ritual, e.g. with the *kowuemba*;

the claim to a fuller initiation ritual when growing up;

the claim to assistance of the clan-members when building his *seraruma*;

being entitled to bear ceremonial titles;

claims to lesser privileges, like e.g. the heads of large fishes caught by members of the clan; wearing a head-covering, which in actual fact is not permitted to the *waribo*; wearing the *damasura* on the occasion of initiation feasts; some kind of homage from his inferiors, especially from younger people who are expected, for instance, to stoop when passing him, so that their head is lower than that of the *serabawa*, and to cede the first place in the prahu to the *serabawa* when rowing. Also in modern times the *serabawa* is likely to perform much less manual labour than the common man who, e.g. when engaged on public works like the building of rest-houses under the orders of the Administration, is always far more active than the *sera*.

These prerogatives are not considered to be all of equal importance. The commoner, for instance, is also allowed to have a full initiation ritual performed for his children, although in most cases he will not go so far. However, the Serakokoi ritual is only performed very rarely. This happened on the occasion of the death of old Ekamai of Apeinawo, but before this it had not been performed for many years. All these prerogatives are reserved for the oldest *serabawa*. In this way Muda, the influential Administrative Chief of the Pedei clan, is not a *serabawa* and his house is therefore not the *seraruma*. For this reason he remains a *homo novus* in the eyes of the villagers, and newly married persons will not offer tobacco to him, but to the much younger Adori, who is merely Deputy Administrative Chief (the so-called wakil).

The prerogatives of the nobility are protected by the belief in the myth concerning their establishment by the ancestors. The ancestors would punish those persons with illness who would presume to arrogate things to themselves which were not in accordance with their station. In actual practice a commoner who would venture to do things of this kind would be made "ashamed". He also has to take the displeasure of the *serabawa* fully into account; in the old days they would e.g. not have hesitated to defend their claim on the slaves who had been caught.

In the marriages we observed, the clan-chief acted as the person who concluded the marriage; our informants told us that this was really very good, but still not necessary. Any other superior, properly married man would also be able to do so. Nor was the clan-chief *ex officio* the leader of the initiation ritual; the initiator is the mother's brother. However, the *dama* had to stand at the side of the *seraruma* and because this *dama* was the clan-sanctuary, as will be shown below, one is at once led to suppose that the *serabawa* was also concerned in some way with the *dama*. This is also indicated by the fact that he was entitled to wear the *damasura* (see p. 155).

The clan-chief undoubtedly possessed a certain dignity and authority; the different prerogatives conferred on him clearly show the negative or taboo character of his dignity.

Now we have to ask whether he was also given several positive powers which clearly show his function. General controlling powers, when necessary supported by external compulsion, are to be looked for in vain as regards the Waropen clan-chief. This is quite likely the reason why writers report again and again that in this society the power of chiefs does not exist. The clan-chief should not believe for a moment that his personal orders would be obeyed; in this respect the members of the clan feel themselves too much members of the family. But on the other hand it would be a mistake to underrate the influence of the clan-chief in matters of everyday life; a clan-chief who is able to become the organ of the *communis opinio* of his fellow-clansmen may quite well acquire great influence.

In general the Waropen are not used to devise measures for their future well-being, but they confine themselves to incidental measures when something goes wrong. Once this has happened, the chances are that the chiefs will exercise their power. The fact that the clan-chief can really interfere is shown by one of the stories I took down. There it is said that a famine started in a certain place, "and thereupon", the story continues, "their clan-chief ordered them to leave" (Texts 178).

However, the clan-chief can do little by himself and therefore he always acts in consultation with other influential men, the so-called *manobawa*, the great men, i.e. the well-known warriors and the leaders of the various important family-branches. In that case the *serabawa* and the *manobawa* together form a kind of council of elders, which usually meets on the porch of the *seraruma*. All important matters are discussed at these meetings and it is a question of personal authority whether a person is able to impose his will on the meeting. It happens, for instance, that a few old men desire a school for the sake of their children and that, even against the wishes of the others, they make their request known to the missionary authorities; usually the others will soon follow the will of these more energetic members and even the clan-chief will have to acquiesce.

In general, the meeting of the chief and the elders should rather be considered as a kind of church-council than as a municipal council. Public opinion is most deeply shocked by violations of religious rules and in such cases the council of elders is the first to act. According to Waropen ideas the laws concerning debt come under the religious rules. In Weinami I was assured that the clan-chief had to feel "ashamed" if one of his people continued to be sued for debt by members of other clans. The reason given for this "shame" was that the ancestors become angry because of such matters and punish people with sickness. As according to the views of the Waropen nearly all cases of bodily harm entail debts which have to be paid, the clan-chief and his elders often interfere also in these matters. However, this interference is only necessary when the damaged party is unable to obtain payment. As far as our information goes, the clan-chief does not concern himself with debts like those arising from the payment of the dowry and the price of the bride. Due to marriage, two parties enter into a continuous exchange of mutual gifts and in that case both parties usually assert their rights. If it should come to blows, it will occur only rarely that whole clans would come to oppose one another, because marriage relations are too intimately intertwined to allow this. On the other hand, the members of the Sawaki clan were ready to fight for their chief when a *mosaba* of their clan had been raped by people from Apeinawo where the girl was about to marry. In this fight hard knocks were given and serious wounds were caused.

Here follow a few examples by way of illustration: The boys and girls of the Apeinawo clan were engaged in their favourite game behind the rest-house, throwing pointed sticks at a target. By accident one of the sticks swerved off and caused a harmless wound in the calf of the leg of a small girl. As by magic the playing-field was completely cleared; in less than no time the angry father of the girl was present, armed with bow and arrow, threatening to shoot the guilty boy, although he was his own brother's son. The other classificatory fathers of the boy left these threats for what they were. The father of the boy had to pay a plate to the father of the girl, thus putting a definite end to the matter. The clan-chief was not concerned with this in any way.

In another case a lover had pulled the hair of his beloved so strongly that an effusion of blood under the scalp was the result, so that she felt ill and painful. Although the boy's family asserted that in this case there was no obligation because quite frequently hurts were caused during love-making, they still had to pay five plates to the injured family who had lodged a plaint with the Administrative Assistant. Also in this case there was little reaction because under the circumstances the guilt had not been irrefutably proved. The clan-chief acted on behalf of the girl.

Again another case: the Administrative Chief of Pedei has married a *mosaba* from the same clan. During a conjugal quarrel he wounded his wife in the upper leg with an iron nail. This affair was considered to be of such importance that the chief of the Sawaki clan conferred on it with many old men, presumably because the *sera*-clan-chief of Pedei was still too young. The old men went in a body to see the guilty man who threatened to leave the village in case of necessity. Thereupon matters quietened down.

Evidently the *sera* consider ill-treatment of a *mosaba* by a commoner to be highly outrageous. The chief of the Sawaki clan was very indignant about the ill-treatment of Ghaisi, the sister of the clan-chief, by an old man of his clan, called Sibuki. The cause of the quarrel was the fact that late at night Sibuki wanted to row into the village with his prahu; he was drunk and when his wife wanted to stop him, he wounded her in the head and in the arm with a knife, whereupon his wife struck him on the back with a piece of firewood. Umesi, the chief of the Sawaki clan and the brother of the maltreated woman, demanded payment of several plates, because he considered it a mitigating circumstance for Sibuki's action that a wife should not interfere in the comings and goings of her husband, especially when the latter is drunk.

It so happens that insanity and drunkenness are considered as mitigating circumstances, unless an intoxicated person kills or grievously hurts another. Formerly, an insane person lived in the village who dragged anyone he could catch as a slave to his house, whence they had to be fetched in the evening with gentle words by the family concerned. In an access of madness this man attacked his brother with a knife; in self-defence the latter wounded him so seriously that the madman died as a result of his wounds. In this case the brother was not held responsible for the death of the insane man and no payment was demanded. When somebody has been really dangerously or mortally wounded, payment is always demanded. A woman, for instance, was mortally struck by a sago-palm which was cut down. The man who had cut down the tree could only save his life by fleeing to his *ruma*. The members of the *ruma* concerned so strongly stood up for the man who had been the cause of the accident that the woman's party were satisfied with not too high a wergeld. When a person is killed by a falling coconut, the owner of the tree is likewise guilty and if needs must be he has to pay with his life.

That it is the duty of the *sera* to see that payment is made in the case of one of the members of his clan losing his life is demonstrated by the following story which relates how the present *ruma* Mamurani seceded from Rumaniowi (Texts 179):

"They possessed one house Rumaniowi (a house of the Pedei clan at Nubuai). Sowoi had fought with Fokei and this had angered Fokei's brother. This brother, called Kapari, thereupon shot and killed Sowoi, his brother-in-law. For this reason Dumoi fled to this place and opened up the dry area upstream (the story is being told at Apeinawo). Dumoi was Sowoi's elder brother. As Dumoi had become angry because of his brother, he went down and cut the *karea-*, the *mandewi-*, the *sagha-*, the *fa-* and the *bina*-trees and established his house there. Formerly this piece of ground was dry, like the gardens hereabouts. There he planted fruit: gourds, sugar-cane and pineapples. Then Namandani, Dowuri and Ghanembiri came to live all three in (the newly established house) Mamurani (at present in the clan Apeinawo). Sowoi's wife, Fokei, had been left behind. Then Fokei gave birth to Sowobini and raised her. Thereupon Angguri showed Sowobini to the people of Mamurani. Angguri said: "You who are living there, look at her!" When they had seen her they approached to fetch her. Fokei had remained with her, in this same Rumaniowi. When they had first accepted the girl, they took counsel concerning the widower Dowui. They conferred with Dowui and said: "Dowui, now you will take Fokei with the little

girl". Then Dowui took Fokei as his wife and also Sowobini and together
they raised her. Some time ago Mamurani was not large. They lived in
a small house which the three of them had established. But when their
children had married, they enlarged the house and lived there. The
chief of their clan was (however, still) angry because of (the murder
of) Sowoi and therefore they wanted to catch Fokei, the wife of Dowui.
Therefore they hid her and they went with her to Rumaniowi, because
they said: "Let this woman (lit. belly) remain alive; then she will be
enabled to bring forth a child to take the place of Sowoi". When they
had taken her there, she was put in custody. Kapari paid for Sowoi.
Thereupon they threw the lime. And Dowui had two daughters; the
oldest was Finai and the youngest Sendini. They died without having
márried. Sowobini thereupon married Raiwowi. After having only given
birth to Seki, she came to lose Raiwowi. Then Sowobini remarried
Ekamai and then she gave birth to Sanggoi and Kaparibini".

This tale, inserted here *in extenso* because it also describes the seces-
sion of the line of a *ruma,* shows that the chief of the clan of the
murdered Sowoi even after a long time remained unsatisfied because
the debt due to this death had remained unpaid. The most frequent
and at the same time the most tragic cause for the action of the clan-
chief for the propitiation of the death of members of his clan is the
"suangi"-murder. According to Waropen ideas, death — and particularly the death of still vigorous persons — is not brought about by
illness or by natural causes, but by the evil doings of certain a-social
elements, *sema,* called "suangi" in the Malay of the Moluccas. When
the first accusations are spread, the *ruma* concerned will usually strongly
defend the accused member of the *ruma,* but when the rumours continue
to grow, the members of the *ruma* wash their hands of the unfortunate
person. Then a conference of the clan-chiefs and a number of old men
takes place. More or less by mutual arrangement the victim is left alone
in the house or in the forest, and then armed men hurl themselves
unexpectedly at the *sema* who is practically always killed. It is not
absolutely necessary that the *sema* be killed by his fellow-clansmen; he
is mostly slain by all those who feel a call to do so, people of his own
clan or of another. The *ruma* concerned will put in a *pro forma* demand
for payment for the member who has been killed as a *sema,* but this
claim is mostly disregarded.

Now it is expected of the clan-chief that he will take the initiative
in organising these so-called "suangi"-murders. That is the reason why
it is often the clan-chiefs who have to be sentenced for these crimes

by the Administration. It is evident that, like Detiger remarks, "returned gaolbirds" are appointed as chief, because it is they who usually are the main instigators. They go to prison for a "suangi"-murder because they were clan-chief, and it is not the other way round that they become chief because they have acquired influence due to their stay in gaol.

My informants could also imagine the possibility that a person would be excommunicated by his *ruma* for other crimes (the *sema* is likewise considered as a criminal), to be killed by the clan-chiefs and their men. However, they were unable to provide me with an actual example, but by way of a theoretical case they mentioned the murder committed some time ago by a man from Mambui on his foster-daughter. This little girl had eaten her foster-father's fish and this had angered the man to such a degree that he simply drowned the child. My informants believed that formerly this man would have been excommunicated by his own *ruma*, to be killed thereupon by the clan-chief and his men.

In short, one may say that the clan-chief, usually together with the other old men of his clan, acts when there are unpaid debts in his clan. Where "suangi" or other excommunicated persons are concerned, blood has usually to be paid with blood. Hence, the clan-chief and his elders came to act also in all cases of serious violation of adat law. As an example, the prohibited marriage of the youthful Seranauri with his (classificatory) mother might be mentioned. According to the old men, the ancestors would punish this transgression by causing the fishing to be unsuccessful. In the old days, I was told, Seranauri would have had to flee to another village in order to prevent his being struck down in his own village. And even now he had to take refuge in the rest-house during the first nights in order to avoid ill-treatment.

Another chapter will be devoted to the function of the clan-chief as leader of the raiding-party, where he attempts to obtain slaves and the property of others, or to revenge the blood lost by his clan by causing other blood to flow. The concern for the acquittal of blooddebts and the leadership in the raids I believe to be the consequences of the more or less religious position of the clan-chief.

The dignity of clan-chief was moved to quite another sphere due to the installation of chiefs appointed by the Administration. In the cases I know, it was in the main clan-chiefs who were appointed Administrative Chiefs, with the not quite correct title "kepala kampong" (Malay: village headman). The main duty of the Administrative Chief is the collection of taxes, for which he receives a certain collecting-fee, and also the supply of a quota of statute-labourers. The Administrative

Chief is also the person to whom travelling higher and lower officials and strangers may turn. He is assisted by one or more "wakil" (Malay: substitute) who, as far as I know, are not officially appointed by the Administration; they are, however, exempted from statute-labour duties. Occasionally clever young men who speak Malay secure the collection of the taxes and promote themselves to "wakil", to the great displeasure of the *serabawa* who prefers to appoint one of his younger relatives as "wakil", often his younger brother. Above the Administrative Chiefs we find the non-Papuan District Chiefs, assisted by Papuan Regional Police and sometimes by a detachment of Field Constabulary. Since 1939 the District Chief for Waropen Kai is established at Waren, formerly at Demba. At Napan there is an Deputy Administrative Assistant in charge of the Administration of this area; the Waropen settlements in Wandamen Bay are within the resort of the Administrative Assistant at Miei. The Administration is trying to bring about a greater unity in the districts under their Administrative Assistants, but for the time being the village population continues to consider the district organisation as an alien institution and the Administrative Assistant as an authority which has been thrust upon them.

As an outward token of their dignity the village chiefs are given a service-uniform which possesses a special significance for the Waropen (see p. 83). Because the Administration tries again and again to make the clan-chief a village chief, they are slightly hesitant about giving every clan-chief a service-uniform. The required number of uniforms is in fact quite considerable. But on the other hand the Waropen themselves do not quite understand why one clan-chief should be given a uniform, whilst another should not get one. This causes discontent which, in view of the latent rivalry between the clans, often leads to a certain amount of dissension between the clan-chiefs themselves.

The position of the clan-chief is not that of a village burgomaster with clearly defined administrative duties. The former *serabawa* was the leader of the raiding-parties, and in general the leader of the potlatch-ritual. With the support of the prominent old men he was quite able to excercise considerable influence. The clan-chief and his men saw to the payment of the wergeld, the execution of "suangi" and the punishment of serious infractions of adat law, because he was the direct, more or less sacred descendant of the ancestors who had once established the moral order and who still now continue to threaten violations of that order with punishment. On the other hand the feeling of solidarity of the clan is so great that the clan-chief need not think for a moment that he might govern his *waribo*; he has no powers at his disposal to enable him to give arbitrary commands or to enforce obedience.

The change from clan-chief to village-chief has made the *serabawa* lose a great deal of his former position. Of course, the village is no longer permitted to pronounce judgement on its own initiative. If the clan-chief were to try to counter the violation of his ancient prerogatives by modern young Papuas with force, he personally would be punished. Because his position is usually not well understood, he is not always treated by the higher officials of the Administration in a way which might restore his heavily damaged authority in the village. On the one hand he is threatened by the dissatisfaction of his superiors, and on the other by the resistance of the members of his clan, so that his position is not always enviable.

Moreover, one forgets only too often that the clan-chief is not the most advanced outpost of modern civilisation, but on the contrary the stronghold of the ancient Papuan civilisation. The clan-chief usually would be the last person to co-operate willingly in executing the very measures which are aimed at changing or abolishing the venerable customs of antiquity.

In general, the chiefs consider their present position in the service of the Administration to be less elevated than it used to be. To explain this they provide two reasons. In the first place, seniority in descent is not strictly considered when appointing Administrative Chiefs, with the result that the *serabawa* not always also becomes Administrative Chief. Secondly, the decisions and the wishes of the meetings of the chiefs and the influential old men are not taken into account. From the standpoint of the Waropen these two reasons are, of course, important enough.

HONORIFIC TITLES OF THE SERABAWA

I wish to devote a special section to the titles because there exists a great deal of misunderstanding on this point. In the Waropen villages one expects to find authorities equipped with certain powers from which they occasionally derive various prerogatives, but one actually finds persons who claim a number of prerogatives, from which they may occasionally derive a certain authority. Furthermore, one expects that these authorities will bear a title which is in accordance with the powers conferred on them, but one actually finds that the title itself is a prerogative. When a Waropen bears some kind of title, this does not imply that he occupies a certain post indicated by the title, but it merely means that in the eyes of his fellow-villagers he has acquired a certain prestige. The title is to be compared with a badge of honour and therefore I use the term honorific titles. Because the clan-chiefs are the persons who enjoy the greatest number of prerogatives, it is mostly they who have acquired a title. However, there are also other men — usually members of the branch to which the clan-chief also belongs — who have obtained a title, whilst it is not always the clan-chief who possesses the highest title. It was therefore possible to find several title-bearers in the old Waropen village, and amongst these usually also the clan-chief. And because people were under the impression that all title-bearers for that very reason were also chiefs, the conclusion was inevitable that in actual fact one could hardly speak of chiefs.

Among the Waropen the usual honorific titles are *sanadi* and *dediawi, korana* and *maiori, kapita* and *dimara*. There was no complete agreement among my informants concerning the exact value of these titles. The general opinion was that *sanadi* was the highest and *dimara* the lowest; for the rest, *sanadi* and *dediawi, korano* and *maiori*, and *kapita* and *dimara* would be of about equal standing. These titles come from the sphere of Ternate and Tidore. D. van Hinloopen Labberton's *Dictionnaire de termes de droit coutumier indonésien* mentions under *sangadji* (Waropen *sanadi*): "chief of a tribe... chef de tribu faisant fonction. — Moluques d'Ambon"; under *majoor* (Waropen *maiori*): "...Moluques d'Ambon; titre distinctif du chef de district (*hukum besar*)"; under *kapitan* (Waropen *kapita*): "Ternatè; title for chiefs; ...Halé-ma-héra, Gorontalo, Moluques d'Ambon, etc."; under *dimo-dimono* (Waropen *dimara*): "les anciens du village, les notables, en To-bélo, Halé-ma-héra. — Circ. de Ternatè"; under *korano*: "chef, en Manokwari — Nouvelle Guinée". According to J. van Baarda's Galelarese vocabulary, the term *korano* occurs in Halmahera with the meaning: "sultan, ruler, king". The Waropen word *dediawi* is probably Tidorese *djodjau*, a term used as the title of the Regent of Tidore, according to the *Encyclopedia of the Netherlands Indies*, IV, p. 335.

As remarked i.a. by Detiger (see above, p. 70), these titles were often given by traders in the old days. However, it would be incorrect to conclude from the unmistakably Moluccan origin of the titles that their importance would therefore have been connected with an appointment by aliens. The custom did actually exist; according to one of the tales, Maseiori, a *serabawa* of Sawaki, was formerly appointed *sanadi* by the well-known Raja Amos. This trader from Ternate was an old acquaintance of the missionaries towards the end of the last century. It is characteristic for the typically mythical turn of Waropen historiography that according to the tale referred to above Raia Amosi was taken to Nubuai by no less a personality than the culture-hero Kuru Pasai. (Texts 49). However, most of the appointments cannot have been made by aliens. Robidé van der Aa says of the Raja of Salawatti, who is said to have exercised great influence on the island of Japen in name of the sultan, that already in the middle of the 19th century the Numfoor seceded from his authority, so that these pretended powers "are based rather on historical memories than on reality".[66]

The acquisition of honorific titles is closely connected with a curious system of friendships which connect individuals from different villages. In the old days, the life and property of travellers were constantly endangered by the raiding parties whose activities will be discussed below. However, persons who stood in the relationship of *kamuki* (friend), spared one another. It was therefore important to establish a wide net of such *kamuki*-relations because in this way one was enabled to move freely in the villages where these *kamuki* were living. The children of a former clan-chief at Napan, called Sopi, remembered, for instance, that their father had stood in a *kamuki*-relationship with a certain Sekuri from Ansus on Japen, with Biriri and Serawoai from Waren, and with Munoi from Nubuai. The informants were, however, neither able to tell me anything about the way in which the partners entered into such a *kamuki*-relationship, nor concerning its obligations and privileges. They only knew that these *kamuki*-relations were usually transmitted from father to son.

One thing is clear, and this is that the *kamuki* maintained some kind of potlatch-relation; they challenged each other. The challenger pledged himself to appear near the village of the person challenged on a certain day, indicated by means of the knotted cord (*ghono*), bringing a quantity of goods, e.g. a prahu, a number of ancient plates, etc., and also a linnen suit. On the day indicated, the challenged man had to be present with a slave he had captured, the latter to be exchanged against the goods brought by the challenger. If the challenged party was able to make good his obligations, he immediately put on the newly acquired suit. The example of such a suit which I was shown

---

[66] P. J. B. C. Robidé van der Aa, *Reizen naar Nederlandsch Nieuw-Guinea* (Travels to N.N.G.), p. 367.

was the remains of an ordinary yellowish tropical suit with a high, closed collar.

Then the people go back to their own village, where a feast is arranged because of the catching of a slave. The challenged person is then allowed to wind a piece of red cloth (*toara*) around his head, whereupon a man who has already acquired a title asks him, which honorific title he would like to choose. Theoretically the new title-bearer is at liberty to select freely one of the titles mentioned above, but in actual practice the man who were to choose too high a title would make himself ridiculous in the eyes of everybody. As the title-bearer may only take off the newly acquired suit during the feast, the celebration on the assumption of such an honorific title is called "feasting for taking off the dress" (*kitaghara saira wa kirua suna*).

There exists not the slightest objection against having oneself given honorific titles by several *kamuki*. According to my informants the old *sanadi* of Woinui in this way acquired titles from Nubuai, Risei, Ansus, Ambai and even from other villages. It is, however, quite possible to stand in a *kamuki*-relation with a person without receiving honorific titles from him.

The acquisition of honorific titles is therefore in actual fact an episode in the ceremonial barter, to which in my opinion in the Waropen area the catching of slaves is likewise to be counted. The slaves who have been captured are primarily destined for the *serabawa* and from this it follows automatically that honorific titles are always found among the *serabawa*. It might be imagined that the "great men" (*manobawa*) would also acquire such titles, just as theoretically everybody might be able to do so, but we found no honorific titles except among members of the *seraruma*. Our informants believed, moreover, that the ancestors would visit a non-nobleman with sickness, if he would arrogate such a title to his person.

Even in modern times this acquisition of honorific titles still continues, although it is, of course, no longer possible to make the stake a slave. During my stay at Nubuai, Munamberi from the Sawaki *seraruma* obtained the title of *korano* from people of Arièpi on Japen, but he simply bought it with sago. Neither did the men from Arièpi give him a suit, but a large prahu, because nowadays the whole ceremony has simply become an exchange. However, on this occasion a raiding party was staged on our behalf by way of demonstration.

The award of honorific titles is therefore distinct from the appointment of chiefs. Haga's report that the Wandamen exercised the supremacy at Weinami because they had appointed one of their people *korano* will therefore simply indicate a *kamuki*-relationship between the people of Napan and those of Wandamen Bay (cf. p. 19). This institution possibly also existed on the island of Biak. W. K. H. Feuilletau de Bruyn says about the *sengadji* of Mambruk "......that of the chiefs who were said to have existed from of old...... several had been elevated to that rank by him (viz. the *sengadji*). As a reward for these services the inhabitants of several villages, led by the newly created chiefs, collected the dowry for his marriage with a woman from Sor".[67]

SUMMARY

The traditional genealogical unit is the mainly patrilineal *ruma,* within which lesser lines attempt to separate under the influence of a notion of status which is based on seniority. Because in these lines it is rather the actual than the traditional relationship which comes to the fore, due to the coherence of the family supported by this feeling of status, the *ruma* is not always strictly unilateral.

It is for this reason that the exclusive cross-cousin marriage in force here has not been elaborated to a traditional connubium-relation between certain groups which receive and those which give wives, as has been observed elsewhere in Indonesia.

In the Kai district there exists furthermore an organisation of mainly local total groups, which co-operate in a ritual; these groups are called *da.* The *da*-relation is also based on a weak sense of family-relationship which is due to the notion of the descent of all members of the *da* from one and the same mythical ancestor.

The ritual on which the arrangement of these *da* in co-operating pairs is based, is losing its total (or clan) character; it is becoming concentrated on those family-lines which grow in importance under the influence of the principle of status. The total ritual is changing into a more economic potlatch-ceremonial in which the rising class of the chiefs, the *sera*, is taking the lead.

---

[67] *Schouten- en Padaido-eilanden,* p. 43.

CHAPTER TWO

MARRIAGE

SEXUAL LIFE

In the Waropen area good breeding does not demand that the relation of the sexes be clothed in an understanding silence. Rather one might speak of an understanding mirth. Stories which undisguisedly mention sexual life are publicly told by young married women, accompanied by shouts of laughter of the listeners of both sexes, including the small children. No attempt is made to keep the nature and the significance of the relation between the sexes somewhat hidden, at least for the younger people, with the result that even the smaller children make gestures and remarks the meaning of which is far from doubtful.

Their dress, at least the old-fashioned pubic belt of the men, does not fully cover the genitals, without this causing the slightest offence. It seems that the penis itself, however, explicitly belongs to the shame, because my assistants who in other respects gave their information without diffidence, were really shamefaced when I tried to note down the words for foreskin, etc. The old women do not cover themselves very carefully either, whilst young girls when having to work e.g. in the mud take off their upper apron and cover themselves more or less with a large leaf. Girls go naked up to approx. eight years of age, boys even longer. Moreover, the arrangement of the houses is such that in this respect nothing is likely to remain hidden. The terms of abuse which men and women apply to each other are nearly always concerned with sexual matters. It is nothing out of the ordinary to hear a woman say to her husband: *"Asua karaba"* (your penis is too long), or the husband: *"Aitasi do"* (your vagina is too deep). An often heard expletive is *ko itasi* (we cohabit in the vagina). It is highly insulting to say to somebody *agho aribingha*, cohabit with your wife. The worst insult imaginable implies an incestuous relationship *agho aghinai*, cohabit with your mother. Still, in these matters some reserve is used; some women talked euphemistically of *mina ri bino* (to be with the wife), instead of the coarser *oko* (to cohabit).

The physiological side of sexual life is therefore far from being shrouded in mystery, nor are the true facts clothed in veils of decorum, especially for the sake of the women. Nobody expects of the Waropen girl that she will behave towards a man like a shy virgin who has to be introduced into this field all confused and blushing. In this society the wife is an independent helpmate in the collection of food, certainly not an inconsiderable task which no man will take from her. In fact, one does not expect womanly weakness in a woman, a weakness to be countered by the man's courtesy. The women are quite able to assert themselves, and also in the house they certainly do not let themselves be pushed from their rightful place. During quarrels they usually create so much noise that the men often are the losers. However, of unmarried young women it is expected that they will not speak so freely about sexual matters as those already married. When a story is told about such things she begins by being indignant, but then she joins in the merry laughter of the others. This is often followed by a joke of one of the men, leaving nothing to the imagination.

Although Waropen possesses a word for shame (*ninggamaro*), this expresses an idea different from ours. In Waropen, shame is rather a loss of prestige than a feeling of extreme embarrassment. A girl therefore becomes ashamed when she is more or less disgraced. Occasionally *ninggamaro* means "to have respect for"; in Waropen Kai, for instance, a girl has to be *ninggamaro* towards the brother of her mother and her mother-in-law. In the myths which discuss the incest between brother and sister, the perpetrator of the incest, i.e. the brother, is said to have been *ninggamaro* so that he committed suicide.

Hence, although there is hardly a question of shyness or diffidence concerning the physiology of the relation between the sexes, this does not imply that the members of both sexes are on terms of easy familiarity. The contrary is true. In public the sexes keep carefully apart; the women always go and sit together and so do the men. Never will a girl engage a boy in conversation in public. However, between members of the same sex the contact is rather intimate. Boys hold each other by the arm or by the neck, and they are romping all the time; here all kinds of sexual horseplay are among the pleasantries that are condoned. In this respect the girls are more reserved. Among men anal sodomy also occurs, but according to my informants this only happens when they are drunk. Although this is not judged to be a direct violation of the manners established by the ancestors, it is

considered highly ridiculous, so that it is a serious insult to say of
somebody that he is a sodomite (*agho rironi*).

Intimacies between members of the two sexes are not permissible
in public. When an older woman had herself massaged by a boy against
rheumatic pains, she was clearly given to understand that this was
not considered innocent. The liberties which people of different sexes
are permitted to take lie within the sphere of a "joking relationship".
Young people who are potential partners in marriage and who are
therefore each other's *firuma*, will commit certain rather rough pleasantries. In the evening, when the sun goes down, the young people

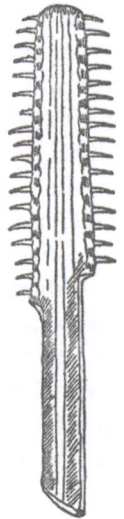 

Fragment of the sword of a swordfish
(length approx. 20 cm.) and leg of a crab,
used to inflict erotic scratches.

come together in the prahus. Then it is the young men's delight to
pursue the boating girls with loud shouts, or to push their prahu over;
then the girls give telling blows. Another pleasantry is often to give
a rather rough pull on the dancing tufts of hair of an unsuspecting
firuma who is permitted to give her tormentor a resounding punch or
knock on the back amidst shouts of laughter.

At the dances such pleasantries go even further. Then the *firuma*,
often egged on by the older spectators, will inflict bloody scratches
with the sharp leg of a crab, or even with the wicked tip of the sword
of the swordfish (*kasura*). Sometimes such scratches go home and
give rise to painful wounds; often, especially among girls, they swell
to considerable scarifications. Such scarifications demonstrate one's
popularity among the other sex and so they are signs of which one is

proud.[68] This custom is unknown in the Napan area. The description on p. 76 proves that in general one may be rather rough in one's behaviour towards the *firuma*.

If the relation between the *firuma* is continued, the erotic scratching and beating belong again to the more intimate love-making. A telling expression for love-making is therefore the term *kisikimundaruko* (these two beat each other). The intertwining of the fingers, bringing together the backs of the hands, is also considered a great intimacy (*kisikikamondaruko*, these two are stuck to each other).[69] Kissing and hugging are only known as funny customs of the Europeans. When the senses have been strongly excited, lovers suck each other's tongue and lips (*kitumararuko*, they suck each other).

When a girl wishes to respond to the attentions of a *firuma*, she receives him surreptitiously at night in her house. Mostly a few girl-friends or sisters are together; they make a younger girl tell their lovers that they are expected. Sometimes the lovers exchange small gifts: some tobacco or a small ring, and then one looks for a quiet place. If the couple have serious intentions, this exchange of small gifts usually is the introduction to a more considerable exchange of gifts between the parties, leading to a marriage. This nocturnal love-making can hardly take place in the crowded room, so that only the central space of the house or the front gallery remain.

Now cohabitation on the front gallery of the house is considered especially dangerous, as is cohabitation somewhere in the forest, e.g. in a garden-hut, where an amorous couple might also be together undisturbed. It is said that the soul (*roséa*) of people who cohabit on the front gallery or in the forest is stolen by spirits (*binano*) who live under water. These water-spirits sit down with the stolen soul in the same attitude as a love-making couple. The intervention of a *ghasaiwin* is necessary to bring back the stolen soul, and even so the soul can only be returned if the child dies that is the outcome of this intercourse.

---

[68] These scarifications have therefore not been burnt in, as reported by the *Encyclopaedie van Nederlandsch-Indië* (Encyclopedia of the Netherlands Indies) (see above, p. 26). According to my informants, no special means are used to develop the scarifications; against the pain the open gashes are sometimes bathed with water. For the rest the development of these scarifications would be a question of hereditary disposition.

[69] When we were still unacquainted with this view, we often had difficulties when my wife put out her hand for support, e.g. when climbing up into a Papuan house. It was only later that we discovered the reason for the laughter which arose when my wife impatiently grabbed the hand of a man who rushed forward to help her, in spite of his attempts to prevent this.

In other cases the soul is so heavy that the *ghasaiwin* cannot take it out of the water so that the guilty party has to die. This belief is only prevalent in Waropen Kai, but young people from Napan who would like to start amorous adventures at Waropen Kai are strictly warned by their parents against the dangerous consequences of cohabiting on the front gallery or in the forest. Furthermore, it is prohibited to step across the head of a sleeping *firuma* because then the *firuma* is forced to marry. It is said that despairing lovers have actually brought unwilling girls to marriage in this way occasionally.

When engaged in amorous dalliance the girl sits on the spread legs of the boy, putting her legs around his hips, an attitude hardly likely to prevent sexual intimacy. Premarital pregnancy rarely occurs in spite of this, and in any case it is condemned. At Weinami the boy is obliged to pay a fine over and above the dowry in case the girl has become pregnant before marriage. Nocturnal love-making is, in fact, not officially permitted by the parents and they put on a show of great indignation when their daugther has been "deceived" (*iokofari*, he deceives her), although actually they were very well informed. Undesired suitors are ignominiously driven away upon discovery.

According to Detiger the Waropen are unacquainted with physiological paternity; he writes: "The Papuas are of the opinion that coition has to be repeated several times before pregnancy can occur. They are unable to establish a correct understanding of the causal relation between coition and pregnancy. Gestation is not conceived as limited to a certain definite period. In this way a Papua will not be astonished in the least at the joyful circumstance that upon his return to his village he finds his wife, who had been left behind, in a pregnant state".[70]

However, the information at my disposal does not indicate ignorance of the significance of physiological paternity. People were unable to say exactly what was the function of the testes, but they did know that a castrated animal was no longer fit for breeding purposes. Pigs are castrated to make them heavy, whilst this is done to a dog for the very purpose of keeping it away from the bitches. In childless marriages the possibility of the husband's incapacity is really taken into account; this would arise if a man had been spit in the face by a large mollusc which lives on the reefs (*nungguba*), as was said to have happened i.a. to the childless Dedui. Pregnancy during a pro-

---

[70] *Adatrechtbundels* (Bundles of Adat Law) XXXIX, p. 431.

longed absence of the husband was only considered possible, if the woman had had intercourse with another man. A Waropen who was in gaol at Manokwari was told with obvious glee that his wife had come down with child, and the rage the prisoner showed on hearing this piece of news caused a great deal of merriment. Although in mythology supernatural fertilisation by contact of an object with the chest or the thigh of a woman and even of a man is repeatedly mentioned, this kind of impregnation was unanimously and definitely denied for the normal and profane world.

However, the men are evidently unacquainted with the processes of pregnancy and birth. Although women speak unrestrainedly about sexual intercourse, they become silent as soon as they are questioned about menstruation, childbirth, etc. I only found one older woman who was willing to give some information, provided her husband and my wife were present at the interview. With this reservation she did not object, however, against the presence of her brother and even of men who were not related to her. From this one can only conclude that for men at least a formal ignorance of these matters is assumed. The information men are able or willing to provide on these subjects is very imperfect. A man who was several times a father pretended not to know that the umbilical cord (*paitokera*) is connected with the placenta. Of the placenta itself he could only tell that it had to do with childbirth. He even maintained that he did not know the word for it (*dembai*). However, another man did know exactly what was the matter. In any case, there exist rather stringent rules of segregation which prohibit the man from being present at the delivery. My informants believed that it would only make him feel sick. Probably we have to think here of phenomena connected with the couvade.

The Waropen do not possess an exact chronology and so they are unable to indicate precisely the duration of the gravidity. Pregnancy is diagnosed by the absence of menstruation and the change in colour of the mamillary region. The pregnant woman is subject to a number of prohibitions, like e.g. the eating of shark (*sokabura*), ikan sembilan (*moa*) and another type of sheath-fish (*foamano*), because otherwise the placenta will not be ejected. Other prohibitions evidently have been deduced from a sympathetic relation: the woman is not allowed to eat mud-jumpers because otherwise her child would have protuberant eyes just like this animal. The state of a pregnant woman is believed to be more or less dangerous, with the result that she wears a variety of amulets; some women wear the tooth of a deceased male relative who has to protect the child against the ancestors and the *dareo*.

Confinement takes place in the partitioned part of the room which serves as privy (*feretei*). The woman squats down on a stick or on a piece of the rib of a sago-leaf and is supported in this position by a mother or a sister in the way support is given to the seriously ill. For a primipara it is assumed that another three days will pass until childbirth, counting from the first labour-pains; in case of a multipara it is expected earlier. When the pains start (*fo ma irofera wa rikugha ra ghero*, she suffers pain towards the coccyx because her child is descending) they may be increased by smearing lime and areca on the belly. The old woman who assists at the childbirth (*ioaiwa sana*, she is sitting at the confinement) knows that it is possible to accelerate the process by breaking the skins (*fafaiana*) and allowing the amniotic fluid (*masino*) to flow out. When the placenta (*dembai*) has also been expelled, the umbilical cord is cut with a bamboo knife (*buranaiwiro*). The only other use for the bamboo knife is for carving the *faia* (Mal.: bawal hitam), a kind of fish which is absolutely taboo for the young child.[71]

The placenta, also called *bawa*, elder brother or sister, is taken to the mangrove forest behind the house where also the children are deposed who have died young. According to some, the placenta of the first-born is buried, but according to more general reports it is put in a basket and placed in a tree, hanging free, in order to prevent the child becoming a cry-baby. The umbilical cord is dried and preserved; if later the child is to go on a long journey, it is to be hung in a tree. If the cord is twisted around the throat of the child, it is permissible to cut it before the placenta has been expelled. Means to expel the placenta are unknown and no effective help can be given when the child lies in the wrong position. According to the Waropen, mortality at birth is not very frequent, but this statement I do not believe to mean much.[72] Although it happens occasionally that the *ghasaiwin* renders assistance, this is not necessarily part of her function. After childbirth women sometimes wear a white porcelain-shell (*korombowi*), attached to the hip-band by a string of beads.

The custom of killing children, e.g. because they are not desired or

---

[71] Would the sound-association between *faia* and *paitokera* perhaps also play a role? *Faia* is also "dowry".

[72] When noting down the genealogies I observed quite a considerable mortality among the first-born; this may perhaps be connected with the youthful age of the young mothers. See on this point M. F. Ashley-Montagu, *Coming into being among Australian aborigines*, p. 246.

because the mother has died or because the child is deformed in a certain way, was said to be unknown. However, this does not mean to say that therefore these children are given the care which would enable them to stay alive. At Weinami a motherless and famished child was given to an old woman who was of course also unable to feed it, so that it died of exhaustion. In those rare cases when a mother is unable to feed her child she gives it to a relative (*raisusi*, my foster-child). The foster-relationship does not constitute an obstacle to marriage, although it is considered unfitting to marry the actual child of one's foster-mother. There is no aversion against twins, provided the mother is capable of caring for them. People were able to mention several adult twins, both of the same and of a different sex. Albinos are likewise left alive; they are rather frequent in the Waropen area.

Abortus is sometimes, especially in cases of premarital pregnancy, provoked by heating the belly, or kneading it, or by the use of external violence. Women believed to be able to prevent pregnancy by stuffing a loin-cloth (*rario*) and a hip-band (*ghono*) into the hole of a certain blackish lobster or prawn with a long tail (*dodoa*) which lives in the mud and which is also occasionally eaten.[73]

A childless marriage is generally considered unhappy and it often leads to polygamy, although incapacity of the husband is also recognised. The unfortunate woman is often heaped with reproaches and she is given all kinds of potions to drink, like the husband. For a young girl pendulous breasts are believed to be an indication of future childlessness. The Waropen ideal of female beauty is a sturdy but supple figure, with a smooth skin, not too dark, and with delicate traits. Their appreciation of beauty does not differ much from ours.

During menstruation (*si erara*, the vagina bleeds), the women use plucked coconut bark; they maintained that it did not give rise to complaints. During this period the woman sleeps apart, but not in a separate little house as is done among the people of the interior. At that time cohabitation is prohibited because otherwise the husband would grow bald.

THE MARRIAGE NUCLEUS

In the chapter on social organisation I explained that there are no fixed connubial relations between the Waropen *ruma*, so that on the

---

[73] At Weinami there are objections against giving the *dodoa* to young people to eat.

occasion of a marriage the *ruma* as a whole never enters into relations with another *ruma*. Within the *ruma* each line has its own interests and objectives which it will defend with force against the other lines if necessary. The marriage agreement therefore includes two lines in a contractual tie which at one time may embrace quite a number of people, where a marriage of important persons is concerned, and rather few at another when the young couple occupy a less prominent position.

The number of marriage contracts which connect the members of one *ruma* with others constitute a complicated set of obligations. With every marriage a number of people are to be found in the *ruma* who remain complete outsiders and who therefore also continue their normal work during the marriage ceremonies. The same peculiarity is to be observed likewise on the occasion of other ceremonies in the *ruma*; the cases where the whole of the *ruma* acts as one single unit are very rare. With initiation and funerary feasts it is always a small group which is concerned; the other members of the *ruma* restrict themselves to the role of more or less interested spectators. Only during the Serakokoi ritual (see p. 193) the whole *ruma* is in a condition of taboo but in that case the taboo is not limited to the *ruma* alone. Therefore the *ruma* is not the unit which accomplishes those "total" ceremonies and in this respect it possesses much less of a religious character than the clan which in any case is still concerned as a whole in the relation of rivalry between the "head" and the "tail".

The group of persons who enter into a contractual relation due to a marriage I shall indicate by the term "marriage nucleus" in the following pages. The marriage nucleus is therefore not a sharply defined social unit, but it simply includes those persons who, on the occasion of a marriage, are willing to concern themselves with the contractual obligations this entails (mainly, therefore, barter-obligations). Just as the marriage contract actually forms the basis for many other barter-obligations — e.g. those of the initiation ceremonies — the marriage nucleus is likewise a combination of persons who stand in a barter-relation, not only on the occasion of one single marriage, but also for many years afterwards.

In its simplest form the marriage nucleus therefore contains a brother with his daughter and a sister with her son. A person possessing several sons or daughters may belong to more than one marriage nucleus, and if the brother-sister relation is taken in the classificatory sense, the extension is still larger.

The relation between brother and sister has especially attracted the attention in Waropen mythology; it is always elaborated to form an incest-motif. Before her marriage the sister is the brother's ally; she tries continuously to obtain goods from the wives of her brothers. However, as soon as she marries, she comes to stand between two parties. Because of her marriage and her residence she belongs to the husband's side (*mambarengga*), due to her descent she remains connected with the woman's side (*bimbarengga*). Her husband is the father of her children; he has to provide goods to the brothers of his wife. But he is also the person who has to try and obtain as many goods as possible from the brothers of his wife in order to transfer these to his own married sisters. The brother is the married woman's official protector and he has to be always ready to vindicate her interests, with force if necessary. On the other hand he is likewise her husband's opponent, continuously attempting to obtain riches from the house into which she has married.

It is especially the wife's brothers who are very aggressive in their attitude towards their sister's husband, who in his turn may be assisted by his own brothers. Hence, it is important for a husband to maintain friendship with his wife's brothers by means of fine gifts. This aggressiveness becomes apparent on numerous occasions, e.g. on the presentation of dowries and on the occasion of the *saira* festival. For when the wife's brothers are not very satisfied about the barter-relations, they are entitled to manhandle the husband's side. This is demonstrated by all kinds of rowdiness, e.g. by agreeing that the whole crowd of dancing participants will keep exact measure in dancing. In normal dancing there is a good deal of stamping and jumping, but there is no danger that the house will be damaged. If, however, all dancers keep time in stamping on the floor, the piles under the house are shaken loose (*kiki rumagha,* the house is shaken loose), so that the house is dislocated. The construction of the Waropen houses is such that in those cases only the central portion (*wundo*) is wrecked, whilst the dwelling-rooms (*arado*) remain undamaged. Sometimes even the beams which support the roof-ridge in the central portion and the foremost of which represents the ancestor, are wrenched free and thrown away somewhere else, of course a very public insult. And occasionally even then the wife's brothers are not satisfied. They penetrate into their brother-in-law's dwelling-room and slash through the four rattan slings in which the fireplace is hung (*awusara*), with the result that this disappears into the mud below the house or in the water at high tide, with all the pots and pans and other household utensils and food which might be on it.

The husband has to be careful to take this treatment as a joke and to put a good face on a bad business by never letting it appear that he has been made deeply "ashamed". The next day the brother of the wife who has provoked this piece of rowdyism will come to him to express his regret at what has happened; by way of excuse he will adduce that he was drunk last night and finally he will magnanimously offer compensation. The husband can do little else but offer a very lavish counter-gift for this "compensation", if he does not want to risk a very serious quarrel with his wife's brothers, with the possibility that his wife goes back to her brothers, resulting in a serious decline of his esteem.

The goods which are generally used in barter are old pottery and sago. Furthermore, from the husband's side to the wife's side there is an exchange of:

dark-blue cotton (*rari*)	
other pieces of cloth	shell-armlets
feathers of the bird of paradise	ancient beads
prahus	beadwork (dancing-aprons etc.)

This division is not painstakingly upheld when exchanging goods, but still it is maintained on the whole.

If a man wants to marry he has to collect a dowry for the wife, consisting of ancient pottery etc. (*ioramuna bingha*).[74] For this dowry he will receive a counter-payment from the side of the wife; there does not exist a technical term for this action, but it is sometimes called *bipora* (where we observe the word *bino,* wife, and *pora,* payment, hence "payment from the wife"). The religious significance of the barter-transactions is expressly demonstrated again by the circumstance that the party which offers sago also always has to present a few bowls of prepared sago.

When children are the result of this marriage, the wife's brothers are the natural initiators of these children. During the initiation ceremonies (*saira*) they are obliged to present old pottery for the sake of their sister's children, receiving again sago in return. Within the marriage nucleus the exchange of gifts therefore takes place in two contrary directions: on the occasion of the marriage from the husband

---

[74] From *ora,* to ward off, and *muna,* to fight?

to the wife (i.e. from sister to brother), and with initiation again from the wife to the husband (i.e. from brother to sister).

By way of a schematical indication:[75]

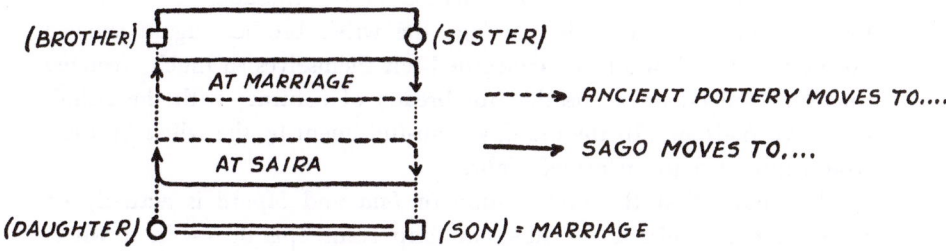

In general the gift of sago is of less value than that of ancient pottery, so that one cannot say that the sago-payment is equivalent to that of the pottery. However, one must not forget that in this exchange of gifts there is a marked tendency to overwhelm the other party with one's present, so as to force it to a large counter-gift. In order to obtain the highest price for one's goods one must not sell them in the open market, but give them away. The contractual value of the objects of these barter-transactions, exceeds their real sales-value which is mainly determined by supply and demand. Giving away an ancient bead or another object to a rich opponent guarantees a rich counter-gift.

Although the return-present in sago may be smaller than the gift of pottery, a certain ratio has still to be observed. When Mui, the senior descendant of the *seraruma* of the Sawaki clan, married his second wife, his family received as *bipora* 29 tumang of sago, among which one very

---

[75] If we may assume that formerly there existed a fixed connubium-relation, this circulation of goods might be exemplified by the following plan:

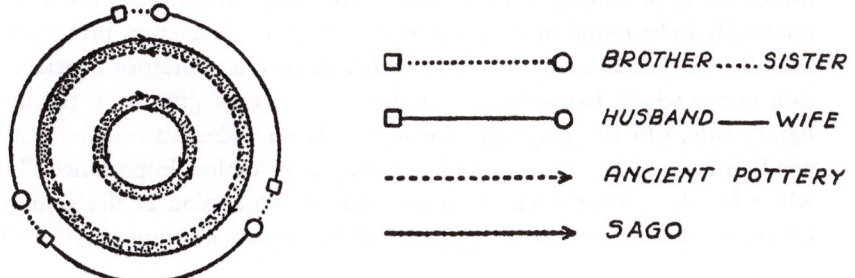

Inner circle: circulation at *saira*; outer circle: at marriage.

large one, the so-called *fi bawa,* as well as three dishes of prepared sago, plus a number of carrying-bags filled with ancient beads and shell-armlets. This payment was carefully noted by means of counting-sticks (*keri*). As a return for the dowry offered by the Sawaki it was considered very meagre and when on the occasion of the initiation ceremony of the first-born child of this marriage the wife's brothers again showed no interest, the Sawaki line concerned felt themselves so much wronged that they lodged an accusation for breach of contract with the Administrative Assistant. In the old days non-fulfilment by the wife's brothers would have led to an armed fight.

This shows that the presentation of *faia* and *bipora* is actually no more than an episode — albeit an important episode — in a whole series of exchanges which extend over a period of years. Marriage is the most important contractual engagement; for the Waropen at least it would be incorrect to consider the presentation of the dowry as an independent act.

We should not forget this when pronouncing an opinion on the dowry; only too often the payment by the husband's family to the wife's family is considered by itself, the payment being taken as the purchase-money for a bought article, and this is contrary to a correct appreciation of the marriage by the Papuas. Indeed, the Waropen does speak about paying for the wife (*iapora ribingha*) when payment of the dowry is meant, but what is furnished is still different from the payment for some bought article, like e.g. a prahu. One might only speak of "buying" if the compensation offered by the wife's side were to lapse.

The word *pora* means "to pay" in a highly pregnant sense; it rather means "requital, retribution". In this sense it is used for the compensation given to ransomed slaves when they have been redeemed by their relatives (see p. 225); likewise for the wergeld for manslaughter (p. 213), or on the occasion of the marriage of a widow (p. 124), or for the fine for breaking an engagement to marry (p. 104). The same word is undoubtedly to be found in the combination *mupora,* to avenge, presumably literally "to kill in requital". The wife's family is therefore considered as a group which has to be conciliated by means of gifts from the husband's side. On the contrary, the *bipora* is an indemnification for the goods presented by the husband and therefore of less importance. The wife's family evidently feels wronged due to the cession of the woman; its aggressiveness has to be propitiated by means of some payment.[76]

---

[76] Would *pora* also be the same word as *por,* the token anciently sent round to related groups in the Vogelkop peninsula to participate in a raid, usually

26, 27. Proud possessors of ancient pottery

28.   Modern coffin

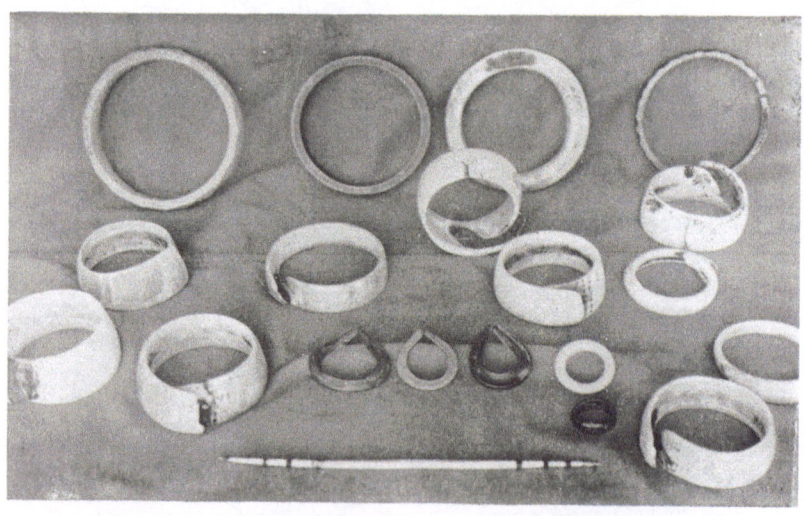

29.   Bands of celluloid and shell, ear-rings and nose-ornament

Etymologically *pora* will be related with Trobriand *mapula*, "being the general term for return gifts, and retributions, economic as well as otherwise" as Malinowski puts it in his well-known description of this specialised exchange of gifts, the *kula*.[77]

In his discussion of this word Mauss brings forward some important views concerning the character of the relations between husband and wife among the Trobriand islanders; of the *mapula* he says: "Il est en général comparé à un 'emplâtre', car il calme la peine et la fatigue du service rendu, compense la perte de l'objet ou du secret donné, du titre et du privilège cédé ".[78]

It happens occasionally that one of the parties is so furiously opposed against the marriage that it refuses to give the dowry or to return a compensation. This has a detrimental effect on the marriage because it does not possess the same contractual ties as another and is not considered quite seriously. When the husband is unable to provide a dowry, he usually goes and lives with his wife. A married woman whose brothers refuse to make the counter-payment, occupies a subordinate position in her husband's house and if her brothers were to take her side, this would not be accepted. They would be told that they had better first improve their sister's position by making a proper exchange of goods possible.

The present conversion to Christianity has caused the lapse of the initiation festivities and the connected exchange of gifts, and this cannot but have an unfavourable influence on the establishment of the dowry, as remarked by a *serabawa* from the Napan region. For if a man had both sons and daughters, his line received ancient pottery; in the first place on the occasion of his daughter's marriage — viz. from the husband's family, and in the second place on the occasion of the initiation of the children of his son, viz. from the family of the son's wife. Now this last possibility of obtaining ancient pottery has been eliminated because of the conversion to Christianity and people therefore try to obtain more when their daughter marries; in other words, the dowry shows a tendency to rise.

---

consisting of a bush-knife or a spear? — I would also refer to the curious expression for one of the *saira*-episodes, viz. *kiweiwa kugha* (see p. 129). This would mean literally "to give together with the child". Would this mean to say that now the wife's family has transferred the child together with the mother to the husband's family, the aggressiveness may be considered as appeased?

[77] B. Malinowski, *Argonauts of the Western Pacific*, pp. 178, 182.
[78] M. Mauss, "Essai sur le don", *Année Sociologique*, N.S. I, p. 82, note 5.

In these exchanges money is hardly ever used in the Waropen area. My informants said: "Money does not stay. It can only be used for buying goods in the store, for areca and tobacco". However, money has a confusing influence because one can easily buy cheap pottery with it and some people are always trying to smuggle this into the exchange transactions as being of equal value with the ancient pottery.

The exchange of gifts on the occasion of a marriage, finally culminating in the presentation of the dowry, starts with a few personal attentions between the lovers, who present each other with some tobacco, a small carrying-bag, a little ring, etc. The village mostly knows before long that the young people are sweet on each other, and once things have been discussed at ' home, if the girl is a suitable *firuma*, industrious and of a good family, and if the boy is able to provide a proper dowry, no great difficulties remain. Then the boy goes to the girl and definitely asks her to marry him. He will say, for instance: "*Aghoai wa rawu aue?*", "Will you wait until I marry you?".[79]

When the girl agrees, a relative of the boy, e.g. his brother or his mother, goes to sound the girl's family who are already informed, of course. The go-between will say e.g.: "*Ruana ariwiamani rawa raiwa-rima*". "I ask whether your girl may go with our boy". When the atmosphere seems favourable, the intermediary leaves a small present, for which a counter-present is given on the occasion of a return-visit. At Weinami no present is given when proposing marriage. If the go-between is going to meet with a refusal, this will be clearly indicated by the lack of enthusiasm with which he is received. A girl without brothers and a boy who is unable to collect a dowry, make a poor match.

Once marriage has been sollicited, relations are officially settled. From that time on there exist prohibitions between the boy and the girl and they have to avoid each other as much as possible. Under those circumstances they amuse themselves with other *firuma*, an intercourse which is not prohibited, but which, on the contrary, seems rather to be encouraged. It is even the rule that on the day before the marriage the boy retires once more into the mangrove forest with his favourite *firuma* and the girl with her admirer, and it is considered as indelicate even to tease the boy or the girl about this.

Now the boy's family starts to collect the dowry, whilst he maintains

---

[79] I could not find out whether this is meant as a formula which serves as the introduction of an engagement, or whether this is simply an example of what may be said on such an occasion.

the friendship by means of small presents. Usually the girl has to move to her parents-in-law a few weeks before the marriage, in order to help her mother-in-law with all kinds of duties, like fetching water, etc. During this time the girl is allowed to take one of her girl-friends with her, whom she has to give a present later. For the time being the boy has to go and live in another house, where his food has to be taken to him.

Parents often do not wait until their child has personally made a choice from among the *firuma,* but already quite early they enter into an agreement for the future marriage of their children. In that case the exchange of gifts extends over a much longer period.

The marriage as such, however, is not concluded at a youthful age, and it would be incorrect to speak here, as is often done in the Moluccas, of child-marriages.

It is considered commendable to present the *faia* and the *bipora* as soon as possible after the conclusion of the marriage.[80]

Some families who are less concerned about their prestige sometimes postpone payment of the *bipora* until children have already been born, because in this way it is possible to some extent to avoid the presentation of gifts for the first *saira.* However, such methods are not conducive to making the position of the young wife more pleasant, although for her livelihood she is never directly dependent on her husband's family. She remains entitled to the sago-property of her own relatives, which her brothers must help her to process. That actually the bride should be solemnly transferred together with the dowry on the occasion of the marriage is clearly shown by the technical term for the presentation of the dowry, *kikaringgiwa bingha,* to load together with the wife, although often the marriage has been in existence for so long that the woman no longer feels like being transported on a festive boating tour.

The presentation of the gifts on the occasion of a marriage is accompanied by some public show. In case of a *faia* the goods to be presented are strikingly exhibited; in the prahu decorated with flags, the pieces of cloth are hung on sticks and the plates are fastened with rattan outside the prahu's gunwales. A small festive procession rows through the village with beating of gongs, whilst girls sing and dance

---

[80] At Weinami it is proper to present the dowry on the day following the conclusion of the marriage.

the *ratara,* the marriage song, in the prahus.[81] It is mainly women who
are concerned with the presentation of these payments. The goods are
spread out at the entrance of the house of the bride, and they are
counted by means of counting-sticks (*kerio*). For large pieces the latter
are bent. Then the *faia* is immediately divided among the women of
the family-line to which the bride belongs. The women who have
brought it, relatives of the bridegroom, are rewarded with some beads
and shell-armlets, and return.

The dowry is always computed in units. One speaks of so-and-so
many *a*, things, objects, in Malay of so-and-so many barang. The objects
are only distinguished as "large" or "small". For the rest, everybody
knows which standard-value is approximately required of a "barang"
to be able to count as a separate unit of exchange. The potlatch character of the whole of this transaction provides a certain guarantee
that people will not try too much to present less valuable objects. In
those transactions which are not based on reciprocity, like payments
for fines, people will occasionally evade difficulties by offering cheap
store-pottery in stead of the commonly accepted pieces of ancient pottery (*mori-mori, katiti, rewanggu,* etc.). However, genuine old pottery
is becoming increasingly scarce in Geelvink Bay, especially because
some thirty years ago this was also often collected as taxes. Some of the
very valuable ancient pieces have a name of their own and their progress through circulation is followed by the attention of the public.

My informants were not acquainted with forms of marriage intended
to avoid payment of the dowry. Even in cases indicated by the name
of "marriage by elopement" or "run-away match", this intention is
not present. According to the teacher at Waren, the so-called service-
marriage is known; here the payment of the dowry is replaced by labour.
However, according to my informants, a marriage without gifts is always
considered as a socially inferior relationship, so that none of the parties
remains satisfied with it. A person whose family is slow in paying the
dowry will, of course, hardly occupy a prominent place in his wife's
house. And if he goes and lives with his wife (which is quite possible,
because slowness in payment usually is the result of dislike of the marriage), the husband will indeed have to work much harder for his
wife's family.

Sometimes the presentation of the *faia* is more noisy, also perhaps
because the wife's family is believed to be more aggressive than the

---

[81] See below, p. 268.

husband's. By way of an example I will describe the payment of the *bipora* by a group of the *ruma* Mamurani to a group of the *ruma* Refasi, both belonging to the Apeinawo clan. Attention should be drawn to a point, to be discussed in more detail below (p. 163), viz. that among the Waropen the formal execution of most ritual actions is determined by the constantly changing conditions. For this reason many of the ritual actions will not be discussed *in abstracto*, but concretely, as we observed them in actual life, because in their formal execution again and again considerable differences may be noticed.

At the presentation of the *bipora* mentioned in the preceding paragraph the wife — who had become a mother in the mean time — was present in bridal array. The different tumang of sago as well as some bowls of prepared sago had been arranged in the central portion of the house of Mamurani. Moreover, there was a *sabakugha*, a tobacco-prahu, which actually belongs to the *saira* ceremonies, so that the payment was evidently also intended more or less for the initiation of the child which had been born meanwhile; when presenting a *bipora* this *sabakugha* is definitely superfluous. To this tobacco-prahu a *sambonama* had been attached, a kind of wooden jumping-jack who holds an oar in his movable arms. Playing children will sometimes attach such a *sambonama* to a stick in their prahu; if a string is attached to the movable arms of the puppet and the other end of the string is held, its arms are pulled at every beat of the oar, so that the manikin also goes through the movements of rowing with its oar. According to my informants the *sambonama* is nothing but a toy which I have also always seen playing children use in this way; attaching a *sambonama* to the ritual *sabakugha* was therefore evidently merely done for fun.

Whilst the sago-stock was being exhibited in the Mamurani house, a number of female relatives of the wife stormed angrily at the members of the Refasi house, dancing in a challenging fashion like men do who want to challenge an opponent, bush-knife in hand. Her father's sister, an old woman, carried on as if she were mad, yelling in the direction of the Refasi house: *"Mapora andiaini ewomo"*, "You have not paid our goods"! This was meant to indicate that this *bipora* was more than ample repayment for the dowry.

Then everything was loaded in the prahus. The bride took her seat on a clothes-chest, an object which in these modern times is used by way of a ceremonial seat, and accompanied by singing and the beating of gongs the procession rowed to Refasi. Here the inhabitants behaved as if the whole affair was none of their concern. The front gallery, where the men usually sit, was completely deserted. The wife's female relatives took their bush-knives and struck at the supporting pillars of the *ruma* Refasi, shouting angrily "*Mimbo maeeee! So raka mueeee*"! "Come on! Are we perhaps afraid of you"!, indicating on their part that the people of Mamurani did not yield to the dowry from Refasi. When the goods were being taken upstairs, an old woman of the husband's family took one of the bowls with prepared sago-mush and danced around triumphantly. Both the *sambonama* and the *sabakugha* were immediately pulled to pieces by the husband's brother. Several men hastily snatched some tobacco from the *sabakugha* whereupon the rest was thrown into the water in front of the house, as a sacrifice to the ancestors, under the words: "*Daida, mindisabaku inekini*", "Fathers, here are bits of tobacco for you". The women who had brought the *bipora* were rewarded with plates, beads, armlets, etc., which were accepted with loud jeers at the stinginess of the Refasi people.

Among the tumang of sago one often observes a very large one, the so-called *fi bawa*, the large tumang; these *fi bawa* are to be met with not only on the occasion of marriages, but also with other festivities. The party which has to provide the sago (hence, for a *saira* the husband's party, and for a marriage the wife's), often has the leaves for packing such a large tumang solemnly brought to its home by the other party, in a decorated ceremonial prahu manned by singing youngsters. Accompanied likewise by singing the sago for the *fi bawa* is thereupon solemnly packed in these leaves by the men and women of the party who have to provide the sago. It is from this *fi bawa* that the sago for the feast is taken; the remnants may be taken home by the party which has provided the leaves. All this is therefore a kind of potlatch challenge: "You have not so much sago that you could fill these leaves", countered by the challenge: "You cannot eat so much sago during a feast as we can pack in these leaves".

Breaking the marriage promise implies repayment of payments already made, and this is particularly true when the girl breaks the promise. Mostly the families on both sides will only acquiesce if the break is irreparable, e.g. because the girl has run away with another man. Besides restoring the goods already received, the girl's family also has to pay a fine to the injured party (*iapora kiriniamaragha*, he pays their shame). The injured party also publicly puts forward its claim for restitution and for payment of the fine by presenting the *mumui*. This is a piece of the rib of a sago-leaf into which a number of small sticks have been stuck, indicating how many "barang" are demanded — the whole usually fastened on a stick or a piece of bamboo. This stick is delivered to the other party and tied to the entrance of the house. The fixing of the *mumui* has to be observed by one of the members of this house, or he has to receive it. This can only be done by an old man, because a youth would run the risk of likewise becoming involved in such an affair later on.

When the girl Sorei from Pedei ran away with Kewowi from Apeinawo, in spite of her engagement to Sewuri from Kai, Sorei's family was obliged to pay a fine of twenty objects, beside restituting the gifts that had already been received. When later Kewowi's dowry was handed over, the *mumui* was also taken along as a sign that the fine had been paid. This *mumui* was therefore given by Sewuri to Sorei and by the latter again to Kewowi, i.e. it was not Sorei herself, but Kewowi who had to pay the fine. It should be noted parenthetically that during the presentation of this dowry one of the girls very demonstratively carried a stick on which silver armlets had been put. The reason for this was that during the quarrel which arose at the first rumours concerning the flight of Sorei and Kewowi, Kewowi's family had been taunted by Sorei's relatives that they were too poor even to be able to present one silver-armlet with the dowry.

When a man breaks the marriage promise, the results are mostly not so far-reaching. Normally, a man will not begin offering presents to one woman, to divert his attentions later to another. In most cases these are flirtations where the girl has taken things more seriously than the boy. In such a case the girl has no official means of compulsion at her disposal, so that she often takes recourse to black magic, usually directed against the first child of the man; in any case, mortality among the first-born is often ascribed to this cause.

Divorce is comparatively rare, because a marriage is only completely liquidated when all payments and counter-payments have been restored. And because the objects have long since re-entered circulation and the sago has been eaten, the relatives on both sides do their utmost to prevent a divorce. Repeated adultery may give the husband cause to give up his wife (*ionarera ribingha*). Usually it is the woman who breaks the marriage-tie by fleeing to her family, because of repeated adultery by the man, and more often because the husband wants to take an additional wife. In case of divorce the woman who runs away has a practical advantage: her husband is not entitled to her sago property which is and which remains for her and her brothers. The dowry has been paid to her family. Moreover, as the wife living in her husband's house, she controls part of the objects which were taken there, e.g. during *saira*-ceremonies by her brothers for her children, and when marrying off women from the *ruma*, these objects by rights belonging to the husband's *ruma*. Mostly, the woman finds an opportunity to take along at least part of these objects on her flight, with the result that the husband's side will do the utmost to undo the separation.[82]

In Waropen society there is no question of married property or of objects which belong to husband and wife together. All transactions run between the wife's family-line and the husband's. Husband and wife personally posses the tools for performing their labour; for the wife these are therefore her sago-beaters, spatulae, cooking-pots, etc. And when in these modern times husband or wife earn wages, they do not attempt either to form a family fortune, but they divide the goods they acquire among all those persons with whom they maintain barter-relations.[83]

CONCLUSION OF MARRIAGES

In the Waropen area the conclusion of a marriage is accompanied by a certain amount of ceremonial, depending on the status of the bride and groom. The remark made above (p. 103) also applies here; when marriages are concluded, there is so much difference in detail, in spite of all similarity, that a generalising description seems less desirable.

By way of example I shall describe the conclusion of the marriage between a man from the *ruma* Rumanioghi of the Pedei clan, and a girl from Sapari (Apeinawo), both of good social standing. According to my informant Adori, the girl was entering into this marriage more or less under pressure from her family, but Adori was not quite impartial, because it was only the firmness of his first wife which had prevented him from taking this girl as a second wife. The girl herself was said not to have been averse from a marriage with Adori, because, as Adori revealed, they had already five times "scratched each other" (i.e. had intercourse).

---

[82] To Christians the marriage regulations apply, mentioned in F. J. F. van Hasselt, as amended since then: *Adatrechtbundels* (Bundles of Adat Law) X, p. 288 ff. See also Texts 117, 118.

[83] From the standpoint of the estate-coolie it is therefore not so stupid as is often assumed, when he converts his wages into presents for all kinds of relatives and friends. Of course, personally he remains as poor as he was before, but under the existing conditions he could hardly have made a better investment.

The marriage was concluded in the wife's house, for the simple reason that a room was available there for the young couple; otherwise it is the rule that the conclusion takes place in the house of the husband. Preparations for the marriage proper start one day before. Then, whilst the *ratara* is being sung, the decorations for the bridal couple are prepared, in particular the *sasa* or ceremonial comb for the bride, and the oil with which she has to be rubbed. In the afternoon, a prahu goes out with festive songs to fetch the sticks and poles for building the *ina*, the marriage scaffolding.[84]

The *ina* is a scaffolding which is erected in the central portion of the house on six poles stuck through the floor into the mud below, one on each corner and one in the centre of each of the long sides. The whole forms a rough kind of platform, about one meter above the floor of the *wundo*, and measuring approximately 1½ by 2 meters.

When towards dusk the *ina* has been erected, the *mauno* or gong is hung up and three or four girls of the husband's family arrange themselves on both sides of the *ina* to sing the *ratara*, making also light dancing movements whilst they sing, interrupted from time to time by the heavy stamping and loud hissing which always belong to the Waropen dance. This celebration of the marriage scaffolding lasts until about midnight; meanwhile one also beats the gong. The *ina* is covered with a small decorative mat (*egharo*); with its gaily coloured decorative stitching it is one of the few objects in the Waropen household which are pleasing to the eye. The *egharo* is, by the way, an imported product from the Barapasi region. On this mat a few pieces of old pottery and some *saparo*, shell-bracelets, are laid.

At midnight, first the bride is carried in on the back of a woman of the husband's family, her legs being supported by another woman, for good manners demand that the bridal couple must not undertake any action whatsoever, so that up to the end of the ceremony they behave with complete apathy. Neither the bride nor the bridegroom adorn themselves beforehand. The bride only wears the loin-cloth of blue cotton, having taken off her apron, like married women usually do, especially when they are working. The bride is placed on the *ina*, where she can only remain sitting with the support of her brother. Some time

---

[84] One will likewise often see such festival prahus starting out on other occasions, e.g. to collect the leaves for packing the ceremonial sago-cakes or the *fi bawa* (see p. 104 above). These prahus are manned by friends and acquaintances who want to make themselves useful in this way; for their troubles they are given some refreshment.

later the bridegroom is likewise carried in by men from the wife's side in the same way. He also wears his daily apparel, without the sarong which young men usually affect. According to some, the party of the person who marries into the house of his marriage-partner have to buy their entry against a small gift from the men who want to keep out intruders on the front gallery. This show of rivalry is necessary because otherwise the partner marrying-in would get to hear later that he (or she) had come into the entrance like a slave.

Now bride and bridegroom are seated on the *ina*, facing the entrance, turning away from each other as if in disgust. Then a man steps on the *ina*; he has to fulfil the requirements of being of good social standing, of being happily married, whilst his first children must have remained alive. On several occasions the clan-chief took this part, although he did not fulfil the last two requirements. At the wedding under discussion it was Gharori, the chief of the Apeinawo clan who acted as the person who concluded the marriage. As such he took a few leaves of the *mamano* — a strong-smelling herb girls use as a perfume at dances and which is also used as "medicine", viz. to promote the health of newly-born children. With these leaves he rubs the sleeping-mat whilst pronouncing the formula: "*Wotitititi! Awudomi ona rawudoma Rari. Wotitititi*"!, "Marry her for always, like I am married for always with Rari" (the latter being the wife of Gharori who concluded the marriage).[85]

Thereupon Gharori rolls a cigaret, pushes the burning cigaret into the bride's hand and brings her hand several times to her mouth. The bride remains apathetic and does not smoke the proffered cigaret. The same manipulations are repeated with the bridegroom with the same negative result. The object is therefore to make the couple smoke one and the same cigaret, but the bride and groom are not allowed to offer their active co-operation for this purpose.

---

[85] At Weinami, where the ceremony is practically identical, except for the exchange of bride and bridegroom discussed below, the person who concludes the marriage pronounces a similar formula with sympathetic effects: "*Tititiaiee! Kinisikiwe ro baba yawe rowenio. Tititiaie*"! "May both fare as well as I do"! My informants were unable to tell me the meaning of the exclamations *wotitititi* or *tititiaie*. In one of the tales *wotitititi* is the sound ascribed to a small snake (*sisaneio*); from the pitch at which the word is pronounced one might conclude that it is intended as an onomatopoeic reproduction of a squeaking sound.

These ceremonies have disappeared in the Christian weddings at Napan; however, the clan-chief there still seems to conclude some kind of civil marriage beside the church ceremony. He joins the hands of the couple and says: "*Asi aribinani tingguo. Awuiengga mimundaru bewara*", "Recognise her as legal wife. Do not fight with each other during your married life".

Now bride and groom let themselves be pushed backwards in a recumbent position on the marriage scaffolding, whereupon four young men who belong to the friends and relatives of the husband, lift the latter, whilst four others lift the woman, all eight of them standing on the *ina*. Now the groom is laid fully stretched on top of the bride, while being transferred into the hands of the young men who had been carrying the bride, whilst she moves into the hands of the groom's party. In this way they are pushed over each other and handed from one party to the other. Bride and bridegroom are exchanged eight times in this fashion, and this is controlled by a number of people present by counting loudly *wosio, woruo, oro, ako, rimo, ono, iko, gharo.*[86]

Every time the body of the bride is laid on top of that of the bridegroom, all present loudly shout *bohooo*. This is called *kikadighanggi,* they are exchanged against each other. It is therefore a demonstration of the equivalance of the parties. After the eighth time the bridegroom is laid again on the *ina* at the side of the bride, whereupon the man is given his neck-support (*runa*) and both are covered with a piece of cloth.

Meanwhile the night has advanced and several members of the wedding-party have laid themselves to sleep here and there, after some of them have done themselves well on palm-wine. The girls resume their singing of the *ratara* and until deep into the night the sound of this monotonous and slightly melancholy song is heard in the village and by the young couple who have to remain lying still in their uncomfortable position.

Towards day-break the bride's sisters and girl-friends begin to adorn her, whilst brothers and friends perform the same task for the husband. The bride's hair is combed until it stands out, some coconut-oil is rubbed into it, it is decorated with four combs (*sasa*) and powdered over with pigeon's feathers, sometimes with flowers. The upper part of her body, like that of her husband, is slightly oiled. Then strings of beads are hung around her neck and sometimes crossed bands of beads (*sare*) are slung across her chest. Arms and legs are adorned with different kinds of armlets and leg-rings. She wears a red dancing-apron, embroidered with figures of bead-work or small shells. Some brides also wear a piece of cloth by way of a shawl, the tips being fastened in the arm-rings. From his brothers the husband gets as many armlets (*ponisi, sarako*) as he can possibly wear. Handkerchiefs are pulled

---

[86] At a wedding the numerals *banggaikena* etc. are never used; the numerals used at weddings are perhaps old-fashioned.

through the arm-rings, a neck-cloth is wound around his throat. Around his waist he wears an apron like the bride, but less beautiful. That means that both are practically dressed in the usual dancing-array; however, on those occasions the girl usually does not wear a *sasa* and the boy no apron, but a sarong, whilst he would be wearing a comb (*sura*).

Next, a coconut is broken over the bridal couple, so that some of the juice is spattered on their heads.[87] Women and girls grab the pieces of the smashed nut from the hands of the man who cut it. Thereupon a sister of the bride brings a dish of sago-mush, of which she tenders eight times a mouthful to the bride and the groom.[88] Even now these two remain passive. The husband's brothers snatch the dish of mush away and finish it.

In the mean time three bamboo poles have been erected, extending across the *ina* to the garret. [89] This ladder (*embo*, also "upper leg", "bow-string") usually is an *araiembo*, an *embo* made of bamboo, but at some *serabawa* marriages it is a *kowuembo*, an *embo* of sugar-cane.[90] At other weddings I observed that the *suna* or slave-block was put at the foot of the ladder. At the place where the poles reach the garret the bark of a coconut is hung, wrapped in rattan.

Now a married man brings in a bow with eight arrows of a small type (*kowai*). He takes the bridegroom by the hand and leads him behind him up to the stakes, the bride following the groom; she is considered to be supporting him. The married man shoots one arrow into the suspended coconut-bark, whereupon he hands the shooting gear to the groom for shooting seven more arrows into it. At every hit (the distance is practically too short to miss it) young and old shout again *bohooo*. This is called *iamba nighaigha*, shooting for a

---

[87] On the island of Dobu a connection is laid between coconut-milk and semen, cf. R. F. Fortune, *The sorcerers of Dobu*, p. 238.

[88] J. L. van Hasselt reports that the Numfoor couples were four times offered a mouthful of sago-mush; this number four the Numfoor consider an admonishment to maintain conjugal fidelity; see his *Gedenkboek van een vijf-en-twintigjarig zendelingsleven op Nieuw-Guinea, 1862—1887* (Memories of 25 years as a missionary on New Guinea), p. 141. Does this number of four refer perhaps to the four primeval clans (*er*)? And would therefore the preference for the number eight among the Waropen be a reminder of eight clans?

[89] The garret is a sacred place, cf. p. 134.

[90] In several cases one of the bamboo poles had a small cross-piece on top, like the bamboo poles used as boat-hooks. Would this mean to imply that in this way the sacred world is hooked, like it is done in the mythology, where the islands are hooked as they come drifting past?

coconut. The object is therefore not merely to hit the coconut, because then one would say *iangha dia nighaigha,* but to acquire the coconut by means of shooting, whatever may be the meaning of the coconut-bark.[91] According to some informants this shooting is also a means of divination, because a missing shot would mean that the marriage would not last. In one case the groom was so nervous that he asked another man to shoot the arrows in his stead. The informants believed that the meaning of all this would be "to see the sun", so that the ceremony would be connected with the important sacred occurrence when night changes into day (see p. 264).[92]

In the mean time it has become bright daylight, and now the bridal couple still have to be rowed ceremonially around the village. They take their place on a clothes-box in a large, decorated prahu, a sister of the wife acting as train-bearer to the bride, carrying the train of the apron in a plate. The morning-sun shines gaily into the smooth water of the river and before long the prahu glides swiftly past the houses. The rowers row with the rapid short beats they use to gather speed, the gong resounds and cheerfully the *soitirano* [93] is sung, ending every time in joyful shouts. First one rows to the spot downstream, close to the village, where the male *anano,* the sea-monster Worodauni, dwells, to whom the sister of the wife offers tobacco. Then tobacco is presented to the chiefs of the five clans (*kiwoiki ma seraruma,* they are rowed to the *seraruma*). When finally the chief of the Sawaki clan has appeared on the front gallery to receive his tobacco, also the female *anano* Gharéondai upstream, close to the hindmost houses of the village, is given tobacco and then one rows home. The ceremony is over. Bride and groom take off their ornaments in so far as these incommode them.

Good manners demand that they still refrain from intercourse during three nights, because the wife is still "angry" at her husband. In the home they are chaperoned by an old woman, and when they go for

---

[91] Perhaps a cut-off head.

[92] At Weinami the families of both sides finally throw the pieces of coconut-bark at each other.

[93] The following is an example of such a *soitirano:*
Anggamo kaisa we mainggo ako anggamo kero
Anggamodoma na Kondirei munino Kondirei kawasa
                         Sieeee!
Anggamo ghiri dangha anggamo kero
Anggamodonma na Kondirei kawaso. Bingha riondai
                         Sieeee!
A translation of this piece of poetry is highly problematical. Perhaps it means in short: We rush into Kondirei, where we will kill the people.

the first time to the sago-gardens, a small brother usually has to go with them. When they have worked there together, the relation between husband and wife has been normalised and their normal married life takes a start.

Other marriages were sometimes concluded much more simply. At the conclusion of the marriage of a fosterchild of the chief of the Sawaki clan no *ina* had been built and hardly any preparations had been made. Bride and groom were not "exchanged against each other" in the central portion of the house (*wundo*), but in the *arado,* the living quarters, after the marriage formula had been pronounced by the clan-chief. That was the end of the ceremony. I did not understand why so little trouble was taken in this case, because usually people will go to greater lengths on the occasion of a marriage. Because similar, extremely simple marriage ceremonies were also observed rather often in the *seraruma* of other clans, the question arises whether perhaps special rules apply to people of lower status (descendants of slaves?), a question to which no clear reply was received.

A general trait, inherent not only to marriage ceremonies but also to others is this, that one can hardly speak of a ceremony in our sense of the word. Everybody loudly shouts his opinion and because there hardly seems to exist complete unanimity on any ceremonial, the whole seems like a confused last rehearsal. The use of palm-wine contributes a good deal to the noisiness, with the result that the co-operation between eight pairs of hands which have to "exchange" bride and groom leaves quite something to be desired. Moreover, this also leads to some confusion in the counting, so that the position of the couple — who are not allowed to co-operate — is not always enviable. Sometimes one hears somebody say at the conclusion of this part of the ceremony that the couple has been "exchanged" more than eight times, or less, a circumstance which would have an unfavourable influence on the marriage.

Here follows also the description of the marriage ceremony as given by District Commissioner L. J. Huizinga, which differs in several points from ours.

"In this whole region child-marriages occur, i.e. with postponement of the period of conjugal cohabitation. Quite soon after the birth of a girl the father of a slightly older boy comes to arrange the future marriage. On this occasion the Waroppen still bring a deep plate which is handed to the girl's parents as a pledge. At this early stage these parents are already given clothes, food, fish, etc. by way of earnest-money, but this has to be repaid by the recipients. When the children grow older, they gradually learn to whom they are to be married. They avoid meeting each other as far as possible, from a sort of sense of shame. All the while further instalments have been paid. The boy's parents gradually have given larger sums, in the form of a prahu (or a slave, in the old days), and all compensations have been requited by the recipients by means of counter-gifts, until the children have reached marriageable age. Then a date is fixed for the beginning of connubial cohabitation. Towards that moment all kinds of food and drink have been prepared for the feast. The wedding does not start in the morning, but at night, towards seven o'clock, when during the day the boy's family have prepared a sleeping-place in his house. The necessary wood has been fetched with songs in a decorated prahu. First the house is passed in all directions, until at last the work is started, i.e. a sleeping-place is built, about 1.90 metre in length and 1 metre wide. In the course of the day bride and groom have separately gone to the rhizo-phore-forest in order to surrender themselves, for the last time and by way of leave-taking from their bachelor's existence, to the game of pukul (Mal., to beat) during which the girls usually beat and scratch the boys on the upper part of the back, either with jungle-grass or with pointed shells.

"Towards evening the bride and the groom are each hidden in a house — this applies to the Waroppen — where they have to be found by their relatives who give ransom to the owner of the house. The way in which the future couple are brought together is not very encouraging; both are shy, the girl often cries and also the boy occasionally shows his aversion — all these are the well-known 'mariages de raison'.[94]

"Bride and groom are taken to the bridal bed, surrounded by six or eight 'virgins' and seated on the back of a relative. They are placed on the bed and the wife's family transfer the girl a number of times to the man's family, who in their turn transfer the boy. During this hand-over the girl is always below, the boy on top. Then follows a rather plastic demonstration and instruction concerning coition and after some time the young people are covered with a new sarong which is only taken away in the morning. During this time relatives and acquaintances have done themselves well on toddy, fish and sago, whilst young girls have enlivened the proceedings with songs and dances.

"Very early next morning a coconut is hung from the ceiling and the husband has to show his skill with a bow and small arrows. If one

---

[94] Note of the Committee for Customary Law: This has, of course to be taken in a different way, e.g. as *rites de passage*; in any case as magico-religious rites.

8

of the eight arrows — which he shoots from a sitting position — goes home, this is a sign that a child will be born within exactly nine months. Thereupon bride and groom are adorned with flowers, beads, etc. and then they have to climb a ladder 'in order to see the sun'. Next they start out in a decorated prahu 'in order to see the sea', and only then they go and present themselves as married husband and wife to the head of the family".[95]

SELECTION OF MARRIAGE PARTNERS

The kinship-terms are the following:

ku, waitéa N., child

warima, waitaimano N., son, boy { *wari bawa,* oldest son

{ *wari kuboma, wari furi,* youngest son

wiama, waitaibino N., daughter, girl { *wiama bawa,* oldest daughter

{ *wiama kuboma, wiama furi,* youngest daughter

*fofo, grandchild......... grandparent [96]

ombo, grandchild, { *omokamanda* N., grandson, *omofi,* *omokabina* N., granddaughter, grandparent, } { *omofimai,* grandfather *omofinai,* grandmother

mambo, mangga N., brother .... *bimbo, bingga* N., sister
bawa, *yosaba N., older brother (sister)
etoku, kokuma N., **yosama* N., younger brother (sister)

*daidai, imai, father, father's brother
*daidaibawa, imaibawa, father's elder brother
*daidai { *kokuma* N., *kuboma,* } or *imai* { *kokuma* N., father's younger brother *kuboma,*

---

[95] *Adatrechtbundels* (Bundles of Adat Law) XXXIX (1937), p. 436.
[96] The terms marked by an asterisk are used in the sphere of the first person. When I speak about my own grandfather (grandchild) I say *fofo;* speaking about another person's grandparent I say *omofi.* Terms without equivalents restricted to the first person are valid for all persons. In Kai therefore *bawa* means both "my oldest brother" (*raibawa*) and "his oldest brother" (*ribawa*), but Napan possesses a separate term for "my oldest brother" (*yosaba*), so that *bawa* occurs only in the sphere of the 2nd and 3rd persons. — N. = Napan.

*mini, inai, mother, mother's sister

*minibawa, inaibawa, mother's elder sister

*mini $\left\{\begin{array}{l} kokuma \text{ N.,} \\ kuboma, \end{array}\right.$ or inai $\left\{\begin{array}{l} kokuma \text{ N., mother's younger sister} \\ kuboma, \end{array}\right.$

---

*amoi N., father's sister

---

nomano, male relative by marriage
nowino, female relative by marriage

---

*amai, ondagha nunggu, arai N., *orawi N., brother- or sister-in-law
*daidairongga, imai iongga, father-in-law, mother's brother
*minarongga, inai iongga, mother-in-law, mother's brother's wife

---

onggamano N., sister's son
onggabina   N., sister's daughter

---

firuma, cross-cousin, friend, girl-friend
doufiruma N., mother's brother's son, father's sister's daughter

---

mano, manda N., husband
bino, binda N., wife

In actual practice it proved rather difficult to note down the terms enumerated above; this is explained by the circumstance that some of these terms have a very wide application. The terms for "father" and "mother" are not only applied also to the classificatory fathers and mothers, but they are likewise the usual terms for "father- and mother-in-law", and their classificatory brothers and sisters, and finally for all venerable persons of an older generation. The same remarks apply to the term for "grandfather". A son born from the marriage with a younger woman even said "father" to an older adoptive son of his father. The terms warima (wiama) have an equally wide scope as they serve to indicate not only sons (daughters), but also brothers (sisters) and fellow-clansmen. The terms mostly used in the Kai district, viz. nomano (nobino) probably contain the element no, fellow-man, companion, and they find a very wide application. The terminology does not distinguish between the daughter of the father's sister, with whom a man cannot marry, and the daughter of the mother's brother, with whom he is allowed to do so, with the result that the word firuma often does not mean much more than friend (girl-friend).

We do not find, therefore, that the terminology in use includes an indication of certain classes of relatives or marriage groups. It is based on the relationships within the family or the marriage nucleus.

It is curious that the Waropen can only realise the relations of the marriage or family nucleus in the terms for "brother" and "sister". The usual type of marriage is always concisely indicated by the formula: "the son of a sister has to marry the daughter of a brother", and when I replaced this by the formula: "a man marries the daughter of his mother's brother", people had to think for a moment until they had reduced this formula to the one they knew better. Unintelligent people even denied that the two formulas were identical. Nobody could realise either that in this system there had to exist a group providing wives and a group receiving wives, beside ego's own group, but the actual structure of the *ruma* being what it is at present, this does not become apparent. However, everybody immediately denied the possibility of exchanging brothers and sisters.

The contrast between brother and sister is also expressed in the terms of relationship. Just as in several East and Central Polynesian systems, the Waropen system counts according to seniority between members of one generation of the same sex. The man therefore speaks of the members of his sex as *bawa,* elder brother, or *etoku,* younger brother. The woman uses the same terms to refer to the members of her generation. However, the difference in sex is expressed in the term *bimbo,* used by the man in referring to his sister, and *mambo* used by the sister to indicate her brother. In these terms the words *bino,* woman, and *mano,* man, are to be recognised. Persons of the same sex cannot refer to each other by means of the terms *mambo* and *bimbo.*[97]

In Waropen society the household does not often come to the fore as a unit. There is no family-property and due to her living in the *ruma,* the wife is very much in the company of her mother-in-law, her sisters-in-law and the wives of her husband's brothers, whilst the husband usually looks for the company of the other men in the *ruma.* A married woman can mostly expect that it will not be her husband who will help her most in the many quarrels she usually has with the other women in her husband's *ruma,* but that the greatest assistance will have to be provided by her own brothers. She therefore remains in close contact with her own family and even if the latter is living in another village, she returns there regularly.

---

[97] See further R. Firth, *We, the Tikopia,* p. 278.

Still, the ties of the *ruma,* would be overestimated if one were to suppose that the household were of little importance. Those lines which in the last instance take sides against each other, e.g. during quarrels, and who also are parties in a marriage nucleus, are only small when all is said and done, and they feel themselves to be one, just like the children and grandchildren of one not yet forgotten couple of parents.

The influence of the wife on the husband is usually still so great that he will not take a second wife against the wishes of his first mate. The couple often go together on a journey, and innumerable times one sees them together in one small prahu fetching something inside or outside the village. The husband has first of all to provide his own household with fish, and although husband and wife do not always go out together during the day-time, they still possess one room, or at least one sleeping-place, and although they do not eat together, they still share the same pot. The tie is strongest when there are children. Waropen fathers are very fond of their children and because the children have to stay with their mother for quite a long time, the father prefers again to be close to his wife.

When educating their children, the Waropen from the beginning bear the future marriage in mind, and the barter-relations which devolve from it. This is probably one of the reasons why even comparatively young couples soon proceed to adopt a child, if they do not soon get children of their own. In a childless marriage the barter-relations of course come to an end, because there are no *saira* for which the brothers of the wife have to provide gifts. If the family has no daughters, it lacks the possibility of obtaining a dowry. A childless couple has no possibilities of maintaining itself in the ceremonial barter-transactions and so in their old age they remain more or less dependent on the willingness of others to assist them. My informants mentioned as the aim of adoption that the parents hope to obtain from their adopted daughters the sago they will need when they have grown so old that their sago will have to be worked by others.

In other cases adoption is not much more than a sign of great friendship between two families. In young households where a daughter had been born already, a boy was also accepted quite soon as an adopted child (*kufi, yaiwaita yakufi* N.), without there being any reason to assume that a boy would no longer be born there.

Brother and sister show a certain preference for adopting each other's children. I was assured at Weinami that the term *imai iongga* which is used in the Kai area for "father-in-law, mother's brother", here meant "adoptive father". The brother who adopts his sister's son therefore accepts the ideal husband for his daughter as his child. Now adoption has rendered this marriage impossible, but marriage remains possible for an adopted child and a girl from the *ruma* of his adoptive father, so that adoption may also be conducive to making the *ruma* endogamous.

Usually children are adopted at an early age so that they are still young enough to be completely absorbed in the household of their adoptive parents. This is not accompanied by any ritual or ceremonial. Neither is any attempt made to have the child forget its own parents. At a more advanced age it distributes its attention equally between the two households. My informants believed it impossible that after a long period of time a father would demand his child back because the adoptive father in that case would demand payment for the care given to the child. Neither did they know of any case of an adopted child running back to its own family.

When an adopted daughter marries, the father and the adoptive father receive part of the dowry; in case of an adopted son, both have to assist in collecting the dowry. The adopted child inherits both from the sago-property of its adoptive father and from that of its natural father. However, it mainly follows its adoptive father who is responsible for the child (*iateina rikufigha*). At Weinami also slaves were adopted; in that case they followed their adoptive father in rank and station. In general it is therefore also in the interest of the child when it is adopted, because it acquires an extensive number of relations due to this. The treatment received by an adopted child shows no difference from that received by a natural child.

Due to the not always strictly unilateral tracing of relationship among the members of a *ruma,* due to the prevalence of adoption and due to polygamy, it is sometimes an almost hopeless task to try and unravel the family relationships. Dusi from the *ruma* Sawaki (clan Sawaki), for instance, has married four women from the *ruma* Manieghasi (clan Kai). At Sawaki it was said that Dusi's mother, Mereni, had come from the *ruma* Sapari (clan Apeinawo). According to the marriage system in force, Mereni should therefore be the sister of the fathers of these four women from Manieghasi. Among Dusi's four wives three were full sisters, daughters of Meni, the fourth (classificatory) sister being a daughter of Ghafai. Now Ghafai is a man from Rumanioghi (clan Pedei) who has come to live in at Manieghasi, although this had already been forgotten in the last-mentioned house. Mereni, however did not occur in the genealogy of Sapari, but in that of Rumanioghi. The confusion arose from the fact that Mereni lived in for some time at Sapari. Mereni and Ghafai are therefore both from Rumanioghi and actual brother and sister. At Manieghasi, however, Ghafai was believed to be a man from Manieghasi, i.e. a brother of the member of his generation Meni from the same house. And for that reason Meni was again a "brother" of Mereni. With the result that Meni's daughters could marry Mereni's sons, just like Ghafai's daughters could.

A son of this selfsame Ghafai married a girl from the *ruma* Pedei who had been adopted into the *ruma* Erari (clan Apeinawo). Because of this adoption Ghafai became her (adoptive) father's sisters's husband, so that she could marry Ghafai's son. The marriage between this couple, however, was not concluded at Manieghasi or at Erari or at Pedei, but in the house of Dusi, mentioned in the foregoing paragraph. After their marriage the young couple went to live for some time in the *ruma* Aibini which forms in fact part of the *ruma* Pedei. And finally they returned to Manieghasi, to which the bridegroom actually belonged.

In this way the potential number of husbands is practically unlimited, because it is always possible to compute the required relationship in one way or another. When I occasionally tried to establish how the relationship had been exactly traced in actual fact, the answer was simply: "We do not know how they are related, but in any case their son and daughter have married each other, and therefore they are brother and sister".

The number of *firuma* among whom a man may select his wife is therefore practically not restricted by the limitations which would seem to be logically connected with a marriage with one's mother's brother's daughter alone. This does not mean to say, however, that a Waropen man or woman would be completely free in their choice of a marriage-partner. On the contrary, there is quite some pressure exerted by the family, who are mainly interested in the barter-relations that have already been established. Once two parties have started to provide each other with presents, they are also committed to each other. The difficulty is not solely that the girl's party feels averse from surrendering the gifts already received, but it would also seem to some extent that they feel themselves committed by these presents.

When discussing the Christian marriage which demands a greater freedom in the choice of partners, the informant Adori remarked that

it was indeed disagreeable to drag a weeping woman to her husband.
When I concluded from this that he therefore would not force his
own daughter, once she was grown up, he replied that this would
depend on the exchange of gifts; he would consider himself to be
committed by the presents accepted.

The woman who is forced into cohabitation usually becomes resigned
in a few days and the marriage in question to all outward appearances
is in no way different from the others. It is possible that also the show
of aversion, demanded by etiquette, exercises a certain influence. It
remains an open question whether it is justified to press for a completely free choice of partners, as is done nowadays. The authority of
the older people in these matters is great and people usually are quite
ready to submit to the decisions of their elders. Moreover, a marriage
concluded without the complete consent of the parents lacks the
guarantee which is connected with the exchange of goods. And nobody
is interested in its maintenance, so that divorce becomes easier in this
way. The advocates of a free choice of marriage-partners forget that
also in Western society there exist all kinds of restrictions, connected
with the wealth, the social position, the religion, the education, etc.
of the partners; one should only think of the sharp conflicts which
may occur if there is a difference in religious sect, let alone in religious
denomination.

Also under the ancient conditions pressure of the elders was not
always accepted. In all those cases, where somebody tries to have his
way there remains the possibility to flee together with the woman
(amokiwa bingha). It is not always a boy and a girl who elope, but
sometimes one of the parties (or both) is married. Elopement is a
rather frequent offence which is counted more heavily against married
persons than against unmarried ones. The fleeing couple have to find
shelter somewhere, usually outside the village, either in a hidden spot
in the forest, or with a friendly family. To run away with a woman is
always an infringement of the existing order, whilst in many cases it
is a breach of contract, i.e. when the woman is already engaged or
married. Although it depends on the circumstances whether the opposition will be very strong, the run-away marriage remains an offence,
in spite of its frequent occurrence. Of course, the Waropen care very
little about the maintenance of law and order in the abstract, but an
elopement causes harm to concrete interests. Persons who receive the
fleeing couple are also guilty and they therefore have to feel strong
enough to face the anger of the injured party.

The violence of the conflicts arising between personal choice and family pressure is demonstrated by the romantic history of Sorei and Kewowi. Sorei was engaged to Sewuri, but she loved Kewowi, and therefore one afternoon Kewowi and Sorei escaped together. In the old days they would have run far away, but in these modern times, when the police has put an end to all attempts at the personal settlement of grudges, people are less quick to take up bow and arrow in order to start pursuit. Kewowi and Sorei therefore kept themselves fearfully hidden in Kewowi's house. When this had been discovered, a great deal of abuse was used and Sorei's father, armed with bow and arrow and a bush-knife, went up to Kewowi's house where he let himself go in a raging torrent of challenges, like the Waropen often do for the most unimportant reasons. The father roared out his diatribe, hacking into a post here and into a prahu there, and finally returned home. He had sufficiently shown Sewuri's family that he was not involved in the plot. When the anger has gradually calmed down, the girl may return to her family, payments are arranged and the marriage may be concluded with the man with whom she eloped.

This happened also in the present case; Kewowi's brothers payed the fine demanded by Sewuri's people for their "shame", and they started to pay the dowry. However, Sorei's family and her mother's sister in particular, continued to withhold their consent. Now Sorei could have simply gone to live with Kewowi, but being the daughter of a nobleman she did, of course, not feel attracted by the idea of delivering herself like a slave without a dowry into the hands of her mother-in-law. She then had recourse to an original and modern means of compulsion, viz. a hunger-strike. However, she had to continue her strike for nine days, until her family had been completely intimidated by public opinion which had turned in favour of the girl. But she had been so much weakened in these nine days that she repeatedly lost consciousness, her family having the greatest trouble to bring her to again. Finally the family was ready to make any concession and Sorei's promise of co-operation and an invigorating potion were sufficient to change her back again into the lively bride who married Kewowi one week later. For the rest, during the ceremonies Sorei and Kewowi were no less "ashamed" and "angry" than all other couples.

Less romantic but therefore not less violent was the struggle the senior representative of the *seraruma* Pedei had to maintain in order to enforce his will. In this case, however, the bride was not a young girl, but a grandmother with eight adult children, who had been a widow for years. Moreover, she was a (classificatory) mother to the bridegroom Seranauri, so that the relation was considered to be incestuous. After her husband's death the woman Fanggadei had gone to live with one of her husband's sisters who had married into the house Mamurani. The members of the *ruma* Pedei maintain relations with Mamurani, so that Seranauri often came there. Whether the old woman went to visit him there (as was said by the *ruma* Pedei), or whether conversely he went to see the old woman (as maintained by her children), the fact of their contact was known and recognised. Due to this the *ruma* Pedei was quite definitely "ashamed", whilst her children were angry.

Although the woman personally would have been satisfied with a fine which the *ruma* Pedei offered to her, her children demanded a marriage, in which matter they had Seranauri on their side, curiously enough. However, according to the *ruma* Pedei marriage was out of the question, firstly because the relation was incestuous, so that the ancestors would send down punishment, consisting e.g. in bad luck when fishing or going on a voyage; secondly, because the marriage would remain childless, and thirdly because intercourse with a widower or a widow would sap a young person's physical powers. In spite of all resistance, Seranauri took the old woman with him into the *seraruma* Pedei. Seranauri's father steadfastly refused to pay a dowry or even to assist Seranauri in collecting his tax-money.

In the same way Fanggadei's sons and sons-in-law refused to hand over Fanggadei's sago-palms, for according to the prevalent ideas the adult children work the sago-palms of the old people, being obliged in return to maintain the old people completely. Now Fanggadei maintained that her palms had not yet been actually inherited by her children, so that she could work them together with Seranauri and his people. To her this was a question of the utmost importance because at the moment she had been practically left to the philanthropy of the *ruma* Pedei which was in none too generous a mood as matters stood. Seranauri had the rather delicate task of preventing the adult sons and sons-in-law from working Fanggadei's sago-palms; in the old days he would probably have paid for his obstinacy with his life, either to the members of his own *ruma,* or to Fanggadei's children. This affair remained unsolved during the time of my research.

It also happens, however, that a man is able to persevere in an incestuous relationship against everybody. In this way I know of at least one marriage between the son of a second wife, who married his father's first wife after the former's death and whose marriage is now fully accepted; it has also produced children. There are rumours that in polygamous families illicit relations between the father's elder sons and his younger wives are not unfrequent and people will even indicate children that have been born from such relations.

Polygynous marriages are not uncommon in the Waropen area, although monogamy is the rule. Only people of an elevated social standing will conclude such marriages, with the result, conversely, that prominent persons sometimes believe it to be due to their station to take more than one wife. In the Waropen region polygyny is the best means of acquiring property. Although one has to possess many goods in order to be able to pay the different dowries, one also has many relations who again have to offer presents. This once led somebody to remark that the ideal marriage policy would be to enter into relations with a woman from each of the *seraruma* in the village. In polygamous families differences in age may be considerable, so that rather old men are found to possess still very young children. Sometimes the women dwell together in one house, but mostly they live with their brothers.

In many cases the women are opposed to their husband marrying an additional wife, and often they make their opinion felt so strongly that the husband prefers to continue the monogamous marriage. In one case an influential man ran away with another man's wife. The man was considered as a specialist in magical practices, so that the wife's family gave up any attempts at reprisal. For the woman the man's knowledge was one of the very reasons why she had eloped, because up to that time she had remained childless and she hoped that his potions would enable her to have children — which she actually had. The two first wives, however, were not to be intimidated and after a terrible village scandal they lodged a complaint with the Administration which punished the offender with emprisonment.

All these figures are to be found again in mythology, both the figure of the sick sister who is maltreated by her brother's heartless wives, and the distressed and indignant wife who runs away when her husband takes another wife. Another figure, also taken from real life, is the unfortunate younger wife who is maltreated by the earlier wives. Especially when the first wife (*binarengga*) is childless and the later wife (*binaghofuri*) is not, a serious animosity exists between them (Texts 109, 110, 116, 117, 118). It then becomes difficult for the husband to maintain the peace and due to the women's mutual jealousy the most frightful quarrels arise again and again.

Polygyny is promoted by the institution of both the sororate and the levirate, which result in a widow marrying a, usually younger, brother of her husband, or a widower the usually younger sister of his wife. However, a widow often continues to live unmarried in her husband's house; sometimes she returns to her own house. Marriages with a widow are considered as being harmful to one's health, except the levirate marriage which on the contrary is recommended. When the widow remarries with another man, she causes shame to her deceased husband; it is for this reason that her new husband has to pay a fine for her, in which all former payments are settled. *Iapora inggoigha,* he pays the ancestors, or *iapora raietokugha niriragha,* he pays for the bodily dirt of my younger brother (see p. 182). At Ambumi this fine is called *iapora bindoferifaia,* he pays the woman's bark-cloth, for during mourning the woman wears a short jacket of beaten tree-bark (see p. 182). The marriage with a widow or a widower is concluded by simply pronouncing the marriage formula; further festivities are omitted.

Married life among the Waropen is generally rather quiet. It is expected of a married woman that she will remain faithful to her husband; adultery on the man's side is less sharply condemned. A young man may occasionally go to the dances where he may continue the relations with his *firuma,* and also a young woman will sometimes go dancing after her marriage, but they have to be careful so as not to give rise to talk. Of a sensible wife it is expected that she will not try too much to keep track of her husband's movements, but in view of the tempestuous scenes witnessed occasionally not all wives are sensible, nor are their husband's movements beyond reproach.

At Rumanioghi there was a young woman who got herself talked about. She was married, but because of bad treatment she had left her husband. Now she had received visits in her own house and the inevitable happened. The young woman's father had not remained unaware of these visits, which is hardly possible in such cases, because the marks of the scratches which lovers inflict on each other are much too eloquent, but he defended himself with the statement that nevertheless he had not approved of these visits and that in any case he had not expected that anybody would "deceive" his daughter.

Had she been an unmarried woman, the solution of the difficulty would have been comparatively easy. Pre-marital pregnancy is often judged rather mildly, and in actual practice several young men only marry when their *firuma* is pregnant; in such cases it is sufficient that the girl reveals the name of her lover. But the situation becomes difficult if she has to mention more than one name, because then the lovers are not so willing to marry and it is sufficient if they pay a fine. Giving birth to a child out of wedlock is considered shameful, so that it is said to occur occasionally that young women kill such children, but that is a question which the family concerned decides by itself. If this were to become known, the woman is not accused of infanticide, but she is made "ashamed" for having borne a child out of wedlock. If the child remains alive, it is always reminded of its origin and people tease it by asking: "*Auari aghani ghoe*", "And where would y o u r father be?"

The case of the young woman from Rumanioghi was complicated, not only because she herself was married, but also because she had to mention two married men. The woman's brother, who according to custom took his sister's part, was a hot-tempered person and he went out repeatedly with his bow and arrow and a bush-knife to raise a shindy in front of the houses of the two men who on their side were quick to join the chorus of those who directed the most violent reproaches against the young woman. The chief of the clan suggested that one of the two men should adopt the child, possibly taking the mother as a second wife. But this solution was resisted by the mother who did not want to lose her child, and also by the brother who did not want to see his sister married as a second wife. In this way the affair remained hanging fire for a year. As the woman was young and good-looking, she then found a man who was willing to accept her with the child. The two men had to pay a fine.

It is difficult to give an opinion on the position of women in Waropen society. A woman has her own important task in supplying food and together with the other women she forms a rather closed block. In general, she seems to be able to fulfil the demands required by society as regards physical and psychical powers of endurance. Because in the social intercourse between men and women the element we call "courtesy" is lacking, many people are of the opinion that the lot of Papuan women is "practically unbearable", as Feuilletau de Bruyn puts it. His verdict concerned the Biak women and it is curious to observe that a Numfoor (who therefore belongs to a tribe which is very closely related to the Biak), said again that Waropen women were in a worse position than the Numfoor women. In support of this statement he did not refer to the fact that the Numfoor men do some work in horticulture, but to the circumstance that the Waropen men wore a sarong, but the Waropen women only an apron.

It must be said that in Waropen married life hard knocks are given at times. In one case an unfortunate blow even caused the death of the wife who was pregnant at the time, but in this case public indignation ran so high that the guilty man will never dare to show himself again in his village. In the old days he would probably have been caught and killed. But on the other hand, the women know how to acquit themselves well and they do not hesitate to give as good as they get with pieces of fire-wood.

Waropen women are quite expert in the use of their most dangerous weapon, a sharp tongue. Frequently one hears the Waropen wife who believes to have good reasons for being dissatisfied pronounce an overloud diatribe against her husband, who repeatedly tries angrily to stem this flow of words, but who knows only too well that for the moment he is the butt of the whole village. The long drawn-out screeches of her accusations are audible everywhere in the stillness of the evening and the listeners indicate by malevolent laughter that they know how to appreciate her ready witticisms. When more persons are concerned with the affair, the controversy starts to rage throughout the village with lightning-like rapidity and comment is voiced on all sides. And it is not easy to bring an angry woman to see reason. A woman whose water-bamboo had been broken by her husband, continued to sit for hours on the front gallery of her house, wailingly addressing the funeral dirge to her deceased father who during his life had always protected her.

When the quarrel assumes more serious proportions, the woman runs through the house as if she were mad, and dances angrily on the front gallery to make the injustice she suffers known to everybody. The extreme to which she may go is to kick the room to pieces in which she and her husband live, and to pull off its roof-covering (*okowa ruma*), demonstrating in this way that she considers the quarrel serious enough to break the marriage-tie. If still unsatisfied, she flings herself on the cooking-utensils which disappear underneath the house, flinging the plates into the mud and kicking a box of pandanus leaves into bits. The husband hardly attempts to calm her down; later, he resignedly collects the lost objects. If at this stage the quarrel is not composed, divorce is the only means which remains, but this step is not taken without very serious reasons.

There are also several marriages, especially of older people, which are evidently very good and where even childlessness does not lead to polygamy. Then one sees husband and wife constantly together; the husband helps his wife at processing the sago and she assists him in fishing. Angry words are heard only rarely, but often jokes and good-natured teasing. The picture of real Waropen happiness in marriage is that of a wife lousing her husband on the front gallery, both keeping a watchful eye on their naked, black little children.

CHAPTER THREE

SAIRA: THE RITUAL OF LIFE

THE CYCLE OF INITIATION CEREMONIES

In this chapter and in the fourth — which, as seen from the title *"Munaba, the ritual of death"*, is conceived as the counterpart of the third — I have collected the material related to initiation and death. The antithetical connection expressed in the designations life and death was not constructed for the sake of providing a convenient arrangement of the material I had brought together, but it was laid by my informant Adori. Speaking about the significance of Christianity, Adori remarked that it rendered impossible the festivities connected with death, the so-called *munaba*, but that the festivities for life, the *saira*, would have to be maintained. The oppositional connection Adori construed between these two series of festivities may perhaps not be conceived equally clearly by all Waropen, but the importance of cultural elements is not determined by the degree or the way in which the bearers of this culture are personally conscious of these elements.

As the *munaba* stands at the end of man's earthly career, so the *saira* stands at the beginning. Male hands initiate him, feminine hands lay him out; at the *saira* it is especially the men who dance, at the *munaba* it is the women. Among many peoples the idea of the ultimate unity of life and death is well-known, although it may not be so strongly alive and although it may at least not have been clearly elaborated in a cosmological system, but they are nevertheless conscious of the mythical oppositional connection between the two sexes and the labour of both. And in any case, the relation between the sexes seen from the mythical point of view is not isolated from the connection between life and death, nor is it detached in a general cosmological sense from the connection between day and night, a connection we shall discuss in more detail in the final section of the sixth chapter.

Before proceeding to discuss a few general characteristics of Waropen ritual, I shall first produce the material which, I believe, justifies my abstracting these characteristics. In doing so, I shall again not give a generalising description of the different ceremonies, but in each individual case I take as my guide one ceremony as it was actually performed in a certain case. I shall start with the first *saira*-ceremony, performed for a child still unborn, to end with the perforation of the nose, which concludes the whole cycle.

Among the Waropen the initiation ceremonies do not mark a transition to a higher age-class because no such grouping in age-classes exists. According to age there are the distinctions between *manakoido* (*binakoido*), very young, still nameless, children, male or female;[98] next *warima*, boy, and *wiama*, girl; *waribo*, young man and *wiamabo*, young woman; *mangguo*, "real" man, already married, and *bingguo*, "real" woman, already married; *manobawa*, "great" old man, and *binabawa*, "great" old woman.

The *saira* cycle contains several episodes, not all of equal importance; of these I mention the following:

1. in the seventh month of pregnancy *kitaghaiwa kiwe rarigha sakora wa ninagha*, there is singing because an apron is tied to her belly.
2. at birth *kitaghaiwa bingha we sana kugha*, there is singing because the woman gives birth to a child.
3. the child recognises its surroundings and so *kitaghaiwa kiwu kugha ma sedogha*, there is singing because the child is taken outside.
4. the child walks well and so *kitaghaiwa kiweiwa kugha*, there is singing because the first leg-rings are put on the child.
5. the child has luxuriant hair and so *kitaghaiwa kugha niworaigha*, there is singing for the hair of the child.
6. the child is taken on its first journey and so *kitaghaiwa kiwoiwa kugha*, there is singing because the child is taken along.
7. the child is given its first clothes and so *kitaghaiwa kituna kugha*, there is singing because the child is clothed.
8. the child is sexually mature and so *kitaghaiwa kika niabogha*, there is singing because its nose is perforated.

The technical term for the celebration of a *saira* is *saghara*, to sing, hence *saghara saira*, to sing an initiation ceremony, or *sagharo rano*, to sing a men's song. In Waropen parlance a *saira* is not "celebrated", or "danced", but "sung", for the sacred, mythical chants are sung. It is curious that a *munaba*, to the contrary, is not "sung", but "danced" (*kikowa munaba*), although on that occasion there is singing and

---

98 The terms *manakoido* and *binakoido* are rarely used; they seem to have some humorous connotation — they may be terms of endearment. From *koido*, frog?

dancing, like for a *saira*. At Napan one says also *weusara kugha,* to initiate the child.

As regards the number of *saira*, it should be remarked that the cycle may be extended if desired. At Waren e.g. it was the custom to celebrate a *saira* when the child was given its first oar (*kitaghaiwa kugha asona rinamagha*). On the other hand, nobody is obliged to have all the ceremonies enumerated here performed for his children, and in actual fact it will only rarely happen that they are all performed for one single individual. Those one does not like to omit are the numbers 2, 4 and 6, whilst number 8 is compulsory for everybody. As to the remainder, one either drops several episodes or one combines these. If one person has episode 8 performed for his children, another will let his child participate, although it is not yet sexually ripe. Occasionally also episodes 3 and 4 are made to coincide. For the eldest child one tries to have the fullest ceremonies possible. For the remainder it is left to the families themselves to determine whether they want to have the ceremony fuller or simpler; it all depends on the family's social status.

RITUAL DURING PREGNANCY

The feast on the occasion of the changing of the apron of the pregnant woman is intended to influence the process of pregnancy and eventually that of childbirth. The ceremony aims at an effect which is similar to that of the action performed. Pouring water on the breasts is to have a favourable effect on lactation. This aspect of the ritual might be called "magical", were it not that the term "magic" is already used generally to indicate a large number of actions, partly religious, partly technical, which aim at an effect similar to the action. By ranging this ceremony among "magic" and by stressing a few vaguely outlined "magical" associations, it is torn from the context in which Waropen religion situates it. For this is an initiation ceremony, a *saira*. After all, any ceremony aims at a certain effect, but it is the categories of the primitive culture itself which are of importance. The "magical" explanations are often arbitrary constructions, conceived by an observer who belongs to a different culture. For this reason it is preferable to avoid using the term "magic", till the time when it will have become a scientific term with a well-defined meaning.

The rite we are discussing at present is only rarely performed with great display, so that during my investigations I was only able to

witness it once, viz. during the beginning of the pregnancy of the wife
of Munamberi, the chief of the Nuwuri clan, in the paternal line actually
the senior representative of the *seraruma* Sawaki (see the genealogy on
p. 141). In other cases this ceremony is celebrated as a very inconspicuous
family episode.

A peculiarity to be observed at all *saira*-festivals is the special decoration
of the prahu which starts out in the afternoon preceding the night in
which the *saira* is to be sung, going to fetch the leaves in which the
sago-cakes are to be packed, and some bamboo stakes and leafy
branches. In the prahu are dancing young men who also beat drums
and blow on triton-shells. The carved decorative pieces have been
placed on bow and stern; flags wave in the wind. The most striking
is the enclosure of blue cotton (the well-known so-called kain chelopan
or *rari*, i.a. generally used for women's aprons) within which the young
men stand and sit when they return from the forest, branches in hand.
Such leafed branches always belong to the activities of the raid.

Like the prahu which fetches the leaves for the *fi bawa* (p. 104), or
the boat which brings the stakes for the *ina* (see p. 107), this prahu
likewise approaches the festive house at great speed, whereupon it is
eight times rowed round in a circle.[99] Just as in the raiding-ritual, two
girls dressed in red dancing-aprons appear on the front gallery, carrying
bow and arrow and dancing in a challenging manner and emitting
the cry *bieee, bieee* (*owaura*, to dance in somebody's way, i.e. to challenge
whilst dancing, in the way in which a Waropen angrily challenges
his opponent). The triumphal branches that have been fetched find
no further use; they are simply placed on the front gallery of the
festive house, where also all available flags are put.

As soon as the sun begins to go down, the rhythmic rumbling of
the drums begins, accompanied by the monotonous singing of the
men's songs (*rano*), at one time rising to a chorus of shouts, at another
diminishing to a dull litany, until a solitary singer carries it up again
with a shrill long-drawn note accompanied by a renewed pounding
of the drums, supported at times by the tearing bellowing of the

---

[99] The prahus mentioned on p. 104 and p. 107 have no enclosure. At the *saira*
of Munamberi one of the youths in the prahus carried a coconut in his hand,
another a pineapple, and a third a bow and arrow; according to my informants
this had no special meaning, for I had considered the possibility that people
had thought of the expression "to go and fetch forest-products", which alludes
to head-hunting among the tribes of the interior. Did perhaps anciently a
prahu which fetched leaves for the ceremonial cakes also go on a raid at the
same time?

triton-shells and the booming sound of the gong. The men dance two by two in a long line which endlessly turns around the pillars of the centre of the house. At times the line is just long enough to enable it to turn always around the foremost central pole of the house; then other men join the dancers and the stamping line winds through the whole length of the house. At Waropen Kai it is only the men who are allowed to dance at a *saira*, and the women sit and look; in the Napan area and on the Moor Islands the wife joins her husband, which is done at Kai only for the *munaba*. The songs which are sung always deal with mythical material, but they are in such an incomprehensible language that I cannot offer a translation (cf. p. 269).

The dance has to be continued during the day-time, but then the urge to dance has been reduced to a flickering flame; usually the revelries are at their height at midnight, also at the *munaba*. Then many dancers retire, whilst others lay themselves to sleep in various places. Towards morning some wake up and shiveringly join those who are continuing with little enthusiasm to dance until morning, either because they belong to the nearest relations, or because they expect a reward. The actual ceremonial actions of the *saira* take place at sunrise, when night changes again into day, an occurrence which is considered from the mythical point of view as the metamorphosis of the bird-fish (see p. 264).

The final song which concludes the night's singing is the *amairano*, the morning-song. According to the myth this is the song of a snake which helps a boy in his trials and tribulations, quite probably the initiation-demon assisting the initiandus.* The snake takes the boy with it into the sea, whence he will soon return to his village accompanied by a wife; the villagers regale the snake and in reward they obtain all manner of wonderful things (Texts 51).

The morning-song given by way of example by the person who told the myth runs:

Maine maino. Siriri pumbo. Ranae sawa. Seri tunie.
Maine maino.

This means perhaps: "Let us go to sea. Catch them — hurray. Night at sea, the clear deep. Let us go to sea". This is apparently a reference to the slave-hunt with which the initiation is connected.

The singing of the sacred songs is always followed towards morning

---

* This Latin term used in the original Dutch text we have kept in the translation. Ed.

30. The "ashamed" bride and groom sitting on the *ina*

31. Rowing round the new prahu

32. The initiandus on the back of his mother's brother (left), the tobacco-prahu (centre), the tobacco Morning-Star (right)

33. Fetching the leaves for wrapping the ceremonial sago-cakes

by the exhibition and the distribution of a quantity (usually not very large) of food, tobacco and areca which is intended partly as a counter-gift for the old plates to be handed over by the brother of the wife, and partly as a reward for the participants.[100]

When during the ceremony we describe here, the last songs were started towards the end of the night, one of the men present started to whittle a bamboo-joint, about one span in length. He made a small hole in the partition at the joints so that by closing one of the holes it could be used as a primitive pipette.[101]

Now the tall bamboo, which had also been fetched the day before in the prahu with the cloth-enclosure, was put up at an angle, so that it reached from the floor of the central part of the house towards the garret, as is done with the *embo* on the occasion of a marriage. Then a small hole was cut in five of its eight joints and into these water was poured, whereupon the small prepared bamboo was fitted into these holes. Finally the water was poured again from the tall bamboo into a plate.

Then the pregnant wife of Munamberi for whom the *saira* was performed entered and sat down on a mat on the floor. Munamberi's mother took the small bamboo and blew through it on the breasts of another pregnant woman. A fourth woman, the sister-in-law of Munamberi's wife, pregnant herself, but having already passed through a fortunate delivery, took an **uwifi,** a sago-cake packed in ubi-leaf, into which a spoon for eating sago-mush had been stuck, and with this cake she lightly tapped the chest, neck and back of Munamberi's wife, saying: "*Aninadowako ro una raninadoni roini*". "May your belly be good like my belly is good". According to some, a little sago-mush

---

[100] B. Malinowski in his *Argonauts of the Western Pacific*, p. 209, supposes that this distribution of food (which in the Waropen area does not occur at the beginning, but at the conclusion of a feast "imposes an obligation on the others to carry through dancing, sports, or games of the season".

[101] Sometimes this instrument is made more carefully. I was shown a bamboo used for the same kind of *saira*, about 40 cm. in length into which a stick had been inserted, the tip of which fitted in a hole in the lower partition. If this bamboo was filled with water and the stick subsequently drawn from the hole, the water came out of the opening in a jet. Utensils for ritual purposes are to a certain extent considered as sacred, as is proved i.a. by the disinclination often shown against the making of models for demonstration, because this is believed to be dangerous for young people, especially for those staying abroad. But this view does not imply any respect; when this squirt was tried out, somebody remarked amidst general laughter: "*iimamisusu*", **"he pisses".**

should be spread on the nipples by means of the sago-spoon. Young girls grouped themselves around the woman and sang the *ratara*.

Later, the sago-spoon is taken from the *uwifi*, tied to a bamboo-sprout and placed in the sago-forest; when this sprout begins to grow, the spoon is taken back to the house, again with a small *saira*-ceremony.

Now the small bamboo instrument and the plate of water were taken. The bamboo was sucked full of water, whereupon one of the pregnant women squirted it in a thin jet on the breasts, the neck, the shoulder-blades and the back, i.e. on those places which had also been tapped with the *uwifi*. According to some, the way in which the water trickles down permits predictions regarding the sex of the child.[102] According to others it serves to satisfy the ancestors, in order that delivery may proceed as smoothly as the water runs down.

After this treatment with water follows the actual changing of the apron (from which the feast takes its name), but this is not done in public.

Finally an inventory is taken of the presents offered by the wife's brothers: a few ancient plates, some strings of old beads, one shell-armlet and four bead-armlets.

Before proceeding to describe a following *saira*-episode we shall first try to establish the meaning of the long bamboo pole, placed in the *wundo* and leading up to the garret.

In mythology, it is always the first ancestor who dwells in the garret,[103] and this is also the place where the skulls of the deceased relatives are kept. The garret is one of the sacred places of the Waropen house, and the bamboo we already observed at the marriage therefore connects the central portion of the house with the sacred space of the garret. According to some it is in fact intended that the woman should climb up to the garret by means of the bamboo, in order "to see the sun". Evidently this ceremony is therefore also linked up with the cosmic crisis of dawn, which in a mythical sense is a metamorphosis of the heavenly bird (see p. 264). Furthermore, it is clearly intended to indicate that the bamboo is filled with water; the relation between

---

[102] The same idea is said to exist among the Numfoor (*sor bin sneri*) by F. J. F. van Hasselt; it is likewise reported on the Banks Islands, cf. W. H. R. Rivers, *The History of Melanesian Society*, I, p. 147.

[103] See p. 281. Among the Makassar the ancestral statuettes are to be found in the garret. The Toradja imagine "the Lord of Heaven as dwelling in a small room in the attic of the house or the temple", according to A. C. Kruyt in *TBG* LXXX, p. 224.

water and heaven is self-evident, because *dora* means both "heaven" and "rain".

On the use of the bamboo we find some other curious remarks in a paper by Van Balen on the funerary feast of the Papuas of Geelvink Bay.[104]

The first passage reports how the spirit of a murdered man is led to the land of the souls through the bamboo: "During a drunken revelry a man had been stabbed dead, and because the spirit which leaves the body due to violent death does not go to the *"suruka"*, but remains drifting in the air, one has to assist the spirit on its way to the *"suruka"* via Roon Islands. A tall bamboo pole decorated with leaves and wild fruit was planted on the drying beach in front of the house of the murdered man; during the whole day no fire was allowed to smoke in the settlement. In the afternoon, when the beach was dry, a large crowd collected around the planted stake, whilst a large fire was made with moist wood so that a heavy cloud of smoke arose. An aged man, who acted as the medium in this case, took hold of the stake with both hands and began to shake it, slowly at first, but then faster and faster, until the shaking became so violent that all decorations dropped of...... Finally the man's extasy came to an end...... The spirit was now on its way to the *"suruka"* because through the bamboo it had entered into the man who shook it, and after having conversed with him, it now went its normal course".[105]

A bamboo pole was likewise used at a funerary feast on Roon Island. "On this day also the tall bamboo was prepared for its specific object. At the bottom-end eight small holes are cut with a knife, at some distance from each other, four on each side, and into these holes small spouts, cut from thin bamboo, are inserted, pointing in the same direction; all sections of the stake have been punched through, except the bottom section". The next day the snake-dance is performed. "The back of the last house [viz. to which the dance proceeds] has been masked by the sail of a prahu. From below this sail the bamboo stake protrudes, sloping down to the beach, so that all the spouts lie flat. Once the dancers have approached the last house, the nature of the dance and the song change. The dance assumes the movements of a snake, winding forward and back, to the left and to the right, whilst the dancers

---

[104] J. A. van Balen. "Iets over het doodenfeest bij de Papoea's aan de Geelvink-baai" (Some notes on the funeral feast of the Papuas on Geelvink Bay), in *TBG* XXXI, p. 556 ff.

[105] *Loc. cit.* p. 565.

sing: 'Aya Wakoei, aya Wosei', 'I am Wakoei, I am Wosei', as people
sometimes sing when they have become drunk, either on arrack or gin
they have obtained from one ship, or on their own toddy. When this
has continued for some time, the inhabitants of the last house and all
the onlookers start to shout and yell, stamping and kicking with all their
might. Simultaneously, from each of the spouts in the bamboo there
issues a jet of water, mixed with red pepper and other pungent fruit.
As soon as the dancers observe this, they rush with loud cries for these
jets and all drink of this tasty drink".[106]

This last description undoubtedly refers to the dramatisation of the
myth known in the Waropen region as the battle between Kirisi-
Aimeri and the snake Ghoiroponggai (Texts 33, 34, 35).

Hence, the bamboo connects heaven and earth.[107] The water running through the bamboo can hardly be anything else but the rain-
water, so that the smoke which rises through it represents quite probably the clouds. The intention would seem to be that the pregnant
woman is surrounded by the dora, heaven, the heavenly water, to
assist her through childbirth which is conceived as a crisis-rite. The
idea of water as a bearer of sacred power we also find in the
roséamasino, the water of the soul (see p. 249). The opinion of some
informants, viz. that the pregnant woman has to climb up to the garret,
just like the bridal couple, "in order to see the sun", is another indication
of the conception that the woman has to be brought into contact
with the sacred forces which gain the cosmic victory of the day
over the night. The ritual of pregnancy is therefore a process of
sacralisation or a ritual promoting life which runs parallel to the
rhythm of life in the cosmos.

CHILDBIRTH AND TAKING THE CHILD OUTSIDE

Pregnancy is a period of crisis which finds its culmination in childbirth. After childbirth a process of de-sacralisation is executed in
stages, enabling mother and child to take their normal place in life.
The whole process is a gradual abolition of a segregation which is
concluded by bringing the child outside.

---

[106] Loc. cit. pp. 567, 571.
[107] For the initiation ritual on Biak a bamboo pole (ambober) is used through
which smoke is sent up. "It is believed that the souls of the ancestors can
enter into contact with this world by means of the ambober"; see
W. K. H. Feuilletau de Bruyn, Schouten- en Padaido-eilanden, p. 105.

As we saw above, childbirth takes place in the screened corner of the room which is also used as a privy (*feretei*). Here the woman has to remain for a number of days after delivery — eight days in fact, but in actual life many women, especially after the birth of later children, do not take this isolation so seriously. As long as the woman remains in the *feretei* her girl-friends and female relatives come to sing the *ratara*, at which no men are allowed to be present. The father has to provide tobacco and sago for these women.

At the conclusion of these eight days the Waropen believe that a special *saira* is necessary to liberate the woman from the isolation of the *feretei* and to extend her liberty of movement to the rooms. But even then she is not allowed to come into the front part of the house. On the occasion of this *saira* the woman has to walk eight times around a pole decorated with leaves; according to others she has to pass underneath it. Another woman, armed with a stick, precedes her. According to the guru at Waren, who has a Waropen mother, one sings on this occasion:

> *Yasoi nda yamaino*
> *Yamaimba sirewino*
> *Sirewino sirewino*
> *Sapi yasoinda mano*
> *Yamaimba sirewino*
> *Sirewino sirewino mano.*[108]

I have never seen this *saira* at Nubuai, although there too many women draw a line of soot above their breasts parallel with the shoulders, some time after having given birth to a child, as the women in Waren are said to do after this *saira*.

The women are obliged to stay indoors for quite some time longer. The pregnancy ritual for Munamberi's wife we described above and which may be estimated to have taken place in the seventh month, was celebrated between the afternoon of 3rd March until the morning of the fifth. The child will have been born in May; it was a boy who was named after one of his famous ancestors, Maseiori. On April 10th

---

[108] I may be permitted to omit the translation. Perhaps some words may be recognised in it: *yasoi ndau*, I throw him outside; *yamaino*, I stay waiting in the prahu; *yamaimba*, I wait for; *sapi*, hibiscus; *sireghi*, a piece of cloth. Does the initiation-demon wait until the child is sufficiently grown to be taken outside (i.e. into the forest) and to be carried in a carrying-cloth?

of the next year the *saira* for taking the child outside was celebrated, so that mother and child had to stay in the stuffy atmosphere of a Waropen room from May till April, i.e. about eleven months. During this time the skin acquires a sallow colouring against which the tattoo marks stand out more clearly; this is considered very beautiful and it is even said that the tattoo marks are applied for this very purpose.

Most mothers keep rather strictly to this seclusion, which is sure to be far from agreeable. Only in case of the most pressing necessity do they leave the house, for instance because the child has become so ill in its original home that it is considered necessary to transfer it to another place. A mother who had been obliged to take this step, preferred to remain all those months in the strange house, even when her child had grown well again, rather than venturing outside prematurely in order to regain her own house. According to one of my informants it is actually insufficient that the woman remains all this time in one room and in the old days she had to stay in the *feretei*, but this is undoubtedly an exaggerated piece of information.

There are a number of criteria to see whether the time for the "taking outside" has already arrived. The woman has to remain in isolation (*ioaina sea* or *seka*) until the child recognises its surroundings (*kugha si iko*, the child recognises us), or until it is able to sit, or until it possesses at least two teeth.

The first time I was able to observe that a child was taken outside it appeared at the same time that there are also people who take this less seriously. It concerned a child which was certainly not much older than one or two months, and the participants showed quite clearly that they considered the whole feast a make-believe performance. And because during the same time there was a *munaba* in which the people were much more interested, the whole feast fizzled out like the proverbial damp squib. The intention of the originator of the feast will have been mainly to reintroduce the woman as quickly as possible to normal life and to have her work the sago.

The *saira* was sung at great length when Munamberi's child was, taken outside, i.e. the child whose mother's pregnancy ritual was described above. Munamberi, the chief of the Nuwuri clan, is a prominent person who belongs to the *serabawa*. He has gone working as a coolie more than once and so he has acquired a certain amount of worldly wisdom, as well as a number of dearly bought trophies of Western civilisation, like linnen suits, pith helmets, umbrellas, etc., all of which have considerably increased his prestige.

We may omit the description of the tour of the prahu with the enclosure; its evolutions are repeated at every *saira*. Let us assume that in the manner described above the *saira* has been sung for a number of days and nights and that we now are present during the latter part of the night before the execution of the actual ritual. We find the child's mother, Munamberi's second wife, in the room, together with the boy beside the fire on a sort of charpoy (*kambo*), hemmed in with blue cotton cloth (*rario*), so that this screened space is none too roomy and airy.[109] As in the case of the enclosure of the prahu it is not intended to close this space to the eyes of unauthorized persons, for it is easy to look over it.

In the chilly morning there appeared before the *kambo* Munamberi's crippled sister, Imberusi, who never married because of her hunchback. She brought two strips of coconut-bark and two choppers, whilst another woman carried a plateful of water. Imberusi stuck each chopper into a piece of bark, dipped this in the plate and pushed the two across the slats of the floor at the place where mother and child would pass. Imberusi, like many of her crippled companions in misfortune an energetic and sharp-tongued creature, with angry mutterings and curses cleared a passage for mother and child, whose way was outlined by the two choppers, like two iron legs with feet of coconut-bark.[110]

On the front gallery Foiwai takes over the child from his (classificatory) sister (see the genealogy on p. 141) and makes it sit in a carrying-cloth on his back.

Then he joins the ranks of the *saira*-dancers who now no longer dance in a simple circle, but assume the special formation indicated by the term *ghoisaira*, snake-*saira*, in which the row of dancers continuously makes the figure 8.

The small child, like all Waropen children accustomed from birth to be in the hands of everybody, lets itself be carried around quite composedly, and so Foiwai with the child on his back arrives at the head of the snake-row in the back part of the house. Here the row stops, amidst increasing blares of the horns and the noise of the gong. The atap roof-covering is slightly pushed aside so that the light of

---

[109] At Weinami it was not customary to enclose the *kambo*.

[110] In Huon Gulf the girls had to walk on pieces of coconut-bark when they came outside at the initiation, as reported by R. Neuhauss, *Deutsch Neu-Guinea*, III, p. 419. Also in this case with intention is probably that mother and child should not come into contact with the profane earth, as is likewise indicated by the use of water.

the morning may shine on the face of the child; a woman wipes the child's face with water from the plate in which Imberusi had dipped the coconut-bark shortly earlier, and Foiwai starts the following song:

> *Ianima Samanie iana paia*
> *Kiwe ieasanie.*
> *Ianima Wororinui iana wusidana*
> *Kiwe ieasanie.*
> *Ianima Pedei iana mundana*
> *Kiwe ieasanie.*
> *Ianima Aiomirewo iana gharadani*
> *Kiwe ieasanggindekio.*
> *Ianima Maniwurirewo iana rorarie*
> *Kiwe ieasanggindesikie.*
> *Ianima Pandori iana wusina*
> *Kiwe ieasangginekinie.*
> *Ianima Roarewo iana watamano*
> *Kiwe ieasangginekinie.*
> *Ianima Maturirewo iana oana*
> *Kiwe ieasangginekinie.*
> *Ianima Birandewo iana bararo*
> *Kiwe ieasangginekinie.*

Although the meaning of this song remains uncertain, it evidently indicates that now the initiandus is looking towards a number of localities and capes (*rewo*), whilst every time the animals are mentioned which he is allowed to eat (*iano*, he eats). The refrain seems to mean: "they are there, his teeth here". Then the leaves are enumerated in which the sago-cakes may be packed which the child is now allowed to eat (*gharaidana*, rattan-leaf; *wusidana*; *mundana*); furthermore *paia* = *faia*, a kind of fish, called "bawal hitam" in Malay (see p. 92); *rora*, a kind of mackerel; *watamano* for *watama*, mud-jumper (?); *oana*, a kind of fish, called "kerapu" in Malay; *wusina* for *wusi*, tree-kangaroo (?); *bararo*, gourd (with palm-wine?).

Even though the translation of this song is not quite definite (neither is it clear to the Waropen, as it seems), one thing is certain, and that is that at each subsequent *saira* the taboos lying on the types of food mentioned are lifted. Only, there exists no unanimity as to which taboo is lifted at each of the episodes.

When the child had been shown the sky from the front part of the

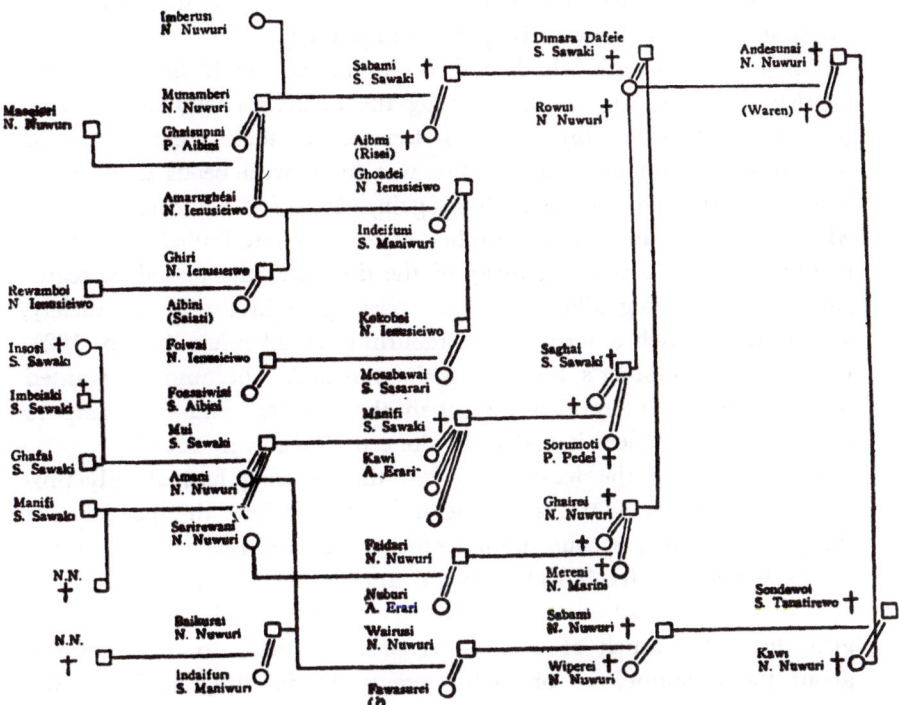

GENEALOGY I

Explanation of the signs:

□	— man	
○	— woman	
——	— descent	
=	— marriage	
†	— deceased	
† N.N.	— died before receiving a name	

Capital letters — indicate the clan; they are followed by the name of the *ruma*; K = Kai, A = Apeinawo etc.

( ) — native of another village.

house, it was taken over by the father, whilst in the mean time an *egharo,* a small decorated mat, had been spread on the floor of the front part as a seat for the mother. She looked tired and pale. For that matter, the women from Munamberi's house showed signs of irritation, and during the singing of the *saira* their cantankerous voices were heard all the time. His sisters were angry because their brother received so little in return from his wife's brothers for his large dowry.

Then the mother took over the child from the father and put him on her lap with its feet in an ancient plate. Several older women between whiles had strung a number of old beads on a string and the old *ghasaiwin* of Pedei told the mother in detail who it was that had given the beads, as these have to be paid in sago later.

When Munamberi's child had been adorned with beads, it was handed to one of the women among the onlookers who immediately put the child at her breast, whilst the mother took a child adopted by Munamberi on her lap and this was hung with beads in its turn. Thereupon she took her own child again, whose feet were now washed with soap and water by the brother of the mother. Probably we have to think here of a washing away of the dirt after the period of segregation, before being allowed to take part again in the life of society, a custom to which a term of the mourning ritual refers (see p. 182). However, the mother's brother made a mistake, because he wanted to wash the right leg first and then the left leg, and according to many protesting voices it had to be done the other way round.

This concluded the feast. However, the morning had already progressed so far, that the sun was quite hot before the participants thought of leaving, because no agreement had been reached concerning the exchange of gifts. To all outward appearances the cultured Waropen man smiles in extenuation of these feminine bickerings about gifts; he is too much of a *grand seigneur* to make so much trouble about the insignificance of such presents. Disdainfully he leaves the accepting and the bringing of the gifts to the women. But behind this lack of care a keen interest is hidden. That is also the reason why the men likewise linger in the front part of the house. For that they have not remained untouched becomes apparent in the evening when under the influence of palm-wine the tongues are loosened; then the men are quite ready to stand up for their interests, with force if needs must be.

CUTTING THE HAIR AND FITTING THE FIRST LEG-RINGS

As an example of this *saira*-ceremony I have chosen the feast which was given for the oldest son of Mui, the senior representative of the house of Sawaki, when two episodes were combined. The prahus with the enclosure, the dance of the men, etc., were like those on the occasions already described. Munamberi, whom I mentioned above, acted as the initiating brother of the mother (see p. 141) and carried the child in the carrying-cloth during the snake-dance (*ghoisaira*); it is here that I start my description.

The little boy held in his hand a decorated carrying-bag, filled with sago-cakes. In the front part of the house presents were exhibited: a few coconuts, about ten plates, large and small, and a tin clothes-trunk. The ceremonial sago-cakes exhibited there deserve special mention. These cakes, the wrapping-leaves for which have been fetched the day before in the prahu with the enclosure, are careless imitations of all kinds of objects. On the occasion of this *saira* there was one cake, also often found at other *saira*, viz. the *ghoifi*, snake-cake, a sausage-like object, coiled on a plate. Without the information provided by my spokesman I would not have been able to conclude from any outward characteristic that this cake was meant to be the likeness of the mythical snake Waisimaimuni, to be discussed in the seventh chapter. There also was a *maghafi*, i.e. a sago-cake, *fi*, modelled roughly after the shape of the sago-beater (*magha*). A flat, round cake, about ten inches in diameter, was called the *ghafafi*, mooncake; another was called *mamboifi*, brother's cake.

Another ritual object to be seen in the hands of one of the dancers in the photograph is the *sabakugha*, the tobacco-prahu, mentioned on p. 103; this is a structure plaited from tobacco rolled in pandanus-leaves, which in fact slightly suggests a prahu. Another object was the *Saparisabaku*, the tobacco morning-star, to be seen on photograph no. 32 as a plaited construction made of such rolls, roughly in the shape of a star.

When the dance was over, Munamberi sat down on the floor of the front part of the house and took the child on his lap, its feet again in an ancient plate. Now the mother's brother Baikorei approached with a decoction of *maiwua*-leaves with which he washed the child's feet. These *maiwua*-leaves do not possess any ritual value, according to my informants; leaves from any other tree in the tidal forest might have been taken for this purpose. Baikorei took a leg-ring and tried to slip it on the child's right leg, but as he was far from skilful, the child began to cry and the ring snapped. As usual everybody started to shout his advice; the father, who was looking on attentively, also interfered, whilst the mother who was singing the *ratara* in the background, together with the other women, observed things with a worried air, without interfering, however.

With every ring Baikorei fixed on the child's leg, the bystanders shouted gaily *iiiooo*. Then the child was adorned with all kinds of ornaments: on its arms bands of bead-work (*koruwangga*) and bands of shell (*saparo*) were fixed, on its legs beside *saparo* also rings of celluloid (*ponisi*), and bands of beads (*korubo*), around its shoulders crossed bands of bead-work (*sare*), around its hips beaded fringes (*foina*) with white porcelain shells (*korombowi*). When the uncles seemed to intend to hang even more decorations on the baby's body, its grandmother, i.e. Mui's mother, a quick-tempered woman, angrily interfered, so that Baikorei shyly restricted himself to cutting a few locks of the child's hair.

In the mean time Ghoadei, a classificatory mother's father, had taken his place beside the child and counted all his victories in war and his great deeds during the raids on the beads the child wore around its waist. Ghoadei was a famous warrior, but the names he enumerated were so many that it seemed hardly likely that he restricted himself to his own deeds of valour.

The relation between the ritual of the *raak* and the *saira* becomes apparent on several occasions. On p. 131 it was stated that the prahu with the enclosure is received in exactly the same manner as a successful raiding party. Formerly the leaves in which the sago-cakes for the feast had been wrapped (*fema*) were preserved for a raid, when they were hung in the trees near the raided village.

THE RETURN FROM A LONG VOYAGE

In the Waropen area this feast belongs to the less important ones, whilst among other tribes, like e.g. the Numfoor, where it is called *mansorandak*, it is considered very important. Usually one waits until a voyage to the Moor Islands is undertaken; occasionally a voyage to Japen is sufficient cause, whilst the chief of the Apeinawo clan waited until his son had the opportunity to join a voyage to Ambon. Seeing new regions (*siafu*, to see for the first time) is a constantly prolonged initiation to the Papua who loves to travel. Whoever once has made the journey to Manokwari is from then on considered an experienced traveller. This love of travel is also one of the motives for entering into a contract as an estate-coolie. The farther the journey to the estate, the greater is the ambition to go and work there.

The few times I was able to witness this ceremony one could hardly speak of a dancing party. During the *saira* the initiandus sits on a decorated festive seat (*kambo*), and leaves it again when the morning-song has been sung. Also at this moment his feet should be prevented from touching the floor, and because he is too big to be carried, he has to step on ancient plates. Then his mother's brother lifts another taboo on food by giving him a piece of lobster (*aifa*) to eat. The exchange of a few presents concludes the ceremony.

In another case no *saira* was sung at all. The mother's brother gave the child a piece of *aifa* to eat, a few presents were exchanged and the ceremony was over. If possible, one tries to have this ceremony coincide with another. Many people manage things without a feast, as is also often done when putting on the child's first clothes. But this is of course anything but dignified.

THE FEAST FOR THE PERFORATION OF THE NOSE

A fuller description should again be given to the ceremony of perforating the nose, the most important episode in the ritual, concluding the whole *saira*, and obligatory to all individuals, men and women. On the conclusion of this *saira* the years of puberty are past and usually the initiandus marries quite soon. Formerly, a pin (*buriniei*) about six inches in length and made out of a ground down shell was worn in the septum, sometimes a flight-feather of a cassowari was used. Nowadays the wearing of this nasal ornament has fallen into disuse, but the perforation of the nose as such is maintained. As is the case for other rituals, one is free to restrict or to elaborate the ceremony, but in Waropen appreciation it continues to be considered as important, and likewise as highly solemn when it culminates in the torch-dance. I shall take as my guide the *saira* as it was performed for two brothers of about fifteen and seventeen years of age, from the *ruma* Bindosano of the Kai clan; they were called Sighai and Tirai.

Nothing more need be said about the introductory ceremonies, the prahu with the enclosure, the dance, etc. I should only mention in addition that on the occasion of this *saira* the prahus of the wife's brothers upon their arrival in front of the house of the host were received with a demonstration of enmity and a rain of harmless projectiles like pieces of the ribs of sago-leaves, fragments of coconut-bark, etc. However, such demonstrations of ceremonial antagonism between the brothers of the wife and those of the husband were also witnessed occasionally during other *saira*-ceremonies and they are therefore not typical characteristics of the episode under discussion.

When towards dawn the snake-dance was danced, each of the initiandi went to stand beside a mother's brother, who put a slip of the carrying-cloth he was wearing around the shoulders of his sister's son, evidently indicating that the initiandus should be carried like a child by the initiator. Also the appearance of a number of sisters of the initiandi was striking. One walked behind each of them and pressed a *patu*, an ornament shaped like a fish-trap, like a crown on their head (see

10

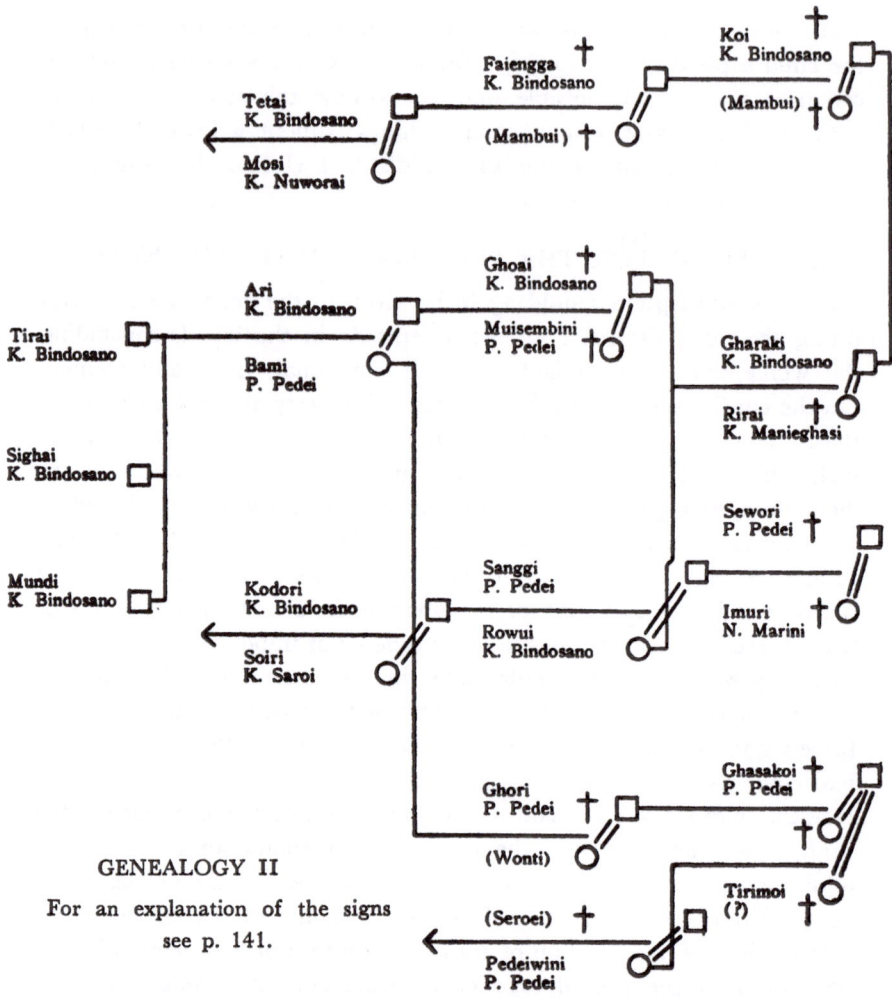

GENEALOGY II

For an explanation of the signs
see p. 141.

p. 158). Another sister carried a bow decorated with beads in front of
the initiandi. One girl carried the *Sapainara*, the torch of the morning-
star, a torch about two feet long and ╪ shaped. Another carried
a ceremonial sago-cake, roughly two feet in length and called *famafi*,
plank-cake, an imitation of the carved plank on which the initiandi
sat formerly. Again another carried the *foiesarafi* (*foisara* is the name
of the strip used for attaching the sieve for washing the sago inside
the rinsing-tub). However, at this *saira* I missed the decorated *fiaisari*,
a stamper used for tamping down the prepared sago in the leaf-

wrappers, which I had observed at other feasts. For that matter it was also the first time during this *saira* that it was not the men, but the sisters of the initiandi who danced around with these objects.

With great pride I was shown a so-called *damafi*, a trunk, with a movable lid, built from the ribs of sago-leaves, and filled with sago-cakes. At Weinami people maintained that formerly *buifi*, triton-shell cake, and *iawaifi*, shield-cake, had been made; such an *iawaifi* was specially destined for the perforation of the nose. First the women rasped the coconuts necessary for the preparation of this cake, accompanied by festive songs; next a scaffolding was erected in the centre of the village where the *iawaifi* was roasted, also with songs. When the women turned the cake, everybody loudly shouted *iiiooo*.

After the conclusion of the dance in the central portion of the house, all the men came together on the front gallery in order to dance the torch-dance. This is one of the few religious ceremonies which visibly move the Waropen who in general hold rather pedestrian views on religious matters. This dance and a certain ceremony in the funerary ritual they themselves repeatedly indicated as being very impressive.

And in fact, this ceremony has a certain fascination for the observers. When coming out of the stuffy house with its irritating smoke and its oppressive exhalation of heavily perspiring dancing men and women rubbed with coconut-oil, the cool wind of the morning is a refreshing comfort. The water races on in the never ending flow of the tides, the sun comes to lie on the rim of the forest, with pastel tints everywhere, in a frame of black-green tidal forest. Now the young men light their morning-star torch, its flame rising and falling above the dark heads of the dancers. Thin and solemn the song sounds through the silent village. The following *rano* is sung, called "the count of the monsters of the sea", *kikeaka ananggigha*:

Umao Saparigha umao.
Kasimbeghare Sapari
  kasimbeghare.
Umao Ombaibai umao.
Kasimbeghare Ombaibai
  kasimbeghare.
Dama rauo saia saiduio.
Saidui saia saimanio.
Iapina sandaso, anano Sowoio,
  iapina sandaso.

Mbarine mbambarine.
Anano Sowoio.
Iapina sandaso, anano Woradauni
  iapina sandaso.
Mbarine mbambarine.
Iapina sandaso, anano Waisimai,
  iapina sandaso.
Mbarine mbambarine.
Anano Waisimai.

Anano Worodauni   *Mbarine mbambarine.*
Dama rauo saia saiduio.   *Anano Ghéarina.*
Saiduio saia saimanio.   *Mbambarine.*
 Iapina sandaso, anano Sareri,
   iapina sandaso.
Mbarine mbambarine.
Anano Sareri.   *Mbarine mbambarine.*
Dama rauo saia saiduio.   *Dama rauo saia aghakie.*
Saiduio saia saimanio.   *Mbarine mbambarine.*
 Iapina sandaso, anano Ghéarina,   *Saiaghakio.*
   iapina sandaso.

The translation of this song confronts us again with many questions, but several things are certain. First the star, *uma*, Sapari, the morning-star, and another star, Ombaibai, are mentioned, of which it is said perhaps that they are red mirrors (*kasino reghare*). Then there is mention of a *dama*, the young men's house, *rauo*, towards the sea. According to my informants one has to think in this case of a mythical *dama*, built on one single pole, which moved towards the sea in the morning and which went back to its place in the village in the evening, all under its own force. This *dama* is again connected with the *dama* of the Pedei clan, which formerly stood on the banks of the river Ghaipono, and the name of which has been retained in the name of the *ruma* Saidui which stands on this spot nowadays. The name Saidui might be connected with a word **saro*, day, and a word *du*, to dive. According to the informants the word *saia* refers to the light of the torch. Perhaps *saimani* also means "morning". Then the song mentions a number of *anano* or *rina*, sea-monsters: Sowoi, Woradauni, Sareri, Ghéarina and Waisimai. Worodauni is the male monster which dwells downstream of the village. Waisimai we find as the snake at the initiation. The *ghéa* is a bird which indicates the change of the seasons; it is said that they emerge, *iapina*, and I would venture to suggest that we have to think again of the metamorphosis of the mythical birdfish which occurs at sunrise. The use of the torch might also indicate this.

When the song of "the count of the monsters of the sea" had been sung, it was possible to proceed to the ceremony proper. For this purpose the two initiandi took their seat on a clothes-chest, painted brown and provided with a lock (*burua*), which in these modern times is used as a festive seat, instead of the old-fashioned plank decorated with carving. At marriages this highly valued sign of opulence also serves as

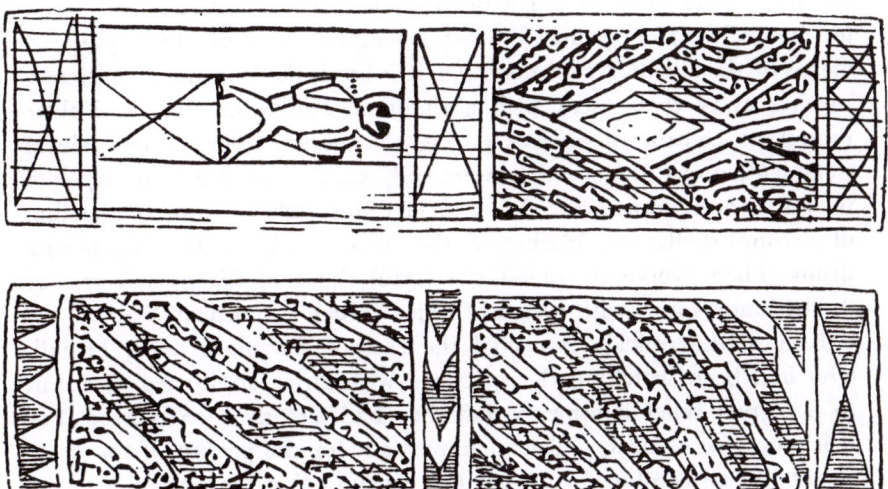

Sitting-board for the initiandus; length about 70 cm.

the seat of honour for the bride and groom, the key often being worn
demonstratively on a string around the neck of the fortunate owner.
The decorated plank was formerly not simply placed on the floor, but
on a ceremonial festive seat (*kambo*). The *kambo* is a kind of square
table, the legs of which — often slightly carved — protude beyond the
table-top. In his manuscript dictionary F. J. F. van Hasselt mentions
Numfoor "*kamboya*, sleeping-place for unmarried young men in a
Papuan house". In fact, young men occasionally sleep on the Waropen
kambo, not because that would be the main purpose of the *kambo*,
but simply because it offers a suitable resting-place. When the *kambo*
is used during ceremonies, a small mat or a mattress is spread on it.

Also at the ceremony described here, the clothes-chest was placed
on a mat; on it sat the two initiandi, staring apathetically in front of
them, like most Waropen do when during a feast they form the target
of public attention. Rari, a young woman from Nuwuri who had married
into this house, stood behind them as a helper. The mother's brother
Kodori was the right person to perform the actual operation of perforating the nose, but he stated that he lacked the necessary cold-
bloodedness and so he restricted himself to sharpening the *kowai*, a
small arrow for bird hunting, right in front of the patients and trying
the point on his thumb.

Therefore he handed over his work to the classificatory father of the boys, Tetei, but to judge by the faces of the initiandi, the latter too worked far from painlessly when attempting to pierce the septum by means of the *kowai*. *Korako, koraka*, it is hard, the bystanders muttered, turning away their faces with a shudder. The young men did their utmost to control themselves during this undoubtedly rather painful treatment, and although they were only partly successful, in any case their smothered cries were drowned in the terrible noise of the beating of coconut-shells, the blaring of the tritons and the banging of the drum. Then everybody raised the joyful shout of *iiiooo*, and one of the bystanders said with a sigh of relief "*Moka*", "finished".

Kodori also quickly inspected the ears of the small sister, Mundi, who had been watching the spectacle with large, frightened eyes, but to her great relief it appears that they had already been pierced. For it is more economical, and least unpleasant for the patiënt, if the operations can be combined. On another occasion first the septum and the ears of a girl were pierced, then the septum of a boy and lastly only the ears of a third, a smaller boy.[111]

After the operation a piece of the *kowai* used is left in the newly made opening, to be carefully replaced by a slightly thicker twig, until at last a hole has been made through which one might easily pass a not too heavy pencil. The patient feels a burning sensation in his nose, and for that reason Tetei blew on it; at another occasion a small decorated fan was used for this purpose, after first having held a piece of glowing coal under the nose of the initiandus.[112]

At the last-mentioned occasion also a coconut was broken over the head of the initiandus; it was hurriedly taken outside, rinsed with water and eaten by the people present. At that time a string of beads had likewise been hung around the neck of the initiandus; later this string was cut and handed to the mother's mother. The latter behaved as if she were very angry, because in her opinion the *saira* for her grandchild had been postponed too long. As already mentioned above, the wife's family always show themselves the most aggressive, because, according to the informants, it is they who are rendered most "ashamed" by a stagnation in the barter-relations. In those cases they do not dare

---

[111] It should be stressed that the women were also initiated, see p. 295.

[112] Among the Biak smoke is blown over the initiandus, see W. K. H. Feuilletau de Bruyn, *Schouten- en Padaido-eilanden*, p. 104. Among the Kiwai the boys are brought into contact with fire when they are allowed to see the sacred *mimia*-stones for the first time, see the *Report of the Torres Straits Expedition*, I, p. 214.

to appear at the feasts of others and this leads those others in their turn to refuse to come to the parties of the former.

Whilst Tirai and Sighai continued to sit on their clothes-chest with faces distorted by pain, another mother's brother appeared with a *manggoti,* a thin stick about half a meter in length, into which a number of white feathers had been stuck. This *manggoti* is a comb such as formerly famous warriors were allowed to wear at the *saira,* showing their victories by the number of the white feathers. This special *manggoti* was an heirloom of a famous warrior, already deceased, and it was with this comb in his hand that Adori, the mother's brother, addressed the onlookers, calling: *"Tinawuaramuo. Ineni...".* "Be silent. This is...", and then followed a string of names of vanquished foes, each feather representing one name. But this was only a mock-performance, because Adori had no longer been able to go on a raid. Because his address was lost in the laughter and the cries, Adori squatted down in some embarrassment and consoled himself with a piece of the *famafi* (sago-cake) which had been reserved for him.

In the mean time one of the relatives had appeared with sago-mush prepared with coconut in a special way, and with some kind of fish (*koipo,* a variety of horse-mackerel). For the initiates these dishes are no longer burdened with a taboo. According to some, also the *rora* (another kind of mackerel), and the *eni* (turtle) had been taboo for the initiandi, but according to others this taboo had already been rescinded earlier (see p. 140).

This is not yet the end of the *saira* because the initiandi still have to remain eight days in isolation, indicated by the curious expression *oai suno,* to dwell in the depths. It would seem that not everybody keeps to this obligation, because several persons, even of noble status, maintained that they had not gone through this experience. At Napan people were of the opinion that this custom had existed formerly, but only, because the initiandi suffer too much to allow them to go outside. In any case, Tirai and Sighai "dwelt in the depths" during the required eight days. Nothing of importance seems to happen during this period, because after much questioning I was only told that sometimes the initiandi receive advice concerning several types of food to be avoided and concerning sexual intercourse, i.a. that immediately after their marriage they should not eat too much mush, as otherwise they would have to urinate a great deal. The only thing which now is still required of the initiandi is that they should remain unnoticed as much as possible; they have to speak in whispers and they must not go out to urinate, but they do so through a bamboo.[113] A brother of their mother, according to others an elder brother, has to look after them and to give them advice.

After eight days the isolation is lifted with a small ceremony, at which I was not present. During the eighth night a number of men sing a *saira,* and when the dance is over they line up in front of the room where the initiandi are. Then the brother or the sister of the father steps out and asks: *"Arinasanai Sarakanie"*? "Are you called Sarakani"? The reply is: *"Io",* "Yes", but this reply is not given by the initiandus personally, but by another person who has hidden himself in the trees behind the house. The choice lies between two names; for male initiandi Sarakani for the elder and Andeimoiwai for the younger brother, and for women initiandae Wisopi for the elder and Winori for the younger sister. In this case Tirai obtained the name Sarakani and Sighai that of Andeimoiwai. As there are only two names, my informants said that it was better never to initiate more than one pair of brothers or sisters at the same time. It is not clear what is intended by these names; however that may be, they are often used in normal life.

NAME-GIVING, FILING THE TEETH AND TATTOOING

Here I want to make a short excursion into the names of the Waropen, although the bestowal of names has not been ritually institutionalised among this tribe as is often the case among other peoples. Every Waropen has at least three names, the first being his real name, *nasana bawa,* great name, by which he is generally known. One attempts as much as possible to retain the names of deceased members of the *ruma,* so that according to many the child should bear the name of the fifth ancestor in the male line. This is, however, not apparent from the genealogies; it is only that in one *ruma* the same name often occurs because one likes to call a child by the name of a brother or sister who had died earlier. According to some informants the father's father is the right person to determine the name of the child at the conclusion of the *saira* before it is taken outside.

Beside the "great name" everybody has two names which are mainly names of endearment used by the mother, but this is done in such a

---

[113] Similar customs, also the confinement of the initiandus, are to be observed among the Biak, see W. K. H. Feuilletau de Bruyn, *Schouten- en Padaido-eilanden,* p. 104 ff; see also J. L. D. van der Roest, "Uit het leven der bevolking van Windessi" (Life among the Windessi), in *TBG* XL, p. 156.

way that the first name of one person may be the second or the third name of another. One of my informants was called Adori Rafeterei Dedui and another Dusi Manupi Makodai. Although often several people have the same first name, one tries to prevent as much as possible that two living persons simultaneously bear the same combination of names. In this way there are several people called Ghafai, but there is only one Ghafai Narowuni and one Ghafai Munamberi. Calling a person by his second name causes embarrassment; this name is especially used by his mother. When Dusi Manupi Makodai was still a small child, his mother sang him to sleep with the words *Manupi ienako,* Manupi should sleep, not using the name Dusi.

Some names are considered as being more stylish than others; in this way high-born women are called Maindidai, Sarafai, Remoki, Bisarai, Sodafoni, or with combinations with the Ternatan word for "princess" (*boki*), Bokironai, Bokiraghai. The Waropen called our Numfoor cook Insosi, a name she did not like very much, because, although this name was quite fitting for a woman of social standing, it was especially used for older women. Some frequently used names are:

Women	*Men*
Akai, atap roof-covering	Bawai, the great
Anderifu, stranger's areca	Fafai, trunk
Aubini, ghost-woman	Ghafai, moon
Dimbarowui, bag with earrings	Ghairoi, lower course of a river
Ghafairari, moon-apron	Ghéai, kind of bird
Ghoabini, woman from the interior	Kambi, goat
Korai, marsupial	Kowui, sugar-cane
Masai, central pillar	Kukui, crumb
Mereni, the soft one	Maghai, sago-beater
Mosabawoai, precious bead of noble-woman	Mambeghai, crow
	Nafai, beach
Sairabini, *saira*-woman	Semai, ghost, suangi
Suai, scrotum	Sireghi, cloth
Suibai, testis	Umesi, house-lizard, chichak
Susibawai, great-breast	Unai, dog
Umai, star	
Unai, dog	

As in daily life it is not possible to call a person by his second or his third name, and because there are many people who have the same "great name", recourse to *soubriquets* and nicknames is unavoidable. When there is a person called Ghafai Munamberi Rakoi and another called Ghafai Narowuni Maninei, they both should be called by their main name, Ghafai, because second and third names are not used in daily intercourse. So, to distinguish one Ghafai from the other, one is given the nickname e.g. of Baitaghi, because at the moment the people had just heard of Batavia, whilst the other is called Roudi, because he had made a journey to the village which bears this name. An old man with the main name Dumoi was often called Dedui after his deceased father, in order to distinguish him from the others. Once people have children they are often addressed as Father or Mother of... (with the name of their first-born attached), like e.g. Insosimai, "Father of Insosi", or Maindidinai, "Mother of Maindidai". Often all these names are still insufficient and then the indications of a person have to be completed by mentioning the membership of the *ruma,* or their nearest relatives, etc., so that for the observer it is often far from easy to find his way through this multiplicity of names. Especially in mythology the great number of names by which the different figures are indicated gives rise to great difficulties. As soon as the Waropen have become Christians and have been baptised, they replace their pagan names by Christian names, often biblical ones, adapted to the exigencies of the Waropen sound-system.

Filing the teeth has not been institutionalised ritually either. Boys or girls who feel inclined to do so may have their teeth ground smooth along the cutting edge by means of a piece of pumice. The person who renders them this service has to be rewarded with a small present. It is said that once the teeth have been filed down, it is to be recommended to chew areca-nut regularly because otherwise one will have tooth-ache. Much ground down, black teeth are considered beautiful, whilst white teeth, so much appreciated by Europeans, are deemed ugly.

Finally we would like to repeat what was said on p. 28, that tattooing also is a non-ritual act. For instance, no ontological myth about this practice was ever encountered.

DAMA, THE YOUNG MEN'S HOUSE

The episodes of the Waropen ritual of life show the classical image of an initiation ritual, which is known from many other civilisations, particularly from this part of Oceania. The role of the mother's brother, the successive removal of the taboos on food, and the myths concerning the initiation-snake to be discussed below are clearly characteristic for the social structure of a society with a clan-system like that of the Waropen. In this ritual we often find that the period of crisis is at the same time a period of seclusion from the profane world, whereupon a process of gradual de-sacralisation again abolishes the seclusion. In this way we observed for childbirth that mother and child have to stay indoors for a considerable period, and that when they come outside special measures have still to be taken to enable them to touch the ground.

Because the initiation ritual is usually closely connected with sexual life, as demonstrated by K. Th. Preuss in his writings, many peoples also observe special measures in order to withdraw the initiandus from contact with the world of women during the initiation ritual. Now, although the Waropen *saira* is sung by dancing men, this group of *saira* dancing men does not form a separate men's organisation which watches carefully that its doings remain secret. Although the girls do not take part in the dance, there exist no objections against their observing everything. Moreover, on the Moor Islands the girls may take part, whilst in one of the *saira* described above the sisters of the initiandi even have a role in the ritual. A clan-sanctuary as described by Rassers in his studies on the theatre, i.e. a secluded, mostly enclosed space outside the village and therefore in the forest or in the mountains, and strictly prohibited for women and children, is unknown here.

On the other hand, during the *saira* described on p. 139, mother and initiandus sit on an e n c l o s e d bench, whilst during the *saira*-ritual the dancers, equipped with leafed branches, appear at the house in a prahu with an e n c l o s u r e. Furthermore, in the ritual the woman more and more takes second place; in the *saira* at childbirth the man still plays hardly any role; when the child is taken outside, the mother still holds it in her arms, but then it is transferred to her brother and she no longer comes to the fore. In view of the theories developed by Rassers, and when we see the enclosed prahu and the enclosed *kambo*, and consider the circumstance that it is only the men who act, the question immediately arises whether the Waropen initiation ritual also knew the isolation of a sacred and enclosed clan-sanctuary, all well-known elsewhere.

A first indication in this direction was provided when I happened to be shown a curious festive comb which, according to my informants, used to be worn by *serabawa* at the *saira*-ceremonies; this was the so-called *damasura*. This *sura* or comb has been made according to the plan of a certain building, *dama*, with one *masa*, central pole, and

four *ekainio,* poles supporting the roof, placed against the central pole
so as to form a cone. At the side where the comb is inserted into the
hair, it had one prong called *bubo,* the back of the house, and at the
other a prong representing the *rengga,* front of the house. The bent

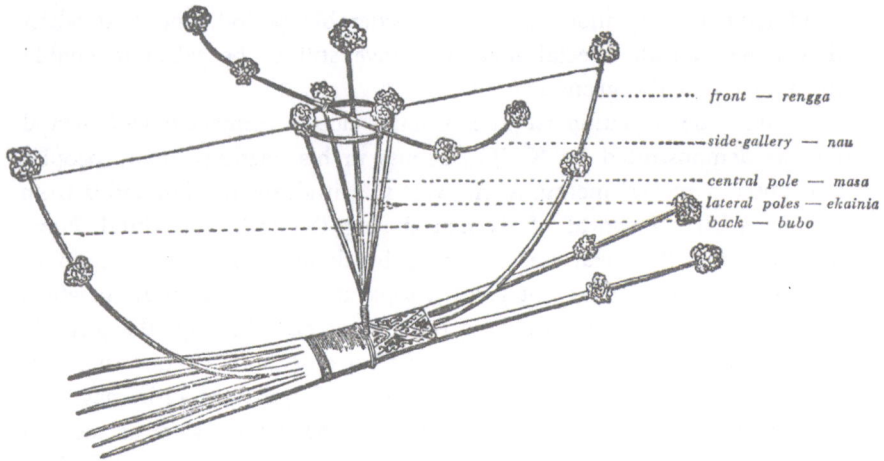

.... *front — rengga*

.... *side-gallery — nau*

.... *central pole — masa*
lateral poles — ekainia
back — bubo

Damasura, festive comb; height 25 cm; length 60 cm; bamboo decorated with
red and blue cotton.

bamboo perpendicular to the longitudinal axis of the comb is the
nau, side gallery. The model of a *dama* construction, provided by the
damasura illustrated here, has to be conceived as standing upside
down. The rosettes of red and blue cotton are therefore placed where
the poles of a real *dama* would be inserted in the ground. Further
questioning elicited that this *dama* was a representation of a small
building which several decades earlier still occurred in quite a few
Waropen villages and which was destined as the living quarters of the
unmarried men.[114]

At Nubuai people remembered that such a *dama* had been constructed beside the *seraruma* of every *da,* except in the Nuwuri clan.
However, according to the people from Mambui, this locality only
possessed one single *dama* for both clans; it stood near the *seraruma*
of Ghopari. At Weinami there were two, one upstream inside the
village and called Saiwini, for the people of Warami, and another,

_____

[114] See the illustration on p. 161.

34. The file of dancing men (Moor Islands)

35. The mother's brother slipping on the first leg-rings

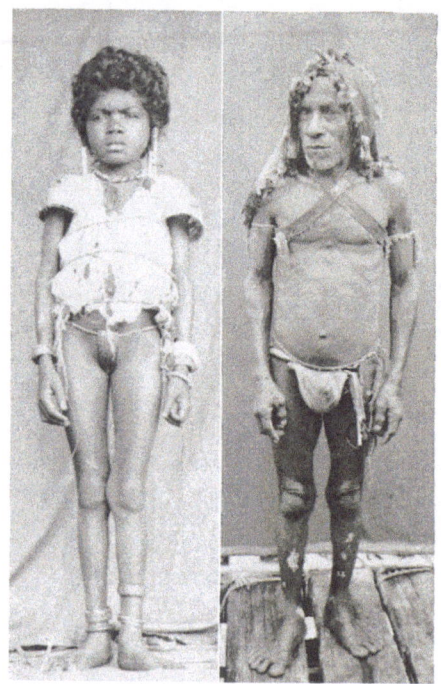

36, 37.　Mourning　apparel

38.　The corpse is taken to the cemetery

called Somari, downstream, for the people of the *da* Waratanoi.[115]
In the old village on the Woisimi river — where for the rest nothing
was known of an organisation in wards — the *ruma* Samberi and the
ruma Wokuai would have possessed a *dama* called Kowei Warami,
whilst the *ruma* Marani had a *dama* called Saiwini.

At Nubuai a *dama* Saidui, stood at the side of the *seraruma* Pedei
which is mentioned i.a. in the song of "the count of the monsters of
the sea" (see p. 147). This *dama* is called after a mythical prototype
which in the morning went to sea and sailed back in the evening under
its own power. The myth which decribes the meeting of men and
women tells how the women dwelling in the forest met the seafaring
men in a *dama* where they were living with a grandfather who dwelt
in the garret. It seems that this *dama* could float, just like the *dama*
Saidui, because the women threw out a rope to pull in this small
building. The *dama* stood on one single pole and for that reason the
pig, which had shown the women the way, did not want to stay
there (Texts 1, 2, 29).

On another occasion we found the *dama* at a *saira* feast in the shape
of a *damafi* (see p. 147). People maintained that this *damafi* should
be cone-shaped, but the object shown proved to consist simply of a
small chest made of the ribs of sago-leaves, with a saddle-shaped
roof, in which there was an opening to take out the sago-cakes.

According to the old people, no feasts were given in the *dama;* for
that matter, the structure was too small to do so. There was no carving,
neither on the poles, nor on the door. It was simply a sleeping-place
for the unmarried young men, where they could also stow away their
drums, and where they could sit and talk on the front gallery and chew
their quid.

Although the reproduction of the *dama* in the *damasura* and the
damafi, and the more or less strict exclusion of women in the whole
arrangement of the *saira*-ritual warrants the supposition that formerly
the *dama* must have been more than just a simple small building, this
still should have to remain a supposition, were it not that we possess

---

[115] These *dama* were still seen and described by F. S. A. de Clercq in 1887. Apart
from the *damau* there were no *rum seram* at Weinami, according to de Clercq,
but in the Waropen region there are said to have been *péré bomau,* in which
widows had to live during their period of isolation. It is not clear what houses
de Clercq means; perhaps the *feretei* or privy. See F. S. A. de Clercq and
J. D. E. Schmeltz, *Ethnographische beschrijving van de West- en Noordkust
van Nederlandsch Nieuw-Guinea* (Ethnographical description of the West
coast and North coast of Netherlands New Guinea), pp. 176, 178.

another telling indication concerning the nature of this structure. For
at Weinami the songs sung during a *saira* are called *damadorano,* i.e.
songs of the interior, *do*, of the *dama*. From this one can hardly conclude anything else but that the initiandus at least dwelt in the enclosed
space of the *dama,* which possesses an opening just large enough to
allow one to creep inside. The prototype of the enclosed space in which
the mother and the initiandus were sitting and of the enclosed prahu
with the cut branches can hardly be anything else but the sacred space
where the initiandus has to stay among other tribes. So, among the Waropen this enclosed space will quite probably have been the *dama,* where
formerly he "dwelt in the depths" (p. 151).

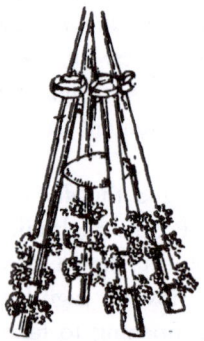

Patu, crown for initiandi.

We may proceed further and devote some attention to the *patuo* [116]
mentioned above on p. 145; this is a piece of decoration, made of four
little sticks tightly fastened around one half of a coconut shell. During
the feast of the perforation of the nasal septum this *patuo* is put on
the head of the initiandus. I presume that this *patuo* is nothing but a
memory of the cone-shaped *dama* with its four lateral poles. Within
the *dama,* the *masa* or central pole represents the ancestor. Putting
the *patuo* on the head of the initiandus implies that he is now sitting in
the space of the *dama* formed by the four *ekaini;* the central pole,
the *inggoro,* is lacking because that is the initiandus himself. For in a
mythical sense, as is also clear from Waropen mythology, the initiandus
equals the initiator; he is one with the *inggoro,* the ancestor, represented
by the central pole of the *dama*.

---

[116] A *patuo* is also a small fish-trap held in the hand. It is about 30 cm. high
and made of thorny rattan. Women place it over the hole in which the
mud-jumper lives in order to catch the animal.

Hence, we reach the conclusion that the *dama* — which according to the myth stood on one single pole — is nothing but a large *inggoro,* represented by this central pole. We therefore have to imagine the *dama* as a large umbrella, the *ekaini* merely serving to keep up the screen. Technically the *dama* was bound to this central pole, and in actual fact the *ekaini* had to be fastened to the ground. The mythological representation of the *dama,* however, suggested the cone-shape, mentioned continuously by my informants, although the models I was shown had not a round but a square ground-plan.

Isolation in the closed room or in the enclosed space at a *saira* has therefore, I believe, come to replace a former isolation inside the *dama,* where the initiandus became one with the ancestor, represented by the central pole. Now the significance of the leafed branches brought in the prahu with the enclosure is no longer subject to doubt. For they represent the forest where the initiandus had to pass through his trials segregated from society. I shall enter further into this question when discussing the mythology. To remove any lingering doubt, one of the myths relates that after humanity had sprung into being from the primeval couple, the "old man" (see p. 281) ordered them to celebrate an initiation ceremony in the sago-forest (Texts 14).

The Waropen *dama* therefore must have been a small building, similar to the Numfoor *rum sram.* [117] And because among the Numfoor the "head"- and the "tail"-clan belong to the two sides of this house, being at the same time the head and the tail of the snake, I conclude that the triton-shell, with the head and the tail of which the component clans of the Waropen are associated, must be the mythical equivalent of the *dama,* the young men's house. Head and tail of one primeval clan = head and tail of the snake = head and tail of the ancestor in the young men's house, otherwise the young men's house itself = head and tail of the triton-shell. The enclosed space in this shell is the enclosed space of the *dama* with its narrow entrance. In mythology the enclosed space in which the initiandus is confined may be equated with any other closed object, like a basket, a jar, etc.

In the literature concerning the young men's houses on New Guinea there repeatedly proves to exist uncertainty on the question whether these young men's houses also possess a religious significance. The sacred nature of the m e n ' s house in the Eastern part of Netherlands New Guinea is rather evident; this building, as Wichmann puts it, is at the

---

[117] *Dama* is probably related etymologically to Numfoor *sram.*

same time church, townhall, local inn, school, barracks and even, when
enemies made prisoner are concerned, gaol.[118] In this case the men's
house is the centre for the more or less organised men. At Tobadi the
men's house and the young men's house (sleeping-place and residence
of the unmarried men) occur side by side. The difficulty is that,
although e.g. the Numfoor *rum sram* was a sacred building in view
of the ancestral images (*mon*) to be found there, and of the special
festivities which the Numfoor used to hold upon the establishment of
such a *rum sram*, no cult was ever observed there, or in any of the
young men's houses occurring elsewhere in the southern part of Geelvink
Bay. G. A. J. van der Sande believes that both the men's house in
eastern New Guinea and the young men's house like the Numfoor *rum
sram* "have their origin in the same fundamental idea, though the
external ceremonies of the service have, in the case of the *rum seram*,
much diminished".[119] However, there is not the slightest trace of a
cult which was observed in these young men's houses in former days.
W. K. H. Feuilletau de Bruyn attempts to show that the young men's
house was connected with a cult of the sun; he seems to believe that
certain things belong to the *rum seram*, e.g. ithyphallic images.[120] But
F. J. F. van Hasselt objects against translating *yaberdares* (according
to F. de Bruyn the original name of the *rum seram*) as "house of the
sun".[121]

When we compare the function of the Numfoor *rum seram* with that
of the *dama*, the difficulties outlined above are quickly solved. There
is no need to look for traces of a former cult in the young men's house,
because this building may quite well have had a religious significance
without this. Along Geelvink Bay the initiandi dwelt in the young men's
house during their period of segregation from society. In this enclosed

---

[118] A. Wichmann, *Nova Guinea*, IV, p. 165, note 1; see also IV, p. 347 ff., and
      III, p. 287 ff.

[119] *Nova Guinea*, III, p. 302.

[120] *Schouten- and Padaido-eilanden*, p. 33. One could rather be inclined to think
      of a relation between the young men's house and the moon, as might be
      deduced from the curious wooden quoit, according to de Clercq only found
      on top of the *rum seram* and indicated by the name *kob*. De Clercq draws
      attention to the fact that *kob* is also the name of a float carved in the shape
      of a bird and used for catching turtles. Now this same bird-shaped float is
      called *ghafa* in Waropen, so that one is led to ask whether the *kob* is not
      intended as a representation of the winged moon (*op. cit.*, plate XXVI,
      fig. 10, and plate XXIX, fig. 4). The same suggestion is raised by the picture
      of a *kob* given by Feuilletau de Bruyn, *op. cit.*, p. 33.

[121] *TBG*, LX, p. 110.

space they were together with the ancestors, or rather, they were inside the ancestor, so that the whole world of ideas is in agreement with what we know from elsewhere in the non-Netherlands part of northern New Guinea. The close relation between the young men's house and the ancestor-initiation-demon explains why it was in the *rum seram*

Dama, the young men's house (model).

that the images of the clan-ancestors had to be found. And as soon as this sacred area is extended and the feasting men perform their dances and execute the initiation ceremonies for the initiandi within its secluded space, the young men's house has become a temple.

This exposition is therefore a confirmation and a further elaboration of the idea propounded by Wilken, who believed this building to be consecrated to the worship of the ancestors of the tribe.[122]

[122] G. A. Wilken, *Verspreide Geschriften* (Collected works), III, p. 212 ff. — On p. 129 I mentioned that the technical term for "performing" a *saira* is curiously enough "singing" a *saira*, although in actual fact there is as much dancing as on the occasion of a *munaba*. If we would be allowed to conclude that dancing a *saira* was a new development, we could quite well conceive that the feasting men originally only sang the *saira* in the narrow space of the *dama*, because there was no room for dancing. The very sacred Serakokoi is not danced either, but only sung. See pp. 189 and 235.

Along Geelvink Bay the young men's house did not become a temple
and the centre of a cult, but it lost its significance when people began
to initiate the initiandi simply at home. The actual ceremonies were
formerly still executed on the beach at Weinami, I was told. In the
mangrove forest the house is practically the only place where there is
room to have a feast. The house possesses, moreover, a *masa*, a central
pole, which represents the ancestor, and it is around this central pole
in the centre of the house that the men dance.[123] For the house likewise
possesses a mythical design which should perhaps be connected with
the prahu, the latter often forming the starting-point for the division of
the tribe.[124] Although in the Napan area the *dama* is not cone-shaped,
but follows the type of the normal dwelling house, as is the case in the
whole of the western part of Geelvink Bay, this does not mean to say
that the young men's house is not built according to a sacred plan,
because even the normal house obeys to such a plan. An exposition given
by Wichmann and the pictures included in his work would even lead
us to conclude that in the period 1775—1827 a young men's house on
Doreh Bay of the *dama*-type changed into the model of the later
rum seram.

DISINTEGRATION OF THE RITUAL

The enclosure of blue cotton cloth is far from excluding the interior
of the enclosed space from the sight of the women and the uninitiated,
who anyway are able freely to observe everything that takes place during
the initiation in the central part of the house.

When we observe the Waropen *saira* it is not only striking that the
segregation is far from strict, but also that the participants do not take
part in the religious ceremonies in a complete, religious surrender. Of
course, there is dancing and singing, and often with great passion, but
the least thing is sufficient cause for the dancers to stop their dance
and their sacred song. Only during the torch-dance something like true
devotion could be observed. The initiation ceremony is of personal

---

[123] According to Feuilletau de Bruyn, among the Biak the initiandus was locked
up in the *kamboi* (cf. the Waropen *kambo* in which the initiandus is placed),
"a kind of cage, screened off with mats and hung on the cross-beams of
the house, so that the women and children see nothing of the young man",
op. cit. p. 104.

[124] Would it be more than only poetic licence when it is said in no. 33 of the
Texts that the open maw of the snake Roponggai (the initiation-demon) was
as large as the centre of the house?

interest only to some of the inhabitants of the *ruma*; the others go about their ordinary daily business. The men collect their fishing gear, the women gather the tools for preparing the sago and both push their way towards the prahus through the persons officiating at the ceremonial. Children quarrel and babies are nursed. And amongst all these sleeping, eating, yawning, walking and dancing people there are the initiandus and the initiator. Even when the people are engaged·in the ceremonial action proper, like e.g. the perforation of the septum, there is no trace of religious respect. Everybody loudly voices his opinion or his humorous remarks. The initiandus tries as well as he can to show an unmoved countenance, although obviously it is highly disagreeable to him to stand at the very centre of public interest. Even the initiator shows nothing of the attitude of an officiant in some ritual, but rather gives an impression of diffidence and clumsiness.

Even more striking than the lack of respect for sacred things among both the participants and the other onlookers is the apparent great lack of decision. There is no question of the consistent execution of a minutely ordered ritual. So much difference is shown in the performance of even one and the same initiation episode, that it is impossible to give a general and generalising description. There are always differences. The one has made sago-cakes different from the other; sometimes there is a mock-battle against the wife's brothers; sometimes the sisters of the initiandus appear in the *saira*. What is a large feast for one person, is hardly noticeable for another. For single individuals never the same number of *saira* is celebrated. Furthermore, at every *saira* there are endless differences of opinion concerning the order or the details of the actions. From all sides one hears the assurance that the ancestors have ordered things in such and such a way and that pretended innovations will cause their anger. Such innovations often concern things which in our eyes are trifles. The use of a clothes-chest by way of a ceremonial seat was one of those innovations which caused a great to-do. In short, a definitely ordered and methodically performed ceremonial is out of the question.[125]

---

[125] These phenomena do not occur among other tribes. P. Wirz writes in his *Die Marind-anim von Holländisch-Süd-Neu-Guinea*, IV, p. 17, "It is curious, how stiffly and ceremoniously the otherwise lively and impulsive Marind behave during the feasts. The people cling to the usual customs with an unbelievable tenacity and a rigid ceremonial nobody would ever think of changing in the slightest detail, or care to enliven its serious and solemn atmosphere by a joke".

This curious phenomenon of uncertainty, combined with little inner surrender I call disintegration. The causes for this phenomenon are not due to a lack of interest among the Waropen. Neither is it necessary to think of the disorganising influence of the contact with Western civilisation, as before 1930 the region of Waropen Kai had hardly any contact with the West. We shall therefore have to look for an internal cause, a cause which also led to the disappearance of the *dama*.

This cause I believe to be a shift in the whole structure of Waropen society. A society in which we could expect a young men's house with an accompanying *saira*-ritual, a sacred enclosure, a forest, etc., would be that of the "total" clan with the initiation ritual well-known in this part of Oceania; Rassers has clearly described this ritual in his studies on the theatre.

The ancient clan-initiation, to which Waropen mythology is still adjusted for a considerable part, was a total ceremony, which resulted i.a. in a certain ritual equivalence of the young members of the clan. All had to pass through the same ritual. The Windessi believe that if somebody's nose has not been perforated during his lifetime, he will have to suffer this operation after his death by the hands of the spirits. We shall meet with a similar belief among the Waropen. The clan-initiation is aimed at an equalisation of the members of the clan in accordance with a scheme provided by tradition; the value of this scheme in itself is of secondary importance.

However, when under the influence of the already existing principle of seniority the oldest lines of the genealogical *ruma* increasingly tend to become a separate layer in society, tendencies start to develop which will eventually lead to a change in character of the initiation ritual. The first tendency is, that all attention is increasingly focussed on the senior descendant of the senior line. The more this moves into the foreground, the more the younger descendants will withdraw into the background. In this way an ever increasing difference between the individuals is brought into being. Whilst for the commoner the ritual becomes more and more simple, that for the oldest children of the senior family-branches develops further and further.

For these high-born children a feast has to be celebrated on an ever growing number of occasions, with the result that finally a nearly endless series of feasts marks all conceivable points in their development. Even the ritual of childbirth with its taboos for the husband becomes involved in the *saira*-ritual. However, the view is still generally prevalent that every individual, especially the man, should have his nasal septum perforated during puberty, even though no extensive *saira* is celebrated. This is a reminiscence of the old compulsory nature of the clan-ritual for all young people.

We also observe that the increasingly elaborate ritual still had not assumed fixed forms; with every ceremonial there is great uncertainty, because the growing ideal of social status has not yet been completely institutionalised. When discussing the mythology we shall see that the mythology in its traditional literary form had not yet adapted itself to this ritual, with the result that in Waropen religion it is difficult to make the myth fit the rites as these exist at present.

A generally accepted ideal personality was likewise unknown in this society; Waropen literature makes only little mention of princes and princesses. The commoner chafes at the presumption of the gods (*sera*) who are, after all, descended from the same ancestor as he. The *sera* in his turn calls himself a god, but he still likes to acknowledge his unity with the lower orders, particularly over against the other *da*.

It is not only in the constantly increasing ritual for certain individuals, the oldest of the senior group, that we discern the influence of a structural change; there exists also a growing tendency towards action in the full light of day. The old clan-ritual sought the loneliness of the forest outside society. That is where the *dama* belonged. There an enclosure kept the uninitiated children and women away.

However, the person who wants to wage a battle of gifts and who wishes to induce his guests, by means of the public humiliation of receiving many presents, to give even larger counter-presents, has little use for all this mystery. The contractual battle of gifts we witnessed on the occasion of the marriage and the *saira* has to be fought in public.

For these reasons the *dama* had to disappear. The ritual should no longer take place in the inaccessible seclusion of the forest, but in the public centres of the houses. An enclosure is still known, here and there, like on Biak, the initiandus is still locked in a cage, but for the rest every woman and child can see the dancing men when their ranks wind around the central pole in the snake-dance. This central pole represents both the ancestor and the old-fashioned *dama* itself.

The new structure has not yet assumed definite forms. The old structure has not yet completely lost its significance. That is the reason why this culture was not concentrated on a definite cultural pattern; it had become disintegrated. In the Waropen area the ritual is more and more beginning to assume the typical characteristics of the potlatch, but it has not yet dissociated itself from the clan-ritual, just as the class-organisation has been unable to free itself from the clan-organisation. The internal tension between the privacy of the men's ceremonial on the one hand and of the public show of status and of priority on the other have resulted in a disintegration of the whole of the initiation ritual.

———

CHAPTER FOUR

MUNABA: THE RITUAL OF DEATH

VIEWS ON DEATH.   THE DISPOSAL OF THE BODY

No more than other peoples have the Waropen learnt from experience to draw the conclusion that man has to die once. They do not consider death as a regular biological phenomenon, but as an irregular interruption of human existence. In Waropen mythology we find a train of thought similar to that described by K. Th. Preuss for religions of a comparable structure. According to mythology, in the beginning the two sexes led a separate existence; at that time human civilisation still stood at a very low level, because in this state of separation both men and women had only inferior or one-sided food. When men and women meet, sexual intercourse leads to procreation, which again must lead to death. And because initiation everywhere also means to obtain entry to ordered sexual intercourse, initiation is connected with death. We find the concept of the dying initiandus in Waropen mythology linked with the idea that due to his death everybody must die.

In general, death is accepted more or less resignedly when very old and decrepit people are concerned; they have obtained what every Waropen considers as his ideal: a very long life. The pagan view of the collection in the Christian church is that thereby one buys a long life. *Raweuri figha kapora doséa,* I collect the sago with which we pay for our souls; in this way the Christians described the aim of church contributions.

The aim of the ritual of death is of course not wholly contrary to the ritual of life and so it does not aim at serving the powers of the nether world. As a ritual, the *munaba*-ritual is meant to promote life as much as the *saira*-ceremonial does. Although there is a certain contrast between the two rituals, in that in the *saira* the men are more active, whilst in the *munaba* it is mainly the women who come to the fore, in the end both *saira* and *munaba* do not aim at opposing, but at identical results.

Saira and *munaba* do not exclude each other, they complete each other. Life after death is a vaguely conceived continuation of life on earth. For this reason the *saira*-ceremonies are actually considered to be as necessary for the dead as for the living. When a boy of about fourteen years of age died suddenly, whilst his nose had not yet been perforated, this operation was performed on him in pantomime by his mother's brother, and ornaments, shell-bracelets and beads were laid on the corpse. In another case first a soul-statuette was made and this was operated on in the place of the dead child. When small children die before having been brought outside, a small *saira* has still to be sung, especially in the case of first-born, my informants told me.

Hence, death is not envisaged as the definite conclusion of life, but rather as its continuation in a changed form, and, especially in case of younger people, as an unnatural change of being. With this conception the view is probably connected that no *munaba* must be held for persons whose heads were lost to the enemy during a raid, or for those who die of a sudden haemorrhage. In these people the principle of life has been maimed so seriously that they will not even be able to continue the shadowy existence of the dead in the hereafter. Their mutilated being constitutes a threat to the living. It will be clear that no *munaba* is held for persons who have been killed as *sema* (Malay: suangi).

The Waropen therefore does not possess any clear ideas on life after death, because it is nothing but a hazy continuation of ordinary life on earth. Death is believed to be brought about by the departure of the *roséa* (secluded sacred interior; soul) which may wander from the body and which according to some is to be compared to a shadow. Ordinary people cannot see this shadow, but a *ghasaiwin* is able to do so. If the *roséa* has gone very far away, one is *fero*, unconscious, dead. Sometimes it is possible to remind the *roséa* of its body, e.g. by beating it with cloths steeped in hot water, or in an emergency by burning it with pieces of glowing coal. Very occasionally the *ghasaiwin*, the medicine-woman, stills knows what to do, especially in the case of young children, because her soul starts out to redeem these children from the *inggoro,* the ancestors, or from the *auo* who have stolen it. I was told that the chief of the Nuwuri clan at Nubuai had already succeeded once in making a person come back to life who was already *fero,* because the *sema* had sent away his soul. But once the *sema* have availed themselves of the absence of one's soul to devour the inside of the body, death follows irrevocably.

What exactly happens to a soul after death, nobody could tell me. According to some there is a heavy tree which blocks the land of the souls and over which the soul of a living person should not climb. Others believe that the dead person has something to do with a snake, but among the pagans no traces of a belief in retribution in the hereafter were found. Although the Waropen were unable to pronounce themselves clearly on this point, I am convinced that they do not imagine the life of the *nunggu fero,* the dead, as being a life of the soul alone. The soul belongs to a human being and according to some it also remains in the vicinity of the places where the dead have been deposed. Sometimes it returns to the house of the deceased, e.g. in the shape of a bat which attaches itself to the roof. Especially when after the burial the necessary presents have not been given, the possibility exists that the dead person will return, manifesting itself by knocking on his clothes-chest.[126]

However, for the rest man after his death becomes an *inggoro* or *nurawa* (N.), a term also used to indicate the soul-statuettes. This does not mean to say that immediately after death one changes into an *inggoro,* for people continue to imagine the dead person as they know him during his life, whilst the *inggoro* are more especially the mythical ancestors who formerly established the world-order and who still jealously look after its maintenance. With the passing of time the dead are associated with the powerful figures of mythical primeval times. Without worrying about the deceased children, they are invoked as *daida* and *mina,* father and mother. Old, influential men after their death quite easily become powerful *inggoro* who are invoked for abundant fishing and whose anger is feared when transgressing their institutions. The *inggoro* live a life of their own in their village (*inggoinu*), according to some situated somewhere in the air, according to others somewhere on the sea or in the vicinity of the burial grounds. At Weinami it was said that the land of the dead was called *Suruka* and that the dead arrived there along the path of the sun, through the hole into which the sun dives in the West; an old woman, called Indoki, was said to be the chief of this village. However, in these representations we may undoubtedly perceive the influence of the tribes of the western part of Geelvink Bay, an influence clearly felt all over the Napan region.

The ancestors are always ready to help their descendants, even when

---

[126] Some Christians believe that the soul remains near the grave until the third day, when a meal is prepared, accompanied by psalm singing. On the third day the soul returns to the house of mourning, where one may see its footprints before the door. It also often makes such footprints around the grave, for which reason several people avoid the grave-yard at night.

the latter want to perform black magic. Their normal task is to assist
the people in their daily occupations. The fisherman invokes them when
casting his line: "*Inggoiee, minda wosoraee! Awe ado bawa natio wu
tasio*", "Ancestors, come close to me! Let a big fish come and bite".
For their benevolence, some palm-wine, some tobacco or some sago-mush
is sacrificed to them. But when the descendants do not please the
ancestors, when they transgress the taboo regulations, when they leave
their debts unpaid or when they neglect things dear to the dead, they
have to fear their spite. Then, in their shadowy and after all rather
dreary world, they are jealous of ordinary men. They punish them with
all kinds of reverses and steal the souls of the little children, which they
lodge as slaves in the banyan houses of the *auo*. When there is a
difference of opinion concerning a certain rule of conduct, a reminder
of the authority of the institutions of the ancients is at the same time
a serious warning for the transgressor, lest they show their anger.
Bawaini kikurasa ku maia, the ancients will be angry at this child. The
ancestors have spoken from beyond the grave. Fortunately they are dead
and in the myth one may make them say what one believes personally.

Both healthy and sick persons take numerous measures and use
various medicines to ensure the much desired long life, but in the end
they face the hard fact of death with a resignation which, although it
certainly does not exclude genuine grief, still gives the impression of a
laconical impassivity. It is often a nearly hopeless undertaking to try
to convince a sick Waropen that his life is in danger, a danger which
might be avoided by medical interference.

As soon as the relatives believe death to be near, they begin to try
to withhold the patient from his intentions by means of loud laments.
The mother takes her dying child on her lap, the son draws his dying
father against him in a half-sitting position (*oai sakuro*, to sit whilst
leaning), and so they continue to sit until they are fully certain.
Raiwarimaeee, my boy, the mother laments; *daidaeee*, my father, the
son keens, uninterruptedly, throughout the night. The mother "weeps
together with her breasts", (*ianikiwa risusigha*), whose milk is the palm-
wine her baby has scorned. *Ariesanieeee*, your palm-wine; with these
words the mother bewails her infant.

Once death has set in, everybody raises a great uproar by yelling and
by beating the prahus.[127] If the deceased was an important personality

---

[127] Although therefore the prahu is used for making noise, an aim to which it
is quite suited, the step towards the invention of the fissure-drum was not
taken here.

and death did not take place at night, immediately and from all sides
prahus start arriving, manned by keening men and women. The women
especially make a great to-do, crying out their grief, despairingly raising
their arms to heaven and then prostrating themselves in the prahu. At
Napan the decease was announced by casting the ashes from the hearth
— and at night the fire — outside the house. This throwing away of
the ashes was repeated there when the corpse was taken to the forest
to be buried. It was done, as I was told, "in order to throw away the
disease", (*kitohara wisio*).

Because people go to absurd lengths in their expectation of a long
life, relatives are usually still taken by surprise when death intervenes.
In some cases the *ghasaiwin* may predict the chances for life, but mostly
her pronouncements only become somewhat more definite when the
final outcome is quite certain. When somebody has died the subject
of the whispered conversation of those present mostly centers around
the possible cause of death. In case of old and decrepit persons one
does not always worry about this problem, but when young people are
concerned one tries to find the actual cause from all kinds of indications.
Sudden death, swelling or change of colour of the corpse are indications
of the activities of a *sema*. For a small child which had drowned the
discoloration of the face was held to be an indication. In this case
people believed that already for some time the child had become a
victim of the *sema* who had already partly devoured it inside, to order
it thereupon to fall into the water, by which means the *sema* attempted
to distract the attention.

Another cause of death, especially for young children, is the activities
of the *auo*, who come to fetch the soul of the child, or of the *inggoro*,
the ancestors, who have been angered by some transgression of the
child's parents, in particular of its father. And then there is the possibility
of black magic and the influence of *dareo* (see p. 216), again in particular for young children.

The deceased is laid in the central portion of the house, the head
towards the front of the house, towards the water, i.e. pointing downstream, and with the feet towards the back, towards the forest, hence
upstream. Thereupon all the openings of the body are closed: the nose
and the ears with pieces of the ribs of a sago-leaf, the mouth is closed
by tying up the lower jaw by means of a strip of blue cloth, the eye-
openings are covered with two small, round shells, a blue loin-cloth is
pulled tightly between the legs of the corpse and tucked in behind
(*kito irogha*). The corpse lies on its back with the arms stretched

alongside the body. At its feet often a small fire is laid, to keep away the flies, it was said. If the dead person is an infant, the mother holds it in her arms until it is buried in the forest.

According to the regulations effective at present, this has to be done inside twenty-four hours. For doing so the people are dependent on the water-level, because only at low tide it is possible to carry the corpse into the forest. In the hours between the relatives squat around the corpse and wail, relieving each other. The closest relatives who sit there and lament (father, mother, brother, sister, etc.) only wail loudly during their turn, to calm down suddenly when they are relieved. The intensity of this public keening is therefore far from indicating the real inner sadness which can hardly be expressed in the noisy wailing and the uproar of the mortuary feasts of the institutionalised funerary ritual. This inner grief is expressed by a certain retirement in ordinary life, in which the mourners are left alone with their sad thoughts by their surroundings.

During the night a number of young men, armed with bow and arrow, stand guard against the *sema* who still would like to feast on the corpse; at low water they should be sitting underneath the house, although I have never observed them there.

The behaviour of the other inhabitants of the *ruma* does not show any particular deference to death; unmoved by the wailing they go about their daily avocations. On the front gallery men sit and talk; on the back gallery women collect around a fire; children walk around and bicker, and a dog who ventures too close to the corpse is given a blow with a piece of wood so that he runs away howling. Still, it is better not to come too near a corpse or a burial-ground if there is no need to do so, because one runs the risk of being punished, e.g. because the dead person inserts a piece of wood into the body of the living, which causes much pain and which can only be removed by the *ghasaiwin* (see p. 251).

When the moment has come that the corpse has to be taken out, one looks for a piece of an old prahu (*ghafema*), or for a small prahu which is also used as a sago rinsing-trough; from this the points are cut. The head of the corpse is placed on a pile of plates and whilst the wailing grows in volume, the head is covered with a piece of blue cotton. Then the corpse is placed in the segment of the prahu and carried out of the back of the house, the feet still directed towards the forest; it is covered with a piece of blue cotton or a mat or a flat

39.   Women in mourning

40.   A valuable plate, smashed on the occasion of mourning

41. Triumphal branches being set out whilst the skull is fetched away

42. Miniature house for the skull

fish-trap.[128] Then the corpse is put in a *sandu,* a prahu made of bark and without outriggers.

Before the corpse is definitely brought to the burial-place (*ferasoa*), an attempt is first made to establish the cause of death. For this purpose the prahu carrying the corpse and the nearest relatives (in the case of a boy of about fourteen years of age these were his mother and her brother, his father and the latter's sister) is rowed to an open space in the tidal forest behind the house of mourning. There the *sandu* is tied up at two punting-poles, one at the stem and one at the stern. Upon the death of the boy mentioned above, the pole at the stem was held hy his father's sister, the one at the stern by his mother's brother. Then the mother's mother crawls into the *sandu* under the mat which covers the corpse and asks in a plaintive, long-drawn tone of voice: "*Kaigha kirisemaigha kimunaue*", "Did the *sema* from the Kai clan kill you?". When putting this question she knocks with her finger against the gunwale of the prahu. If the deceased answers in the affirmative, he makes the prahu roll (*rika gha,* to shake the prahu, to divine by means of the prahu). However, the *sandu* remains still. Then she asks after the *sema* of other clans and at last the few relatives who watch in fearful tension see that the prahu upon the mention of the Pedei clan begins to move, first very little, but then lurching with increasing violence, so that the vessel nearly ships water. The same method is used to establish to which *ruma* the *sema* belongs, and finally who is the person.[129]

Another method of divination is practised in the case of small children. The segment of the prahu in which the corpse lies, is suspended freely

---

[128] According to W. H. R. Rivers, *Reports of the Torres Straits Expedition,* I, p. 286, blue impresses the Papuas as a sombre colour; they are more sensitive to red. Blue cotton is used for the women's aprons, for the enclosure of the *kambo* and the prahu, and for covering the dead. Red cotton is the material for the pubic belts of the men and the dancing-aprons of the girls, and also of the aprons worn by bride and groom.

[129] It is, of course, easy enough to make the unstable *sandu* rock. However, according to the Waropen it is not to be assumed that the persons who hold the poles would do so on purpose, and in support of this view they adduce the far from idle argument that nobody would run the risk of being suspected as a *sema* by purposely making the prahu rock. It is possible that here we are faced with some occult phenomenon. The well-known missionary among the Numfoor, F. J. F. van Hasselt, once told me how he personally had seen a prahu nearly rock itself to pieces by being merely touched at the stem by the captain, who had wanted to find out in this way whether the prahu permitted the voyage he intended to undertake. The Numfoor maintain that their soul-statuettes also reply to questions by moving or nodding independently. See also the remarks concerning the ancestral image Kowei, in chapter VII.

in two rattan slings from the beams of the garret in the central portion of the house. Then the prahu is touched at the head with a *kowai,* a small arrow used for hunting birds, whereupon the dead person replies as described above, viz. by making his resting-place sway. However, I did not see this method being applied.[130]

In the Napan region another method was said to be known; there the right hand is oiled and a lock of hair of the corpse is grasped. The dead person is questioned whilst the hair is pulled. As soon as the reply is in the affirmative, the hair comes off the skin.

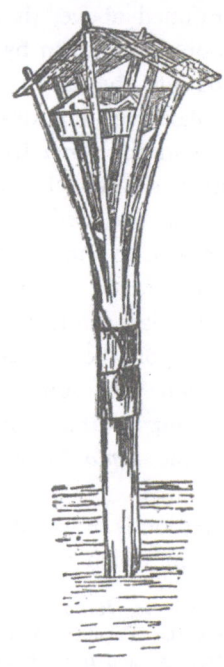

Corpse of a new-born child, coffined in a split bamboo.

When the cause of death has been divined, deposition of the corpse follows. If it is a small child, the corpse is put on its back with the legs drawn up, and placed in a *faisi,* a small case made of pandanus-leaves. Then one takes a long and strong piece of bamboo, of which the final joint is split; half a coconut-shell is pressed down into this section, so that the strips stand out like a fan. Into this holder which looks like

---

[130] Often divination is omitted because of the belief that it might be harmful to relatives absent from the village. For the same reason people sometimes refuse to tell myths or to describe ceremonies.

MUNABA: THE RITUAL OF DEATH

an inverted fish-trap the *faisi* is placed and the structure is provided with a roof. This stand (*saura*) is erected at the side of the house (*kikering-giwi*, they erect together with him) after decorating the bamboo pole with a shell and some beads, and there it remains until the corpse has decayed, whereupon the pole is pulled out of the ground. According to some, a small *saira* is still needed, presumably to free the mother from her isolation. Then a *munaba* has to follow. In many cases, however, the corpses of very young children are simply deposed in the tidal forest behind the house, without any further ceremonial.

If the deceased person is an old man or woman, many people already gather around him when he is passing away. In spite of his helpless protests the sufferer is drawn up into a semi-sitting position; he is pressed to state which *sema* it is that is killing him and with great lamentation one tries to keep him from dying. After the decease the dead person is covered with the blue shroud and a number of shell-bracelets are broken over the corpse. His personal tools are destroyed: bow, carrying-bag and sero-fishtrap for a man, sago-beater and hand-fishtrap for a woman. Only few of the relatives are permitted to assist at the divination by means of the prahu, because the search for the guilty *sema* is an extremely serious business.

Then the sons of the deceased's sisters come together to carry the corpse out, because his own children are not allowed to do so, just as the dead child is carried out by its mother's brother and not by its father.[131] Many mourners gather around the prahu with the dead *manobawa*. The women arrive on the back gallery weeping loudly and beseech the deceased with imploring gestures to return to the house. Others stand straight in the following prahus, despairingly waving in the direction of the *ferasoa*, the burial-places, and the deserted house with plates and mats and *ghafirasaruma*, the carved decorations of the prahus. Helpless in their unlimited grief the women fling themselves prostrate in the prahus, whilst the men look unmoved with tight-set faces.

Arrived at the *ferasoa* the sisters' sons hurry into the tidal forest and soon one hears the sound of bush-knives, as they quickly prepare a scaffolding high enough in the trees to prevent the crocodiles from dragging the corpse away. As soon as the scaffolding is ready, many men take hold of the plank on which the corpse lies, and stumbling and shouting they clear their way with their burden across the slippery

---

[131] The same happens on Dobu, see R. F. Fortune, *Sorcerers of Dobu*, p. 170.

mangrove roots. The sisters' sons have to hurry in doing so, because
they must avoid to seem averse in doing their duty. The other relatives
of the deceased, however, his sons and daughters and wives, etc. resist
the bearers, beating them on the back with their fists and throwing
imaginary spears. Then they fling themselves again prostrate into their
prahus, as if overpowered by grief.

The corpse is put on the scaffolding with the head pointing "down-
ward", i.e. downstream, and the face looking upward. At Napan I was
assured that the dead person "lies looking at the sky and when the sun
goes down, he follows behind" (*iyana iyambora dora. Oro iyamosa wea
iyamokada ora*). Around the scaffolding there are hung sero, mats for
fishing, and other mats, and in the trees around the spot sometimes
shells, plates, pots, etc. The relatives then break a few plates and
suspend broken bows or smashed baskets in the tree, depending on the
sex of the deceased. Then everybody returns to the village.

The *ferasoa*, the places for the deposal of the dead, are usually
situated downstream of the village. At Nubuai, however, some inhabitants of the "upper part" were accustomed to bury their dead upstream
of the village, but according to some informants, this is only permissible
for very young infants.At Nubuai these had to be deposed on Cape
Andegharegha, where, it was said, also the skulls of the conquered
enemies were exhibited as long as the flesh of the skull was not yet fully
decayed. At Nubuai the main place of deposal is the cape at the river-
mouth, Adausanarewo; this is likewise said to be the dwelling place of
the *manduko*, the sea-eagle, who changes into a mythical sea-creature
at every sun-down. Another *ferasoa*, where the clans Sawaki and
Apeinawo usually bring their dead, is called *Begha ienana rewo*, the
cape where the crow sleeps. The Kai, Pedei and Nuwuri have a *ferasoa*
called Saghaidei, closely behind the houses furthest downstream at Kai,
near the creeks Madai and Ghaibo.

Special mention should be made of the curious difference in behaviour
between those who carry the corpse and the others. Whilst the latter are
conspicuous by their great show of grief, the carriers seem to be in a
mood of slightly nervous gaiety. This is perhaps an expression of the
ceremonial relationship between mother's brother and sister's child, or
of a certain fear for the *ferasoa*, which is usually more or less avoided.[132]

----

[132] The Christians are burried in a coffin, according to Ambonese custom covered
with black cotton and decorated with threads of white cotton (*bungga*, Malay:
bunga) and brass tacks. In the Kai district with its marsh-belt of several miles
in width, the burial-place is rather far upstream.

In Waropen Kai a custom was practised, unknown at Napan, viz. that of drying out the corpse over a small fire.[133] In the Napan region after about four days the corpse was deposed in a prahu on a scaffolding in the forest. In Waropen Kai, however, the corpse was placed on a wooden scaffolding in the central part of the house, the scaffolding being hung to the rafters of the garret by means of branches with a rectangular bend (we sara no ruma, to suspend in the house). The corpse rests on its back, but a piece of wood has been pushed underneath the knees, so that these are drawn up; the head rests on a piece of wood.[134] A few old women (binano) have to keep a low fire going, roasting the corpse (kiwe sasoi). When after some time the skin starts to come off, it is removed by the binano (kikewoi, they pluck him). After three days (five days, according to others) the corpse has been completely skinned. The pieces of skin are collected in a plate and thrown away. The corpse now has a whitish colour, which is evidently considered very beautiful and important, for every time this subject came up for discussion, this point was treated most extensively. After about one month's time the corpse is said to have been completely dried out. What happened to the corpse then is not quite clear. Perhaps it was wrapped in mats and kept in the garret until the next munaba. Then probably a parcel was made of the skull, at least of important individuals, whilst the rest of the remains were placed in the places of deposal.[135]

According to the mythology this custom was introduced by Kirisi Aimeri who killed the snake Ghoiroponggai with hot stones and hot water, in this way saving the Waropen from destruction. Kirisi Aimeri went from the Woisimi to the East with the first dead, Aruifono. It was only at Waren that the people wanted to accept the dead man.

---

[133] This custom was likewise known in the southern part of Japen, see *Reports of the Torres Straits Expedition*, I, p. 334. It was abolished under the influence of the Administration.

[134] See the photograph facing p. 237 of *Verslag van de militaire exploratie van Nederlandsch-Nieuw-Guinea* (Report of the military exploration of Netherlands New Guinea), *1907—1915*.

[135] In general, it is difficult to obtain a clear picture of a certain practice from the descriptions given by the informants alone. They are of course quite well informed, but they only describe what is most striking to them personally. The things an ethnographer wants to know they usually consider far too "ordinary" and self-evident to be worth mentioning. And often they pay so little attention to actions which are usual to them, that, although they are able to demonstrate these if asked, they have the greatest difficulty in telling about them in an oral report.

That is the reason why only the inhabitants of the Waropen Kai district and those of Japen since that time have the custom of mummifying their dead (Texts 33, 34, 35, 37, 38, 39).

THE LAW OF SUCCESSION

The division of the property left by a deceased person produces only little difficulty; the personal tools — bows, arrows, neck-support, etc. for a man, the tools for working the sago in case of a woman — are destroyed or burnt by an outsider who receives a small reward. Also the man's carrying-bag (*rowu*) is considered as strictly personal property; the Waropen can safely leave their bag, in which they keep a small comb, a mirror, a knife, tobacco and areca-nut, unattended, without anybody taking advantage of it. When the man leaves his carrying-bag at home, his wife is personally responsible. In one of the stories it is told how a man suspected his wife of adultery solely because she had his bag carried to him by his younger brother (Texts 102). Usually the Waropen has no further personal property, because the old beads and the ancient porcelain which his wife keeps in the *faisi*, the chests of pandanus-leaf, have been acquired with the co-operation of the whole family-line. These constitute the capital of the line, constantly needed by the members for investment in marriages and *saira*-ceremonies, which also constantly supplement it. Only on the occasion of a death valuable pieces of this property are sometimes destroyed or damaged.

In the Napan region the most important sago-property belongs to the *ruma*. E.g. men and women of the house Warami may go and beat sago in an area called Sakererei, after a creek in the vicinity. When a woman from Warami marries, she retains the right to the sago-area of her own *ruma*. Her husband is also entitled to use it, with the permission of the elders of his wife's house, and later of her children. When the children become older, or when the mother comes to die, this right lapses, also for the husband. As far as people knew, alienation of sago-property had never taken place, but according to the general opinion this would only be possible with the consent of all the important members of the *ruma*. The children are mainly entitled to that part of the sago-area which has been worked by their father before them, or where he has planted shoots. Difficulties often arise because shoots of sago-palms come up in the neighbouring area of another *ruma*. Fruit-trees, coconut-palms etc., which have been planted and cared for by a man, are more especially his property and they are not transferred to the whole *ruma,* but only to his own children. So long as a widow does not remarry, she may be permitted to continue the joint use of the sago-property of her deceased husband. The ancient pottery etc. she acquired during her marriage she is not allowed to take along when remarrying; it continues to be the property of the husband's *ruma.*

In the Kai district, however, a common sago-area of the *ruma* does not exist, but it may still be assumed that here also the system of inheritance followed the unilateral organisation of relationship. For the sago-property has to be worked by the brothers and sisters together. Hence, when the sisters obtain through marriage the right of use of their husband's sago-property, it is not in the first place the husband who helps them to work it, but their brothers who often have to undertake rather long journeys to do so, if their sister has married outside their own village.

After the father's death it is mainly the oldest child, boy as well as girl, who takes his place. Because working the sago can, after all, only be done in co-operation with the other brothers and sisters, the younger brothers and sisters are therefore not excluded from the sago-property. All the same, conflicts are still of frequent occurrence, because the oldest child later contests the right of the married younger brothers and sisters to work the inherited sago-property. So long as the children are minors, the property is worked by the mother and her brothers. When the children marry, the mother loses the right to work the sago of her deceased husband. According to some the mother has to be paid for her management with a sago-palm. For so long as she does not remarry the widow also here in the Kai district retains the right to go and beat sago in the area of her deceased husband. At Nubuai a serious quarrel arose because a man denied the married daughter of his brother the right to beat sago in the area of his deceased brother on behalf of the old widow of this brother. The man maintained that a widow is only entitled to her husband's sago as long as her children are unmarried. Public opinion, however, took the old woman's part.

In the village there is no impartial organ to settle conflicts of this kind and therefore they are usually decided by means of armed force. However, public opinion has a considerable restraining influence in these matters, whilst in the villages barter-relations are so intricately intermingled that it is advisable not to go to extremes.

In spite of the preponderantly patrilineal grouping of relationship in the Waropen Kai region, my informants maintained that a child still had to inherit the sago-property of its mother also, and not only that of its father. However, the mother's brothers do not always wish to recognise the rights of their sister's children, and this sometimes leads to violent conflicts.

It is noteworthy that the right of inheritance which is determined patrilineally at Napan, is not exclusively computed through the father but also via the mother in the Kai district, although in general the social organisation in this district would seem to be more old-fashioned than that obtaining in the Napan region. Probably we have to think here of the influence which a fixed connubial relationship between groups of relatives must have exerted on the right of inheritance. When a man always takes his wife from his mother's group, he will of course always obtain sago from his mother's sago-area, also after his marriage, because then he is again allowed to continue to beat sago in his mother's area. The objection old people raise against marriage with a person other than one's mother's brother's daughter is therefore that this leads to a splitting up of the sago-property. The method which is used nowadays to compute the relationship with the mother's brother's daughter is so wide that this form of marriage in its present state will hardly be able to prevent the dispersion of the sago-property. This rule of the ancients will therefore partly be a reminiscence of earlier and more stable connubial relations between exogamous groups.

Another question which arises is whether the computation of the right of inheritance along both lines should perhaps be connected with the imperfect unilateral tracing of relationship of the *ruma*.

MOURNING CUSTOMS

In the Kai district the nearest relatives of the deceased are in heavy mourning during the first eight days. They are not allowed to work or to leave the house; this applies in particular to the widower or the widow who have to remain during these eight days in a screened-off corner of the room (*ioaina sea, ioaina seadogha*). In the period of heavy mourning the mourners are only allowed to eat small sago-cakes wrapped in leaves and small fishes.

In former days when the corpse was dried out over the fire, four or five relatives of the wife's side or sister's sons had to keep watch over the corpse. It was said that these watchers were also called *binano*, like the old women who were to be found willing to dry the corpse against payment. Probably this work of mummification was also the task of the relatives by marriage. At present, when the corpse has to be taken out inside twenty-four hours, the *binano* occasionally still keep watch by a fragment of the prahu in which the corpse has been taken away, or, if it was a woman, at the *enggana*, the mat used by the woman when beating sago to prevent the marrow from squirting. Probably the *binano* formerly had to maintain their watch till the first *munaba*, i.e. during several months after the decease, for it is for this period that the *binano* even nowadays have to receive presents of sago (*kikanduki binano*), like the other relatives who carried the corpse into the tidal forest. This obligation also counteracts the negligence of relatives who would take their time in holding a *munaba*. For the widow or the widower the heavy mourning is raised after the period of isolation by leading her (or him) out of the back of the house and rowing him from behind the houses to the front of the house. Some days after the deposal all relatives and all participants receive a small parcel with some sago-cakes and a fish, for which they have to pay some slight compensation. This distribution of food is evidently destined for the deceased person; it is called *kiwe ma nunggu fero*, to give in the direction of the dead.

In the Napan area the isolation formerly did not last only a few days, but a whole month, not only for the marriage-partner, but also for the closest relatives (parents and children, brothers and sisters). During this period the mourners had to assume the position of a corpse and to sit with the knees drawn up; young people who might be unable to maintain this position sometimes had their legs tied. Only in the evening they were allowed outside to relieve nature; this was called *kiwe safuo* or *kiwe haido*. As is shown by the word *hai* (cf. p. 238) they have come into contact with the dangerous side of the sacred world. Whilst they dwelt in isolation, the other relatives (especially the sister's sons) went out to catch fish which was then distributed in the village, together with sago. At Napan this was called *kikanduki* or *kikanggusara haido*. Here too the distribution of food was intended for the dead person who is invoked to come and enjoy the proffered food.

At the beginning of mourning a friend or relative puts the signs of mourning on the mourners (*wuda nunggu fero*, to assume the signs of mourning for the sake of the dead). Instead of the metal bracelets on the upper and lower arm, the men wear finely plaited bands (*aisa*), which they also put around their ankles; on the upper arm they moreover put a round band of beads. Around the throat they wear a similar band, and on their shoulders the crossed bands of beads (*sare*), decorated with pieces of red cotton.

The woman put around their head a beaded head-band with a fringe likewise made of beads (*fefa*). At present this fringe only covers the temples and the forehead, but possibly it also hung down across the face in former days (see also p. 191). The upper part of the body they cover with a short jacket of beaten bark (*furefa*) tied around with a string adorned with beads; in modern times it is often replaced by a piece of cloth. To the *furefa* the same figures are applied as have also been tattooed on the chest of the woman. Mythology relates that one of the Aighei, the primeval women, made the *furefa* from the bark of the central pole which supported the house of the snake Simundopendi; she laid the bark-dress in the sun and in this way the designs, which are still reproduced on it, came up by themselves (Texts 118, 119, 120).

When the mourners are allowed to break their isolation, they have to cover the head with a mat; the men occasionally use a pointed cap of red cotton. In modern times the men do not always keep to this rule.

The signs of mourning have to be worn until they are completely worn out; then one rows out into the sea at night and the worn clothing is cast off. Some shave the hair which has not been combed during all this time. This is called with a pregnant expression *kito kinirira,* they throw away the dirt of their body; in the Napan region it is called *erinohara mahumbea,* to wash away the heaviness, the same expression as the Numfoor *san merbak,* to throw away the heaviness. Usually a brother of the wife or a sister's son comes to row the mourner out, for which he is rewarded with a small present. In ancient times prominent men could only take off their cap of mourning when they had taken part again in a raid.

For so long as the first *munaba* had not yet been danced, i.e. during a period of about six months, all inhabitants of the house, children as well as adults, were obliged to play at cat's cradle with strings. This was not accompanied by sayings or by singing, except for one figure when one caught with the string one of those present as a "slave" with the words *miroro sapairoro.* Although some people maintained that playing this string-game without just cause might lead to death, it is still played very often simply to pass the time. It is curious that the string-figures do not refer to the production of sago.[136]

The Waropen idiom knows several words to indicate which relative

---

[136] See the figures on p. 362 ff.

one has lost through death. In several cases in stead of the real name
of the mourner a *wudakanasano* is used, i.e. a mourning name, in which
often the name of the departed is also mentioned.

A woman who has lost her father or her mother is a *tobai* or a *woifi*;
she is called Tobai Ghairamoti or Tobai Wororumbai.

A man who has lost his father or his mother is a *tuwofi*; he is called
Sireirewani or Numadamai.

A woman who has lost a brother or a sister is a *raroi* or a *rarowai*;
she is called Sumbaraghani, Fasinanusai or Unaiwiama.

A man who has lost a brother or a sister is an *etawai*; [137] he is called
Etawai, followed by the name of the deceased person, e.g. Etawai
Ghafai.

An (older) man whose first child has died is a *manapa*; he is called
Manapa, followed by the name of the deceased person, e.g. Manapa
Kiriwui. A woman to whom this had happened is a *binapa*. A person
one of whose older children has died is an *atuwai*.

A widower is a *maniwa* or a *maisani*.

A childless young widow is a *sasui* or a *sasuiwuro*. An (older) widow
with children is a *bindo*.

MUNABA

The ritual for the dead is not completed when the corpse has been
carried away, even according to the practices of the present. The mummified body probably remained in the house of mourning till the first
munaba was celebrated, perhaps half a year after the decease. The
skull stayed in the house also after the *munaba,* or it was fetched from
the place of deposal when the corpse began to fall to pieces. And even
when the custom of mummification had fallen into disuse, the skull
was in many cases returned to the house after some time. On one
occasion which I was able to check this happened about three months
after death.

The Waropen were not very communicative on the subject of funerary
customs, because the Administration has taken action against these.
They have been unable to reconcile themselves with the new customs
which in their eyes derogate from the love one owes to the deceased.
Formerly people may have been too quick in drawing conclusions from
the presence of skulls in the houses, which were immediately considered
as direct proof of head-hunting, without considering the possibility that

[137] At Napan an *etawai* is a person whose father has died.

these might simply be the skulls of relatives. The report on the military exploration of Netherlands New Guinea rightly warns against these over-hasty conclusions.[138]

In several cases these skulls were removed on the order of the authorities, although among these there must have been quite a number of the people's own relatives. The first time I was taken along to fetch a skull, the people did not omit to inform me that it concerned the skull of a very small child, so that nobody would have any objections. For the rest, several *serabawa* from the Kai district and from the Napan region assured me that in principle it was only the skulls of the *serabawa* which were entitled to this special care.[139] In the Napan area the custom prevailed which is also known among the other tribes in the western part of Geelvink Bay, viz. of preserving the skulls in the house inside a soul-statuette. In the Kai district such statuettes were hardly ever used. Only once a father made a soul-statuette for his uninitiated son; its nose was perforated in lieu of the child's nose. Evidently no special value was attached to the image itself, because shortly after the decease the father offered it for sale.

Fetching the skull from the place where the body has been deposed is not accompanied by any ceremonial; according to some, it has to be done at sunrise, but in actual fact the people are dependent on the tides. The skull of the young child, mentioned above, was taken by the sisters of the mother, without much ado from the place behind the house where the child had been deposed. The bones were placed in a case made of leaves, whilst the skull was taken along in a new carrying-bag, the chest with the bones being left on a small piece of scaffolding at the original place, after four leafed branches had been stuck in a row into the ground, a few smashed baskets being attached to these (cf. p. 176). If the dead child had been a boy, a bow and arrows would have been hung there. The wife's sisters thereupon sewed the skull in a piece of red cloth; for the services rendered the father had to pay some sago, for which he received again some beads and shells in exchange. The skull really ought to be suspended from the roof-tree

---

[138] *Verslag van de militaire exploratie van Nederlandsch-Nieuw-Guinea, 1907—1915,* p. 252. Also in other respects conclusions, which gave a wrong impression of the Papuas, have sometimes been drawn from certain "proofs". On the same page the author of the report mentions the existence of the slave-blocks as "proof" of "all kinds of cruelties" which the slaves on the Waropen coast were said to suffer.

[139] In Polynesia the head of the most prominent persons was considered sacred, see W. H. R. Rivers, *The history of Melanesian Society,* II, p. 261.

in the garret of the central portion of the house, but in the case under
discussion it was hung near the fire-place for convenience's sake.

Once the first *munaba* has been danced, the skull may be brought
back again. It is remarkable that nobody seems to show the slightest
diffidence or respect when the skull is put away; hardly any interest is
paid to the object. The first skull I saw being brought back after the
munaba was that of a young man. A (classificatory) father climbed
up into the garret, wrapped the skull in a new piece of cloth and then
packed it in a small linnen bag, decorated with a few beads. Then he
hung the skull again for a moment on a piece of string in the midst of
the male and female dancers who had danced the *munaba* during the
night. Some beads, a shell band, a *korombowi* or white porcelain-shell,
and a new carrying-bag were fastened to the parcel.

The line of dancers, mostly sisters of the deceased person, wound
around the skull a few more times; a few women from the house
quickly put a dress on some small girls to have them take part for the
first time in dancing the *munaba,* but one of the girls became shy and
stood against the wall, pouting whilst everybody laughed. The contents
of the decorated bag hanging motionless on a string from the garret
seemed to be completely unnoticed.

In the mean time the fathers of the dead man had sat down on the
front gallery to make a small roof of atap. Then the skull was placed
in the cover of a leaf-case. A number of prahus had collected in front
of the house of mourning and other boats joined these when the fathers
rowed away towards a creek somewhat upstream from the place of
deposal. Quickly a small scaffolding was fastened to a mangrove tree
and the skull was suspended freely under the roof which had been
prepared earlier. The people say that now the dead man is in his *ruma*,
his house, as the skull-shelter is called.

Meanwhile the fathers of the dead man had cut a number of *resa*,
mangrove branches with leaves, in the forest. Jumping into the water
they stuck these triumphal branches with violent gestures into the mud
in rows perpendicular to the shore, loudly shouting the names of the
slaves they had caught. "*Ineni maisa Nubai*", "this one stands for
Nubai", "*ineni maisa Sibai*", "this one stands for Sibai". Similar shouts
resounded on all sides.[140] The younger people looked on with sneers,
but also with embarrassment, for they had never participated in a raid

---

[140] In the Napan area it was believed to be highly auspicious, if a slave could
be caught whilst the corpse was being carried to the place of deposal. See
also W. K. H. Feuilletau de Bruyn, *Schouten- en Padaido-eilanden,* p. 116.

and therefore they were unable to join the old men in mentioning the names of the slaves they had caught.

In the branches of the surrounding trees some baskets damaged by cutting and some pottery were left for the dead man and then the people returned to the village, almost with a certain gaiety, like we observed earlier, when the corpse was taken to the place of deposal. There was only one man whose quiet attitude was already conspicuous in the house of mourning; here also he just sat gazing, wrapped in his own thoughts. Everybody left him alone and nobody offered him tobacco or areca-nut. From time to time he wiped his eyes. This man was the father of the dead youth. His grief was genuine.

When the skull is taken away, or in other cases about six months after the decease, when sufficient sago and fish has been collected, "the *munaba* is danced" (*kikowa munaba*). Although the *munaba*, different from the *saira*, is not "sung", but "danced", the word *muna* by itself means "song", the sacred chant which the *munaba* dancers sing.[141] *Munaba* therefore is "the great song". During the *munaba* an old woman usually acts as the leading singer; this is not a special function, but often it is the *ghasaiwin* who have the greatest knowledge of the ancient chants. At the *munaba* this leader is usually to be found squatting behind the singers near a fire, where she "weeps the death-song", (*ianiko muna*).

Although the word "dance" suggests the idea that the *muna* would be performed, there is as little question of a dramatisation of the mythical material during the feast of the dead as there is on the occasion of a *saira*. At a *munaba* the young men and women simply come together and form pairs, the group of the men remaining separate from that of the women. In principle a married man and an older woman are not allowed to join in the dance, but the young married men in particular often transgress this rule, risking a matrimonial quarrel.

In front usually a few young men dance with the *siwa*, the drum; at a normal *munaba* neither gong nor triton-shell are used. The girls likewise often play the drum, but then it is mostly the short, vase-shaped model, the *siwabuino*, the half-drum. The drums are treated with the greatest care; anciently they were preserved in the young men's house. The drum-skin, on which a few discs of resin have been stuck in order to increase its tension, is constantly adjusted over a low fire. The connois-

---

[141] It is also possible that we should have to think of a word *muna*, to strike, to kill. In Numfoor there exists a word *mun*, meaning "song", "verse", "part".

seurs are very exacting as regards the sound; it should resound well
and harmonise with the sound of the other drums. Two different sizes
with a different pitch are customary. The sound enables a connoisseur
to recognise a drum already at a great distance. Most drums have a
name of their own and are called e.g. Korare, Komambai, Ingganuwini,
Manieghasi, etc. According to the specialists, beating the drum also
requires a certain skill which everybody is far from possessing to the
same extent.

The dance itself is simply a tripping progress, just as for the *saira*;
the upper part of the body is maintained stiffly erect, whilst the dancing
movements are accentuated by the hips, especially by the girls. In one
hand, which hangs straight down, coquettish girls like to carry a piece
of cloth, a mirror, a small tin, etc. The fingers of the other hand are
interlaced with those of the other partner, the backs of the hands
touching each other. Only the drummers are somewhat more free in
their movements; they occasionally make a few steps which remind
one of a mock fight. The other dancers, male and female, trip behind
them, singing endlessly, only from time to time stamping on the ground
as hard as they can, or jumping up and coming down on both feet with
a stamp. When doing so, the partners look behind them on the ground,
as if something were to be seen there, and they make a hissing noise
(*kiwe sisisi*). According to my informants this means that the participants are setting into their stride and are beginning to dance vigorously.
Continuously they encourage each other with the words *korako*, *korako*,
hard, hard, and usually it does not take long before the oiled bodies
are wet with perspiration. Both sexes are very proud of the fact that
the Waropen are able to keep dancing for so long.

According to my informants the stamping and hissing is only meant
as an encouragement; my suggestion that it might refer to the hissing
and wriggling of a snake was received with great hilarity. It does not
seem likely that we should have to think of the word *si*, vagina, although
the sound-association undoubtedly does not escape the Waropen.

A female dancer in festive dress greatly resembles a bride, without
the combs, however. Often she wears a long string of beads hanging
from the forehead over her face. To the beaded shoulder-bands balls
of cotton decorated with beads (*seghano*) have been attached which
jump up and down during the dance. Similar decorations (*mosara*)
have been attached to the beaded loin-band. On her arms she pushes
as many shell or porcelain or glass bracelets as she can. Around her
waist she ties a red dancing-apron with a gay pattern of small shells

and beads or strips of white cotton. The young man also adorns himself as well as he can with armlets of celluloid, brass or silver, jingling bells (*kerekere*) on his belt and a fine sarong around his waist, a gay comb in his hair, a handkerchief or a towel around his throat and other pieces of cloth pulled through the armlets.

It is striking that during the *munaba* the potential marriage-partners i.e. boy—mother's brother's daughter, and *vice versa*) are allowed to scratch each other (*rira*), being even encouraged to do so by the bystanders. "*Aede arira warimagha*", "Come on, scratch that boy" is said to the bashful girls. Most girls carry hidden in their shell-bracelet the sharp leg of a crab, together with the *mamano*-leaf, a strong-smelling herb to which certain properties are attributed. However, one must be careful not to scratch a brother or sister in the flickering light of the bamboo torches, for this is considered to be most humiliating. The custom is also partly erotic in origin; it belongs to the crude form of courtship and the "joking relationship". I already mentioned the case of a boy who pulled the hair of his beloved in so hardhanded a fashion that she was painfully hurt. Perhaps this public erotic scratching is a reminder of the former sexual licence during the *munaba*; this licence is no longer allowed at the present *munaba* and it is not right to conclude to sexual licence from the attitudes or the way of dressing, which strike the Westerner as peculiar. That both sexes try to please each other as much as possible in the Waropen way during a *munaba*, goes without saying.

The *munaba* ought to last without interruption for five days and five nights, but usually matters practically come to a standstill towards noon. Towards night the lovers of the dance come together again in large numbers and quite soon the row of dancers winds again ceaselessly around the central pillar which represents the *inggoro*, the ancestor, the movement of the row being counter-clockwise. The song-leader sings a couplet, usually consisting of two lines, and this is taken over by a few girls. When everybody knows the words the chorus of the others joins in. Every couplet is repeated during five circuits and things are arranged in such a way that the foremost couples have arrived again before the song-leader when another couplet has to be started. Especially the female relatives of the dead person have to show great zeal in dancing and in the later part of the night the men often leave the dancing completely to the women.

Usually the *munaba* is not danced once, but several times, with irregular intervals. At Napan this feast is called *ratisara*, but there, differently from the Kai district, it was only celebrated once after the decease. At Kai the *munaba* is the most popular and also the most frequent feast. When the drum resounds in the village it is mostly because of a *munaba*.

The *munaba* is not the dramatic representation of a myth, but only the chanting of mythical material in a language which is mostly incomprehensible, even to the participants. Whilst at a *saira* it is rather separate and unconnected songs that are sung, the *muna* forms an often endless song on a certain mythical theme. Nearly the whole of the mythology has been elaborated in *muna*-form: the theme of the super-natural fertilization, the origin of the sago, the appearance of Kirisi Aimeri or of Manieghasi, or of the fighting brothers, etc. The *muna* consists of endless repetitions with constantly recurring final words, which nobody understands, or at least which nobody can translate. Still, the Waropen is able to conclude from the words that are used what is the subject of the song.

Therefore, although the actual dramatisation of the myth is lacking (with one exception, to be discussed below), the *munaba* with its youthful dancers in elegant array and with its sacred song cast in a solemn and obscure language is probably meant as a re-enactment of those mythical primeval times when the ancestors, whose actions are sung, lived a longer and happier life than their present descendants. The *munaba* is intended as a periodical repetition of the contact with this happier past, when mankind suffered less from death.

SERAKOKOI

Differences in status become apparent at death mainly by the larger scope and the greater frequency of the *munaba* of the high-born. Apart from the treatment of the skull, on which the informants did not agree (p. 184), there exists no difference in the ritual for the commoner and that for the more prominent. It is, however, remarkable that on the occasion of the death of a *sera* of the Kai clan, who had also claimed a certain superiority due to other privileges, the eating of a number of animals was prohibited until the first *munaba*; these animals were enumerated in a song which was sung during the wake, but not danced, although it has the appearance of a *muna*, a song for the dead. This song follows here, it may likewise serve as an example of this type of literature.

Ambudi ghanano Gheri rasawe
Ambudi nawonano Biri sawe
Ambudi nawana nara Rumasi rasawe
Ambudi nawana nono Kokofi rasawe
Ambudi nawano Sariki rasawe
Ambudi nawano Aramori rasawe
Ambudi nawano ano Sareri rasawe
Ambudi nawano Taiwandini raghewe
Ambudi nawano Wawi raghewe
Ambudi anano Sowi raghewe
Ambudi anano Worodaun daghewe
        Fowore fowore sairone

Ambuidi terere sisawore	*Ambuidi terere mafowore*
Ambuidi terere kuiwore	*Ambuidi terere ghamawore*
Ambuidi terere manggaumbore	*Ambuidi terere ghanawore*
Ambuidi terere opeirumbore	*Ambuidi terere tirewore*
Ambuidi terere samaiwore	*Ambuidi terere susaiwore*
Ambuidi terere nutukawore	*Ambuidi terere sineiwore*
Ambuidi terere kokofiwore	*Ambuidi terere mandukawore*
Ambuidi terere miwore	*Ambuidi terere tusiwore*
Ambuidi terere aghawore	*Ambuidi terere mandarawore*
Ambuidi terere ghegheriwore	*Ambuidi terere toimbowore*
Ambuidi terere bafuwore	*Ambuidi terere sawawore*
Ambuidi terere konandewore	*Ambuidi terere pesaruanawore*
Ambuidi terere kaporawore	*Ambuidi terere aifamandewore*
Ambuidi terere koiwore	*Ambuidi terere foghawore*
Ambuidi terere ghaiawore	*Ambuidi terere aumaniwore*

This song is typical for the *muna* with its endless repetition of the first or the last words of a line with a change in syllables based on assonance: *rasawe — reghawe; ghanano — nawonano — nawonanono — nawano — anano*. The Waropen quite likely recognise some word or another in this play on syllables, but they are hardly able to translate this type of chant, so that we have to depend on conjectures.

The death of a *serabawa* is accompanied by a special ritual; it is rarely performed because it is only proper to have it when really influential men die, old men, the senior representatives of the senior line of the *seraruma* of their clan. According to my informants the procedure was as follows:

43. Dressed for the *munaba*

44. The file of female dancers

45. Women from the Moor Islands

46. The drummer

47. Dancing attire on the Moor Islands

Immediately after the death of old Ekamai — for whom this ritual was performed for the last time at Nubuai — one of his female relatives took a rhomboid piece of bead-work (*aimura garo*), which she tied over her eyes. This particular piece is made of valuable old beads (*woaro*); originally it belonged to Pedei and at present it is among the possessions of Munoi, an old *sera* of the Sawaki clan. Now two women go in front of her drumming, whilst others follow her, guiding her by her hip-band, because nobody must stumble or sneeze, which would be highly inauspicious. The blindfolded woman, standing at the stairs at the front of the house, scoops water into a bamboo, after another woman had first drawn a cross with a bush-knife at the spot where the other was to draw water. The bamboo filled with water is then wrapped in rattan and suspended near the head of the corpse. When drawing the water one sings:

> *Anggedei Inuri masina sinai*
> *We ghonggena ne ghonggena ne*

The first line means: "We fetch the water of Inuri"; the second is obscure. One also sings occasionally:

> *Inuri inuri masina sina*
> *Piamo piamo moiwa renai*

The word *piamo* may refer perhaps to the name Piami Uri which is used on Japen to indicate Uri, the divine trickster.

After this, everything proceeds like at an ordinary death. The relatives collect large stocks of food for a great *munaba*, but this is preceded by a ceremony called *we risura*, to treat his comb. In this case the comb is a piece of the rib of a sago-leaf, into which a number of small white feathers have been stuck, equal in number to the slaves Ekamai had caught. This comb is therefore the *manggoti*, the comb which warriors are entitled to wear. With this object in her hand first a *mosaba*, a noble woman, from the dead person's clan dances, in this case from Apeinawo; then successively a woman from Kai, from Pedei, from Nuwuri and finally from Sawaki. The ceremony is therefore a tribute to Ekamai's brave deeds from all women of the whole village. The Serakokoi ritual therefore is quite obviously a total ceremony in which all clans of the community are concerned. This is expressed again by the ritual mock-fight described below. This *munaba* lasts for six days; then a six day's period of rest follows.

Then the dancing of a *munaba* is started again. After the dance of the first night, the next morning at sunrise an old man enters the house, having a carrying-bag on his shoulder. This old man takes away a piece of sago which had been put out for him beforehand; thereupon all present begin to shout: "*O, Manieghasi wumana fio*", "O, Manieghasi steals sago", Manieghasi being the mythical bird in the river downstream, who puts off his submarine form at sunrise to become a bird.

Imitation weapons, made of the midribs of sago-palm leaves, and used at the mock-fight during the *Inurisaira*.

In the afternoon, towards the time that the sun begins to go down, the so-called *Inurisaira* follows, when the same song is sung as during the fetching of the water. The dancing is not as usual, but the dancers, male and female, remain standing where they are, grasping the heel-tendon of their foreman between the toes of their left foot and stamping on the ground with the right foot. Then follows a mock-fight in which both men and women participate. The weapons used are made of the ribs of sago-leaves and it is remarkable to observe that beside imitation bush-knives, spears, etc., they also include an imitation *bono*, a stone axe, and the *ragharo*, the shield, both of which have long since fallen into disuse. After the singing of the Serakokoi song the *ragharo* must not remain in the house, but it has to be taken to the forest.

The *Inurisaira* is followed by an interval, until the sun has gone down completely. One waits until the children have gone to sleep and then the song of Serakokoi is sung;[142] on this occasion the people present do not dance, but they have to sit down quietly in the central portion of the house to join in the singing. This singing of the Serakokoi chant is considered to be the most sacred ceremony known. As long as it is being sung, a severe taboo lies on the house and its vicinity. Once the leading woman has begun with the words *Tinawuaramuo*, be silent, no more noise must be made, nobody may leave his place, and even rowing in the neighbourhood of the house is not allowed.

The poetic version of the Serakokoi myth was considered as especially sacred, so that the informants were rather reluctant to tell about it, or about the ritual described above. The usual argument in these cases is, that it is dangerous for relatives who are staying outside the village, when without just cause sacred chants are recited or fragments of ritual are performed. Here follows another example of the *muna*, the song for the dead, as reported by a Christian woman; it probably contains numerous gaps. This woman told me:

"The wife of a snake is called Mimiandeisei. It went to...... (here follow a great number of places which the snake visited on its travels, as is told of many mythical figures). When a *serabawa* has died they sing the Serakokoi as follows:

O Kokoi Serakokoi, aghararomae.
Aghasi mitore mitore.
Kokoi Serakokoi, awo Sariandeisei.
Koko Serakokoi.
Kokoi Serakokoi, aghara aroma we Ghaisiri,
Koko Serakokoi.

Koko, awo Maisori,	*Awoi Mimiandeisei.*
Koko, awo Mumbai,	*Awoi Mimiandeisei.*
Koko, awo Woifamai,	*Awoi Mimiandeisei.*
Koko, awo Kobati,	*Awoi Mimiandeisei.*

Koko Serakokoi, awo Rauni na Andeio,
Seghado, awoi Mimiandeisei.
Awimbinuimanie, Kokoi Serakokoi,
Koko, awo Raungharimisi, Maighado, awoi Awimbinghae.

---

[142] In the Napan region the song of Serakokoi is not sung; it was stated that here it was especially Kuru Paisei and Kirisi Aimeri who were known.

Koko, awo Papairumi,	*Awo Nuriwaranderie.*
Koko, awo Ghaiduri,	*Awo Saranaighaie, Koko Serakokoi.*
Koko, awo Papairumi,	*Awo Supiaighei.*
Koko, awo Bebai,	*Awo Fafanaighei.*
Koko, awo Rofoni,	*Awo Mimiandeisei.*

This passage relates how Serakokoi visits several places, accompanied by his wife Mimiandeisei, who is indicated as an Aighei by numerous other names. Awimbinui quite likely refers to the name of the phantom-woman (*auo*) who took the place of Mimiandeisei.

Maighado, awo Awimbinghae, Kokoi Serakokoi.
Numanimosi, awo awe dae, Awo awe nuae.
Asoinde, amaina nusani, nusani Nufori,
Seramanimosi, awo awe dae, awo awe nuae.
Asoinde, amaina nusani, nusani Nuwuri,
Koko Serakokoi, awo awe dae, awo awe nuae.
Asoinde, amaina nusani, nusani Parakiki,
Aighei Saranaighei, awo awe dae, awo awe nuae.
Asoinde, amaina nusani, nusani Kobai,
Kokoi Serakokoi, awo awe dae, awo awe nuae.
Asoinde, amaina nusani, nusani Sabakani,
Seramanumosi, awo awe dae, awo awe nuae.
Asoinde, amaina nusani, nusani Rubasai,
Nuakokoi, awo awe dae, awo awe nuae.
Asoinde, amaina nusani, nusani Fafeai,
Koko Serakokoi, awo awe dae, awo awe nuae.
Asoinde, amaina nusani, nusani Arasie,
Aighei Kukuraighei, awo awe dae, awo awe nuae.
Asoinde, amaina nusani, nusani Nuburi,
Aighei Fafanaighei, awo awe dae, awo awe nuae.
Asoinde, amaina nusani, nusani Kasi,
Aighei Inuriwanderi, awo awe dae, awo awe nuae.
Asoinde, amaina nusani, nusani Kareio.

This passage relates how Serakokoi rows ashore on numerous islands, stays there, catches slaves and trades (or makes clans and villages?). He is addressed by the name of his younger brother Numanimosi. He himself is evidently also the Aighei, called by many names, and therefore essentially identical with Mimiandeisei. Conspicuous among these names are Kukuraighei and Inuriwanderi in which one finds again other names for the divine trickster.

Aigheisari Andeisei iede iani Neawarinafa ietuwao.
Aighei Mimiandeisei iede iani Kaidawarinafa ietuwao.
Aighei Mimiandeisei iede iani Sisinggininafa ietuwao.
Aighei Mimiandeisei iede iani Ruwurinafa ietuwao.
Aighei Mimiandeisei iede iani Aimaninafa ietuwao.
Aighei Mimiandeisei iede iani Ronarinafa ietuwao.
Aighei Sariandeisei iede iani Maindinafa ietuwao.
Aighei Mimiandeisei iede iani Kondeinafa ietuwao.
Aighei Mimiandeisei iede ianisa Kowereinafa ietuwao.

This episode relates how the deserted Mimiandeisei walks weeping along numerous beaches (*nafa*).

Aighei raka nosae, sofosorana weririo eririo,
Aighei raka nosae.
Aighei iepa nosa, sofosorana weririo eririo,
Aighei iepa nosa.
Aighei awoeindi nosa, sofosarana weririmbare eririmbare,
Aighei awoeindi nosa.
Aighei fara magha, sofosorana weririo eririo,
Aighei fara magha.
Aighei raka erai, sofosorana weririmba eririmba,
Aighei raka erai.
Aighei raka gharata, sofosorana weririmba eririmba,
Aighei raka gharata.
Aighei suka gharatai, sofosorana weririmba eririmba,
Aighei suka gharatai.
Aighei iawara gharatai, sofosorana weririmba eririmba,
Aighei iawara gharatai.
Aighei iera rona sofosorana weririmba eririmba,
Aighei iera rona.
Aighei ramaka erai, sofosorana weririmba eririmba,
Aighei ramaka erai.
Aighei rira posarei, sofosorana weririmba eririmba,
Aighei rira posarei.
Aighei fafaka fiei, sofosorana weririmba eririmba,
Aighei fafaka fiei.
Aighei wu kafiai, sofosorana weririmba eririmba,
Aighei wu kafiai.
Aighei iura fiei, sofosorana weririmba eririmba,
Aighei iura fiei.

Thereupon the song relates: Aighei cuts down the sago-palm, she notches the palm, she splits the palm into two, she tightens the sago-beater, she beats the sago-pulp, she cleans the rinsing-trough of marrow, she carries the rinsing-trough on her back, she digs a hole (for the water for rinsing), she kneeds the sago-pulp, she whittles down the band for the strainer, she washes the sago, she takes out the flour, she ties it in a carrying-bag around her forehead.

. This means to say that she performs the whole work of processing the sago, including cutting down and splitting the tree, which is men's work.

Rura topendi wina duere topendi.
Wingha iafa: Aunae, aghafae? Topendi topendi wimamure.
Serasukarori Winanatauri kitutumie regharie.
Sera Serainaniwari nata kitutumie kireghari.
Serasukarori ioditauri kitutumi kireghari.

Mimiandeisei has come back to her husband from her wanderings. The phantom-woman asks: "Did you say something?". The door is opened. Serasukarori and his wife suck each other's lips and make love.

One myth relates how Serasukarori revived during the mummification of his corpse (Texts 121).

Sarawo sarawo Sareri sarawo
Roremaroro, rorema roro, Fifirandomuni rorema roro.
Sarawo sarawo Bambandi sarawo
Rorema roro, rorema roro, firasigha sira orema.
Sarawo sarawo Nighari risarawo
Rorema roro, rorema roro, Fifirandomuni orema roro.
Sarawo sarawo Sareri risarawo
Sarawo sarawo Soteri risarawo
Orema roro, orema roro, firasigha sina orema.
Sarawo sarawo Bubui risarawo
Orema roro Fifiandomuni roremani.
Sarawo sarawo Oandinei risarawo
Sarawo sarawo Doroi risarawo.

The informants were unable to enlighten me on this part. Does Sareri refer to the nickname Mansarareri (Numfoor: liar) given to Uri at Nau? Fifirandomuni is the man who sees the shadow of two women in the water; he conquers them by simulating death (Texts 68).

The connection between myth and ritual is not quite clear in this case. The appearance of Uri, the divine trickster and the mediating figure between life and death, is not inexplicable in the mourning ritual. Even the fact that this funerary ritual includes an *Inurisaira* is not wholly incomprehensible, because in the end the *munaba* can have no other object but to promote life. The possible meaning of the appearance of the bird of heaven, the mock-fight, and finally of the whole of the Serakokoi song we shall try to establish in the seventh chapter.

CHAPTER FIVE

RAIDING AND TRADING

RAIDING EXPEDITIONS

In Waropen a raiding expedition is called *da,* whilst "to hold a raid" is *we da.* In the Dutch literature concerning Northern New Guinea and in the Malay language as current in these regions the customary term is "raak", from the Numfoor *rak,* which again is probably related etymologically with Waropen *da. Da* is not merely the raid itself, but also the band of raiders, "enemy", and finally "clan". One says: "*Kaigha kiwe da*", "the people of the Kai clan have undertaken a raiding expedition", and one speaks of a *Kaida,* "a raiding band from the Kai clan". The word *da* is less frequently used with the meaning "enemy"; it then does not mean opponents in general, but especially the opponents in the raid, the raiders of another group being of course nearly always one's potential enemies.[143]

The Papuas of Geelvink Bay have acquired a certain notoriety as head-hunters and slavers. From the beginning, the reports of the missionaries regularly mention attacks to which the villages in their area were exposed and from which no exhortation could make the Papuas desist. These venerable missionaries intrepidly tried to counter these activities, year in year out, a dangerous enough undertaking before the permanent establishment of the Administration. Only with the conversion to Christianity a change has come about, the punishments meted out by the Administration having also had their effect, of course.

During long years the Waropen had a special reputation in this respect. It was due to their very bellicosity that they remained outside the sphere of the Administration, so that, at least in the Kai district, most of the older men had actively participated in the raids. Till a few years ago it was due to this that missionary activities in Kai were restricted by a passport system.

---

[143] Like the word "clan", *da,* the word "raid", *da,* needs the plural pronominal particle in verbal forms, hence not *dagha ra* in the singular, but *dagha kira,* in the plural, "the raid is coming".

I do not know when the last organised raid was held, but it is quite certain that they no longer occurred since the establishment of the administrative post at Demba in 1926, and perhaps even since 1918, when the first chiefs were appointed. However, it appears from the diary for 1913 of one of the Controllers, that in this year the inhabitants of southern Japen still complained about attacks by the Waropen, who saw their village of Nubuai burnt in the same year by a punitive expedition. One of the older people was still able to give a lively description of this occurrence (Texts 184). The exploration-detachment of Captain Ten Klooster at that time had to deal with "pirates".

This predilection for raiding is all the more remarkable, because the Waropen in general do not distinguish themselves by extraordinary valour; nobody would call them doughty warriors. Old men who had the name of being redoubtable raiders stated quite naively that they had to muster courage by drinking palm-wine. One of these formerly much feared warriors said: "We drank palm-wine in order to become intrepid" (*angguna esa wa aninado kakio*). When the attacked put up some resistance, the often much stronger attackers immediately withdrew, to be heaped with scorn by the women in their own village. For sudden attacks on sleeping villages, unprotected travellers or rowing fishermen and women, courage was not the first requirement. At most it was expected of the very brave that they would defy with a certain frenzy the showers of arrows shot by the opponents. Neither were any great demands put on the ability of the leaders, because of the traditional tactics of falling upon an unsuspecting enemy with a greatly superior number of prahus, sometimes as much as sixty. The time just previous to sunrise was considered the most favourable moment for attacking a village, preferably after a rainy night. According to my informants one did not so much consider the special mythical importance of the hour of sunrise, but trusted with good reason that at that time a deep sleep would have eliminated both watchfulness and preparedness, whilst the prahus full of rainwater would not be ready for the confused flight.

In view of this lack of ingenuity in the attack the defenders would have had no great difficulty in robbing the attack at least of its surprise by means of some slight degree of watchfulness. But even this was beyond their powers of organisation. The defence was restricted to the erection of a kind of palissade of branches placed in all directions *(aisa),* leaving an opening for the prahus and barred under water with pointed bamboos, but this only gave some slight covering to the first few houses. For the rest, it is not so easy to erect an effective bar in the usually rather wide creeks with a current that is swift due to the change of the tides. Evidently the raiders never tried a flanking movement in order to attack the village upstream; they were probably afraid of having no way of escape left in case they would have to flee.

Once the attack had started, or when a raid had been reported, the women and children hurriedly ran for safety into the tidal forest behind the house, taking their valuables along if possible. The able-bodied men remained behind to defend themselves. The attackers were usually satisfied if they could catch a few persons, regardless of age or sex, and rob some objects of value, nets, etc. In some cases they succeeded in driving out all the defenders so that they could completely plunder the village, destroying or burning everything they could not carry with them. It was of course the smaller villages which mostly had to suffer these attacks; in this way Woinui village was shifted three times within living memory, due to the attacks of the men of Nubuai. But that much should be said for the Papuas, they also attacked large villages, like Nubuai itself, albeit that for such an undertaking they had to call in the raiding parties from numerous other villages, *kikonéa nuigha*, to summon the villages for a raid; this was done by means of the knotted cord (*ghono*).

Because of the difficulties for victualling, a large raiding-party could not venture far from home. The raids from Kai usually did not extend further than South Japen and the Moor Islands, whilst on the other hand the men from Weinami, supported by the men from the Haarlem and Moor Islands, did not go further East than the Kai district. Sometimes the fleet of prahus on their outward voyage would first establish a sago-depot somewhere in the uninhabited sago-forests of the extensive East coast of Geelvink Bay, enabling them to stay away somewhat longer without risk. However, usually the raiders stayed close to home. The men of Kai always attempted attacks on the Ambai Islands and on the villages on the opposite coast of Japen, or on the tribes of the interior settled on the Binataboa. A permanent danger of attacks was prevalent in this area; on the one hand because to the attacked party it was essential to avenge themselves; on the other hand because, if they failed to do so, the rumours of their proved timidity would invite further attacks. On the other hand this also obliged people to have as many *kamuki* (see p. 83) as possible in the different villages, because otherwise all barter of indispensable goods would be practically out of the question.

In the first instance the raid is not so much a question of individuals as of clans, as is already apparent from the word *da*, slave-hunt, being used to indicate the clan. Therefore every member of the clan was responsible for raiding activities undertaken by other members. If somebody from clan A, even if he was known by name, had murdered a member of clan B, the vengeance of clan B need not be directed against this very person, but any member of clan A could be struck by it. In actions against neighbouring villages, people usually knew which clan was attacking them. In case of the more distant villages it was sufficient if any member of the whole village was caught, not necessarily a member of the clan concerned. By the same token, the other clans of a village did not consider themselves responsible for the actions of one clan. The inhabitants of Nubuai were amazed and indignant when a punitive expedition, sent due to activities of members of the Kai clan, did not restrict its punishment to this clan, but directed itself against the whole village. It therefore happened as well that raids were made against clans in one's own village, but probably not against the clan of one's own "shell"; only in this case one had to be careful that there were no casualties. At Nubuai I was told of a raid by the Apeinawo clan against the small clan of Nuwuri, and once of the Sawaki against this same clan.

The *auctor intellectualis* of the raids is the clan-chief. He buys large prahus and mans these with his companions, being also responsible for their victualling. When members of the clan had suffered an attack somewhere, they could report to the chief, who then had the duty to stand up for his clan-members and to organise a raid, c.q. to attend to the organisation of a group which could ransom the prisoners. After a successful raid the clan-chief arranges a feast in his house to regale the valiant warriors and to reward them for their services. In return, the slaves that have been caught have to be handed to the *seraruma,* the chief's house, so that the clan-chief also receives the ransom, of which he again hands part to the important *waribo.* Next to the clan-chief, prisoners may also be assigned to important members of the clan, who then receive the ransom, often, however, after having first bought the prisoner from the clan-chief by means of a present. Members of the *seraruma* often also bought prahus which they put up at the different *ruma* of their own clan. Whoever received such a boat, thereby accepted the obligation to deliver all slaves and goods captured during an expedition directly to the chief of his own clan. In return the keeper had this large prahu at his free disposal, to equip and man as he wished

for making trading voyages etc. The following story mentions this custom
and it also demonstrates that slaves were caught inside one's own village
(Texts 192).

"Moi lived in (the house) Erari (of the Apeinawo clan) with Ekamai.
Once Moi and Ekamai went out fishing together and then brought
their fish to father Saremuni in (the house) Ruatakurei. Then Moi
came to words with father and quarrelled with him. Father had said:
"You must never go rowing at night".

"When they started to quarrel, (my father) Korowi (likewise from
Ruatakurei) angrily took Saremuni's part. Thereupon Maresi (the clan-
chief, from the house Erari) shot at the people in Ruatakurei, and
Fewai (from Erari) hit Serawoai (the elder brother of Korowi from
Ruatakurei) in his hand, whereupon Serawoai shot in the hand of Fai
(from Erari).

"My father was angry that Fewai had hit Serawoai with an arrow
and therefore he smashed the prahu (which the members of the seraruma Erari had put up at Ruatakurei). The sera became angry about
this prahu and then Adori and Nakiroi (from Erari) in their rage about
this prahu caught the boy of Saumbeki (the daughter of Serawoai who
was married at Risei).

"When they had caught him, my father came here to cut open this
piece of ground where they have placed this house. And then they let
go the people who lived downstream (close beside Erari), to move
here and built Koghi. That is the new house they founded. But the
upstream part (which therefore did not lie close to Erari) remained
in Ruatakurei. Thereupon they gave goods and they ransomed the boy
Forifori, the son of Saumbeki.

"After a long time they bought a new prahu at the command of their
sera, putting it up at Korowi's. When it had been put up at his place,
they saw men of Nuwuri with nets, rowing into the creek Woiako. The
men of Korowi went down through the tidal forest and caught Sabisai
and his father Ranui. When they had gone into the prahu, they had
them with them in the prahu and rowed upstream with them and
made them stay with their clan-chief Maresi and Sanadi Pari (of
Erari).

"Thereupon the men of Nuwuri gave another boy in stead of the
father, and the boy was ransomed by the chief of Panduamin (on
Japen). And they took the boy, who had taken the place of the father,
to the people of Waren, who bought him".

It was not strictly necessary that the serabawa all personally accompanied the raiding party; when they were too old, they stayed at home. If the *sera* did go along, that did not yet make him the commander, but at most the *primus inter pares*, for it did not lie in the nature of his position that he could expect his *waribo* to execute correctly the orders he gave. At most he could try to establish the will of the ancestors concerning the results of the raid by means of divination, or to give the sign for the attack.

There were always enough reasons for holding a raiding expedition. A *sera* whose wife had died, had to wear a mourning-cap until it fell to pieces or until he had taken part in a raid. A *sera* who wished to acquire a title had to catch a slave. When a *seraruma* was built, a slave was caught and, if possible, the attempt was made to obtain a skull, to be tied to the first central pillar. And in the old days a man had to present a slave with his dowry, if he really wanted to be reckoned with.

All kinds of attempts were made to irritate and to challenge the opponents. In the trees close to their village the leaves were hung which had served as wrappings for the ceremonial sago-cakes for the *saira* festival, as if to indicate that valuable objects were expected from them to serve as payment for the sago. Pieces of wood which had been left when building the front of a new house were likewise tied to the trees near the village of the opponents, indicating that in due time their skull would adorn the central pillar in the new house. Also a piece of rattan was simply hung in a tree, as if to say that such a rattan would serve before long to tie the cut-off heads of the villagers to the central poles of the houses of their attackers.

The successful warrior enjoyed great prestige. During the *saira* the brave old men enumerated proudly the names of those they had vanquished (p. 144); this was done likewise in the mourning ritual (p. 185) and in the Serakokoi episode (p. 191).

The youths whose initiation had been completed by the perforation of the nasal septum will probably have needed but little encouragement to join a raiding expedition which might provide them with a slave, so much desired for their dowry, or at least to make the hardly grown-up child a much fêted lover in the eyes of the women. Under the applause of the women for a successful raid or under their scorn for an unsuccessful one, many a youth will have become firmly resolved also to grow into a famous warrior, feared everywhere and entitled to the names of many vanquished foes.

The equipment of the raider included different types of lances, such as *matane, ghamburu*, pointed bamboos of different thickness; *sara-raiwo*, pointed sticks; *ada*, bamboo spears; *naiwirada*, iron-tipped lances; then *kana*, bows of palmwood with a rattan bow-string (*embo*) and usually a reserve string tied against the front, and arrows of all sizes, like *donggo*, small bamboo-tipped arrows; *buramui*, larger bamboo arrows for shooting at greater distances; *safawa*, tipped with nibung-wood; *repati*, with two toothed, wooden points. Of some arrows the head

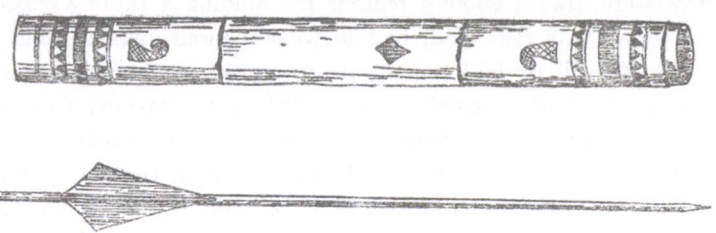

Quiver and arrow for shooting birds.

was provided with a piece of kangaroo bone whittled down to a point or the sting of a ray; these broke into splinters inside the wound and caused protracted ulcerations. For very long distances smaller bows were used, with *kowai*, little arrows made from the rib of a sago-leaf and also used for bird-hunting. The larger arrows were carried in a *wawara*, a bamboo quiver, which was thrown away during flight, often

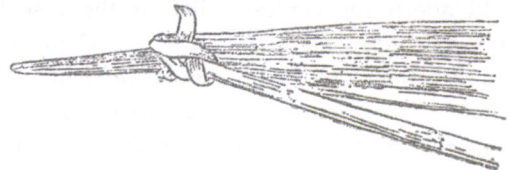

Manner of attaching the cord to the bow.

in order to put the pursuers on the wrong track. In spite of the large number of arrows shot, their effect was not very great; at least during raids the number of victims was relatively small. In addition, the Waropen skill in shooting was not so very great either.

A striking weapon with which the Waropen are extraordinarily fami-liar is the *naibawa* (*sasu* N.), the bush-knife of Biak origin. The axe (*mano*) was little used during raids, although the old-fashioned stone axe (*bono*) was used for this purpose, as became apparent in the mock fight during the *Inurisaira* (p. 192). Means of defence are unknown; at most the oars are used to ward off arrowshots, thrusts or cuts, whilst the sharp blade of the oar should also not be underestimated as weapon of offense. Formerly, however, the shield (*ragharo*) was used, as was proved at the same *Inurisaira*, whilst at Weinami people said that they remembered the use of the shield.

Before the band of raiders start out, the weapons to be used are first anointed with a medicine which ensures success; usually a decoc-

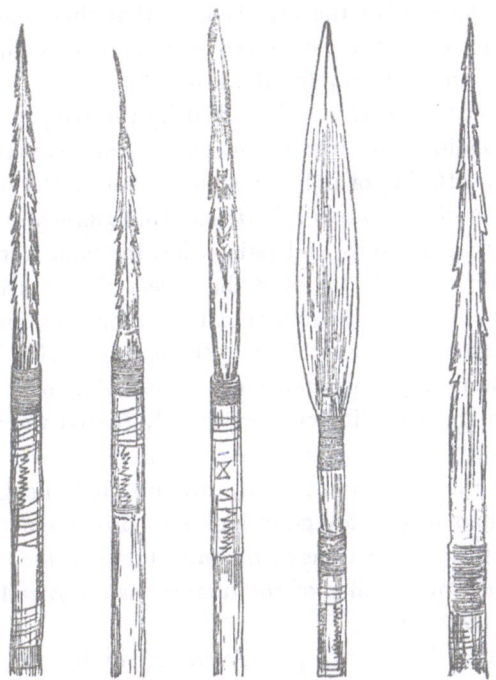

Various types of arrow-heads.

tion of leaves; then a number of auspicious pieces of cloth, beads, etc., are attached to the weapons. At Napan the medicine is called *hesoaiwo*. Arms used for raiding are preferably not used for any other purpose. The night before they sailed the raiders drank a medicine, whilst an augury was taken concerning the result of the raid; this was done by heating a leaf (see p. 245). If one of the participants dreamed that a slave was caught, this was considered an auspicious omen. The people were convinced that the whole action was supervised by the ancestors.

At Weinami, the two clans there each possessed an ancestor, represented by an ancestral statue. The *nurawa* (N.), the ancestor, of the Waratanoi clan was called Kowei [144] and he was the eldest; the ancestor of the younger clan, Warami, was called Warari. Because the village

------

[144] For the figure of Kowei see further p. 282.

of Weinami usually equipped only one raiding party for both clans, it
was the custom to present a sacrifice to Kowei before starting out, the
offering consisting of sago-mush with coconut. When this food had been
presented to Kowei, it was eaten by the gathering of the companions.
It was assumed that Kowei preceded the party and showed them the
way, blinding the eyes of the attacked, so that they would not observe
the attackers; he rendered their bows weak and then flung them out of
their prahus as an easy prey for the attackers.

At Kai prognostics were usually taken by throwing a bamboo, decorated with porcelain shells, into the water (*adumanggusara*). The ancestors received an offering of palm-wine and tobacco, the *serabawa* saying:
"*Inggoini, mindisabaku ghado mindiesawako. Mimbe dora muna ghero*",
"Ancestors, here is tobacco and palm-wine for you. Let the rain come
down". At Nubuai it was believed that every time a slave was caught,
Worodauni, the local representative of the submarine initiation monster,
lifted its head above the water. During the voyage divinations concerning
the outcome were regularly repeated. When the men from Weinami
made a raid in the Kai district, they usually passed the first night near
the river Binataboa, where they took prognostics concerning the result.
During a raid the participants hung around their neck small wooden
figures which had to provide good luck and protection. These figurines
are nothing but representations of the ancestors like these are recognised
in the foremost central pillar of the house; these are called *aiwo*, medicine, Numfoor *hesoaiwo*.

At Nubuai a raid was once performed for my benefit; this happened
when Munamberi, one of the *sera* of the Sawaki clan, had obtained an
honorific title from the men of Arièpi on Japen.[145] For this occasion
Munamberi provided a fine new prahu, which in compliance with the
demands of custom was first rowed to the wife's family, where a mock
fight was staged, including the flinging of all kinds of innocent missiles
to the bow of the prahu, such as pieces of the rib of sago-leaves, coconut-
bark, etc. The presentation of a slave belongs to the acquisition of an
honorific title and so the people were of the opinion that on this occasion
a slave-hunt had to be organised *pro forma*.

The new raiding prahu had been beautifully rigged out with all the
decorative pieces on stem and stern and with different flags, i.a. a small

---

[145] I was told that the whole show had been staged spontaneously for my
instruction, but afterwards I have occasionally wondered whether after all
it was not a moderate form of reality which was performed, because in actual
fact the catching of slaves was considered an indispensable condition for the
lawful acquisition of an honorific title.

pennant, said to have been bestowed by the sultan of Tidore. The bow of the prahu ought to have been festooned with strings of *korombowi*, the white porcelain shells, repeatedly mentioned in mythology and often worn on festive occasions, especially by women, although in other respects no special value is attached to them.

The large prahu was fully manned and first went to hide somewhere in a creek, downstream from Nubuai village. In another smaller prahu there were two fishermen who had to act as the unsuspecting victims. At a certain moment the large prahu shot forward, numerous oars giving it speed, whilst the triton-shell which always has to lie in the bows blared frighteningly. And when soon it became apparent that the small prahu with its two men would not be able to escape, the war-cry was shouted at them (*we dorasa*), the long-drawn and strident *iuai iubiiio*. The excited pursuers started to dance, standing straight in their prahu, yelling in a challenging fashion *u, u, u* (*wororo*); others incited each other with the cry *siriripumbo*. In vain the pursued tried to escape into the tidal forest. Arrows zoomed around them, and although the tips on this occasion had been provided with plugs, they were dangerous enough, to my mind. The pursuers caught them and with a really realistic show of violence they were dragged into the raiding prahu with deafening yells and shouts.

After the victory leafed branches were cut from the trees in the tidal forest, the *resa* we likewise observed in the enclosed prahu at the *saira* ritual and also when the slaves were enumerated as the skull of a deceased person was being carried away. One particularly large branch was put upright and held in this position only by the men who had taken part in the actual catching of the slave. The man who had been the first to reach him, placed his hand the highest on this triumphal pole (*adakainamo*), those who had come later had to put their hand below his.

Those who had stayed behind knew from the blaring of the triton-shell, that the raid had been successful.[146] Then *ghomindano*, slaving-songs are sung; the tune, like that of all other men's songs, resembles that of the normal alternating songs, sung in the prahu. An example of these is:

Bosoi rara karadia awo wiarepawa
Sinawo anggeio bosoi rara karadia awo
Simuniawa rara karadia awo.

---

[146] At Napan the term for "success in the raid" is *heso; kiwe heso,* they have been successful. Cf. *hesoaiwo,* medicine for success in raiding.

I shall not venture to translate this, but I can only remark that *karadia awo* means "to fetch forest products", from Malay kerja, a secret expression, indicating that heads of the people of the interior have been hunted. Another example of a *ghomindano* is:

> *Seruai seruai karaino rarimerano*
> *Serama sewasa Erari*
> *Seruai seruai karaino,*

which mentions the train of the red aprons of the women who dance on the front gallery of the house Erari when the warriors come home.

Those who hold the triumphal pole have their face painted with the raiding-paint (*ghamesaro*) of lime, red sirih-spittle and blacking.[147] Whilst they row into the village in this fashion, the jubilant women come together on the front gallery. The raiders row up to the *seraruma* and throw the triumphal branches they have carried along at the front of this house. At Napan it was customary to stick these branches in the ground in front of the *seraruma* as was done at Kai when the skull was carried away.[148] The triumphal pole was erected on the front gallery of the house to which the slave was awarded.

Whilst the raiding prahu rows eight times around in a circle in front of the *seraruma*, in the same way as on the occasion of a *saira*, two women from this house appear on the front gallery, adorned with red aprons and holding a bundle of arrows in their hands; they dance, challenging the raiding party (*owaura*). Under loud cheers of *ieeewo, ieeewo, ieeewo* the slaves who have been caught are led up the notched tree-trunk which serves as a staircase, whereupon the girls take them by the hand. As remarked above, taking each other by the hand is considered as an intimacy which is only permitted to lovers. This gesture renders the slaves practically inviolable. If a man has to flee during a fight, he may become similarly inviolable if he succeeds in grasping a woman by her apron from behind and in tearing it off, if possible.

Then the prisoners are taken inside and in order to prevent their

---

[147] It is not clear whether this was also done before the attack.

[148] This casting of spears or stones also occurs elsewhere. Among the Windessi, upon the successful completion of a raid the triumphal branches were flung at the *rum sram* (J. L. D. van der Roest in *TBG*, XL, p. 158). Among the Roro-speaking tribes the throws were likewise aimed at the club-house (C. G. Seligmann, *The Melanesians of British New Guinea*, p. 298). Among the Mafulu they were thrown towards the house of the chief when the latter had had a child (R. W. Williamson, *The Mafulu*, p. 157). In the *kula*-barter among the inhabitants of the Trobriand Islands sometimes stones and spears were thrown at the house of the chief (B. Malinowski, *Argonauts of the Western Pacific*, p. 486).

escape, one of their feet is locked in the *suna,* a heavy wooden block shaped like a crocodile, and when there seem to be good reasons to do so, one hand is inserted in the *kaiwa,* a similar, but much smaller block which is worn on a piece of string around the neck. Such blocks are still preserved in several *ruma;* they have a name of their own, being called after the ancestors of the house, or after one of the great warriors in the wife's family. The *ruma* Sawaki possessed two of these, Munamberi and Dafaiei, called after two famous warriors of the past.

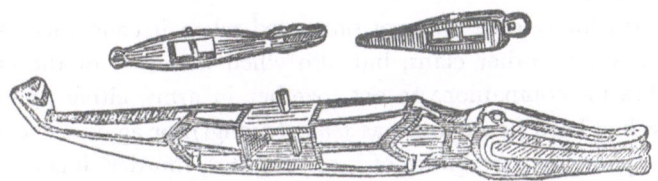

Slave-blocks for legs and hands (*suna* and *kaiwa*); wood, painted in red and white; length approx. 1 m. 80 and 60 cm.

Now the *serabawa* was obliged to present a feast to his companions, lasting three nights (*ghominama; womuna* N.). During this feast only the men danced, just like on the occasion of a *saira.* The guests had to be amply provided with sago and tobacco, and according to the myth, the burden this obligation laid upon the wives of the chief of the clan brought about that also the younger brother of the *serabawa* was given a share in the obligations towards the feasting companions, whereupon the clan was split up into the "head" and the "tail" of the triton-shell (Texts 176, 177). At the feast slaving-songs, *ghomindano,* were sung.

At Weinami it proved possible to obtain more information concerning these *womuna.* The feast was evidently intended for the ancestors, the *nurawa.* If the party returned with cut-off heads, these were arranged behind the first central pole of the house. The skull itself was for the "elder brother", the *ruma* Waratanoi, the lower jaw for the "younger brother", the *ruma* Warami. When the skull had been placed behind the central pole, the image of Kowei, the *nurawa* of Waratanoi, was put behind the skull. And behind Kowei again triton-shells were laid, the first, Warimadai ("boy's body"?) was considered as male, and the second, Rarimeramboai ("new red dancing-apron"?) as female. It was said that the sound of these shells was so clear that it could be heard at Weinami when Kowei blew them on the island of Nau, at least thirty miles away. Closely against the freshly cut head a couple of crocodile skulls were put, so that these held the head tightly between them, as it

14

were. To the right and to the left as many crocodile skulls were put in
a row in the longitudinal axis of the house as it was possible to collect.[149]

However, if the raiding party ended in disaster, the prahus came
rowing surreptitiously into the village, their decorative pieces stowed
away. And the raiders could be glad if no dead were to be mourned,
because the anger of the relatives of the casualties was often directed
against the men who in their eyes had not watched over the life of
the dead man.

The term for raiding was not only used when it concerned the hunt
for slaves among other clans, but also when the chief of the clan had
to mobilise his companions to act together in arms, either against persons who had been pointed out as *sema* (suangi), or against excommunicated people who were guilty of manslaughter, murder, incest, or other
serious violations of customary law. In those cases the raid was not only
aimed at outsiders, but equally against members of the clan. The action
against the persons considered as *sema* is one of the darkest pages in
the cultural history of the Waropen, and all the more regrettable, because the Waropen themselves, and the *sema* as well, were fully convinced of the justice of these actions. Once the whispered accusations
against a certain person, man, woman or child, had arrived at an open
formulation by the influential old men and women, a raiding party
would eventually come together, to attack the unfortunate person unexpectedly, either in his or her own home which had usually been stealthily
evacuated beforehand by his family and relatives, or somewhere in one
of the silent creeks of the tidal forest, where the mutilated body was
left behind without more ado.

Matters wore a somewhat more favourable complexion, when the raid
was directed against an excommunicated fellow-member of the clan who
in anger or through inadvertence had committed manslaughter. In
these cases the offender tried to reach his own *ruma* where he could
count on the assistance of his own family-branch. If he was unable to
do so, or if he believed his family to be too weak for serious resistance,
he took flight to another village, or he retired temporarily into the tidal
forest. Thereupon one or more raids were undertaken against him, the
show of more or less activity depending on the importance of the man

---

[149] There is therefore evidently a close connection between the raid and the
crocodile, as is also apparent from the shape of the slave-blocks. However,
I was only able to note one single story, which told that the crocodile and
the iguana fought for the triton-shell (Texts 151). On the shell the traces
of their claws are still recognisable.

they pursued. Here follows the translation of a story which relates such an occurrence (Texts 190).

"Sawuri. — When after drinking palm-wine they had started to quarrel, Sawuri gave Sighomi a blow. And in the evening Sawuri beat the drum, to have his sisters give him sago to eat. Sighomi became angry and said: "Why are you beating the drum here in the evening?"

"Thereupon he ran upstairs and started to fight with him again. When Sawuri had killed Sighomi, he fell on the ground and then he climbed upstairs and slashed him with many cuts. And he fled away from the house and Sighomi remained lying alone in the house.

"Then a raiding party set out to kill Sawuri. Sawuri had fled into the forest. After having stayed in the forest for some time, he kidnapped his sister and his wife and took them to the centre of the forest, where he became hungry. Due to his hunger he killed a boar during the day-time. But only in the night the fires flared up. He was afraid that the raiding party would come to look for him. During the day he cooked nothing. Only during the night they roasted the boar. During the day-time his wife and his sister remained inside the house, but the man went to look for the raiding party and only returned when it was night. He continued to live all the time in the same forest. And when he became hungry, he went outside, cut down a sago-palm and fetched sago.

"When the raiding party came to look for him, he took them (viz. the women) along to another house, where they continued to live. During the day they went to beat sago and when they had gone to prepare this, he killed a boar for them which they took into the forest and prepared at night. When they had finished with the sago from the palm, they made a prahu from the bark and smeared it with earth and soot. And to this prahu of palm-bark he attached a bow like that of a *soado*. Then they took it outside to the water.

"And during the night he left his wife and his sister at home, but he went downstream in order to spy upon the houses whilst he drifted along. And when the people of the village had put up sero-fishtraps, he laid himself alongside and took away their fish and put these in his prahu. When he had taken them all away, the people of the village came to fetch their fish and then they became angry with the men who had been lying there with their prahus and said: "Where are the fishes?" And then they said: "I put the fishes in this prahu. They must have disappeared".

"When Sawuri had backed away, he remained lying there to spy upon them as they looked for the fish. And when he had done spying, he took the fish upstream, jumped on land, and took them again to his house there, where his wife and his sister fried them, and they ate them.

"During day-time he pushed the prahu of palm-bark under water, and at night he came outside, bailed it out and went downstream to spy upon the houses. When he had caught some fish, he put them down and fried them only at night. During darkness he went to the houses downstream and in the morning he went back to his forest there.

"After some time the raiding party from Sanggei came to kill him. Then he took his bows and went to the open water to shoot arrows at them from there; he stole their palm-wine and then went to sit in the top of a tree. The men of Sanggei said: "Perhaps a raiding party has passed here and tapped our palm-wine".

"But he had climbed into the top of the tree, from where he spied upon them, seeing his own people lie there. Whilst they were talking there, he broke off something and threw it at them from the tree, hitting their *sanduo*-prahu. When they turned around and saw him, he said: "If there were a raid we could hold, that would outweigh the fact that I have fought with you. To-morrow you should bring a raiding party to advance on me and to fight me".

"But he cheated them and he did not remain in the creek Donggori, but stayed at the upper end of the creek Desi. And when that next day according to the agreement the party came to look for him, he shot at them with his arrows and he remained waiting for them high up in the banyan-tree. The raiders shouted in a challenging manner, but they became afraid and fled away from him and he went again to the place where he was living. He reappeared and rowed towards the houses and he met the people who were working in the sago-forest and they all rowed away.

"However, Sawuri's sister and his brother-in-law wept over him and went to look for him in the forest. They were unable to find him, but their dog did. Thereupon Sawuri tied a communication to the neck of the dog. And this communication he carried to the house of his brother-in-law. When the dog appeared, it went around howling with this communication. And when they had felt all over the dog, there was the communication on the neck of the dog. And when the agreed time had come, the brother-in-law went with the sister to look for him. They found him in the forest and they retained him. They remained with him and he gave them of the boar's meat and of the sago.

"Then they took him outside to the river-bank, behind the brother-in-law. And the brother-in-law said: "Let me go downstream first to notify the village, and then you go down to throw lime".

"The next day Sawuri went away and fetched his possessions and left these in the forest and he went to the village and threw lime. When he went rowing downstream, the villagers came rowing up to shoot arrows at him in a fight. Although Sawuri hit the villagers with his arrows, none of the injuries were mortal.

"The sister and his wife put on their ornaments, pushed on their shell-bands, put in their ear-rings and hung their beads around their throat. And when he had finished throwing lime at them, he gave shell-bands, plates, blue cotton and pieces of cloth. And when he had finished paying them, he continued to live in the village".

THE FUNCTION OF THE RAID

In the Waropen area the raid is an institution which possesses several functions that have not been integrated into one definite cultural trait. Here the raid was, for instance, not a purely ritual matter, to which all members of the clan had to submit. Whoever wished to remain at home because he was afraid, need not fear immediate supernatural corrections.

On the other hand, the image of the redoubtable raider stood as a shining example before the eyes of the ambitious youth. Applause and admiration was his share when he returned to his village from a successful raiding expedition accompanied by triumphal songs.

A long list of slaves and a warrior's comb with many feathers perpetuate the names even of warriors long since dead. Their martial deeds are transmitted to the children; a quiver, a tooth or an amulet which once belonged to them are handed down from one generation to the next. At Napan the chief calls his brave companions *raiheso bawa,* my great successes in war. In the *seraruma* of Sawaki the people preserve two ropes that are tied together, one with forty-nine, the other with fourteen knots. These knots represent all the persons who were vanquished by Maseiori, a famous chief of the Sawaki clan, long since deceased; the forty-nine knots are 49 slaves, whilst the fourteen knots represent 14 *sema* who are still all known by name. It needs no further argument that all men, chiefs as well as companions, were jealous of such a fame.

The raid was therefore the institutionalized method for a young man to distinguish himself. It provided him with a standard for his ideals and at the same time a socially approved model of a real man. It also rendered him freer in his movements, because he was able to permit himself infractions of etiquette and liberties with women which would not be so readily accepted of a less redoubtable man. The terrible Munoi from Sawaki was not only able to make half the village tremble with fear due to his threats, but he could even permit himself the liberty of handling his young wife's breasts in public. And people very carefully waited until he was far away before they dared to give vent to their indignation for this "impertinence".

Moreover, the relation between the raid and religion is also still clearly apparent. We only need to remember that the raid was an affair of the clan, under the direct guidance of the ancestors to whom in fact the whole feast of the catching of slaves was dedicated, as was the case at Weinami. It is therefore quite probable that formerly the holding of raids formed part of the inititation of the young men, for the slave-festival was also a men's ceremony, whilst moreover on various occasions during the *saira* the fame of the great warriors was proclaimed. It should also be noted that at Windessi the triumphal branches — which certainly represent the forest into which the initiandi withdrew — were thrown at the young men's house. During his visit to Weinami in 1887 de Clercq found near the *dama* "an open stand made of bamboo and likewise built on poles in the sea, on which *Codiaeum*-leaves were spread out at the time of my visit; after a successful raid the weapons — which are then decorated with flowers and small bits of cloth — are exhibited there".[150]

One of the important things during a raid was to acquire a skull. When a successful party had returned with a skull, in some villages the new head was dried out over a slow fire during the slave-feast, the soft parts being removed; thereupon the skull was tied to the central pole of the front part of the house. Elsewhere the skull was simply placed in the tidal forest until the fleshy parts were decayed. At Nubuai they were hung in the trees upstream of the village, near Andegharewo, probably because there it was less easy to steal them than downstream.

---

[150] F. S. A. de Clercq and J. D. H. Schmeltz, *Ethnographische beschrijving van de West- en Noordkust van Nederlandsch Nieuw-Guinea* (Ethnographical description of the West coast and North coast of Netherlands New Guinea), p. 178.

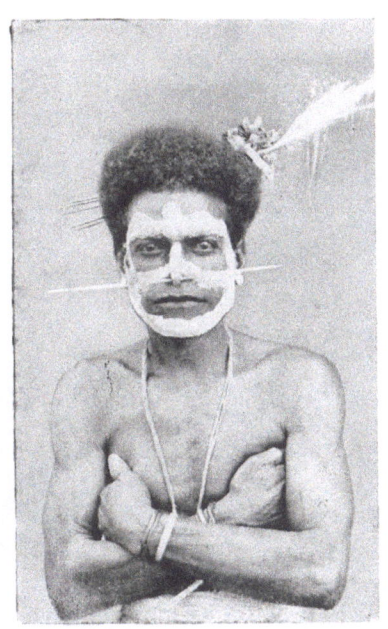

48. War-paint for the raid

49. Archery

50. Bow of the travelling prahu

51. Rinsing the sago

Thereupon the skull was exhibited on the central pillar, if necessary with due regards for the rights of the "head"-clan and the "tail"-clan to the upper part and to the lower jaw. Also at present it is still customary to hang the skulls of crocodiles, dugong, turtle-skins and the heads or the tails of very large fishes on this central pillar, for show.

It is clear that the skull, which presumably represents the whole person as a *pars pro toto*, is meant as an offering of the dead man to the ancestors under whose supervision all raiding activities take place. It is in honour of the ancestors that when the skull of a relative is taken out of the house the names of the conquered slaves are commemorated by planting an equal number of triumphal branches. As we shall discuss in more detail on p. 250, for a sick child whose soul was believed to have been taken by the ancestors, a doll was made which was the slave of the sick child, and this was presented to the ancestors in its stead.

However, this does not mean to say that in our opinion the person who has been killed is destinated to serve the ancestors as a slave in the hereafter; at least, this is not the case among the Waropen. For in the Waropen area the *ghomino* is not a slave or a servant in the sense which these words usually have among us. The Waropen have few tasks which they could leave to servants and for that reason they have not made use of slave-labour, e.g. for agricultural work, as does seem to have been the case among the Numfoor. A *ghomino* is not a servant who belongs to his master and who has to perform all services demanded from him, but rather a person who has had the misfortune to be cut off from his own clan, to fall into the hands of another clan. Everybody trusts his ancestors to help him during the raid, defeat being imputed to one's having neglected or overlooked the inauspicious signs which the ancestors had sent to their descendants to warn them. Raiding is a ritual in which human beings are partly the officiants and partly the victims.[151]

From the reception of the *ghomino* by the women, as this was customary at Kai, one would be inclined to conclude that it was intended to take them into the clan in some way or another. In spite of their remarks to the contrary, it is certain that also among the Waropen

---

[151] Among the Toradja it was believed that one could only be conquered if one owed a debt. How literally this was taken is demonstrated by the fact that the attackers gave payment for everything they took in the alien area. They even did not drink water without throwing a small coin into the river. See N. Adriani, *Verspreide Geschriften* (Collected works), III, p. 317.

slaves simply entered into the clan. Slaves were not exposed to maltreatment if it was not believed .that there was a reason to do so. Neither were the woman deliberately manhandled or prostituted, I was told.

Members of the clan who fall into the hands of a raiding party and who are not ransomed, are simply considered to be dead, without the people feeling particularly "ashamed" about this. In such cases it usually concerns young children or women who do not possess sufficient relatives who care about their fate. On the other hand, people are very sad about the lot of those whose head has been taken; they are believed to be unable to join the other ancestors and therefore they become *dare*. In general *dare* was described as "spilt blood, especially of relatives". Here we must not think of blood which flows e.g. from an unimportant wound, because this causes no special fear, but rather of a mortal haemorrhage.[152] The places where the blood was spilt are avoided from then on, particularly by the relatives. When a prahu lands in a region where somebody fears *dare*, this person will usually refuse to leave the prahu at all; it is dangerous to drink water at such a place, and if it is unavoidable, a cross has to be drawn through the water with a knife. When somebody is struck by the *dare*, his teeth will drop out and his belly will swell until death follows.

However, the *dare* does not become the servant of the person who struck the mortal blow either; he remains a continuous threat to his own clansmen in particular. The *dare* appear preferably at sun-down. Around that time there may be a sudden outburst of shouting somewhere in the village; someone, usually the *ghasaiwin*, believed she noticed a *dare* attempting to approach the small children without being observed and now he is driven away with shouts. Small arrows are shot at him, whilst the *ghasaiwin* defends herself with a knife. Action against the *dare* is one of the most important duties of the *ghasaiwin*, as she is able to see him arrive, a thing others are unable to do. In which way she actually sees him, she was, however, unable to describe.[153]

---

[152] The death of a Numfoor girl was ascribed to the fact that she had been in contact with the blood of her mother who had died when giving birth to a child.

[153] The *dare* may be compared with e.g. the so-called "strong dead" (*dilikini*) among the people of Tobelo, i.e. the spirits of persons who died a violent death while still being in possession of their full mana (*gurumini*). Different from the *dare*, the *dilikini* are protective figures. See A. Hueting, "De Tobeloreezen in hun denken en doen" (Ideas and activities of the people of Tobelo), in *BTLV*, LXXVIII, p. 163.

It is the rule that the shedding of blood has to be compensated by the spilling of new blood. The fact that the shedding of blood in and by itself was sufficient, shows how closely this approaches the sphere of the sacrifice.[154] It is therefore not absolutely necessary that it is the blood of the guilty person or of his clan that should be shed, but it could even be one's own blood. In the tale of Sawuri, quoted above, it is the killer himself who causes another wound in the ensuing, more or less ceremonial, propitiatory fight (p. 213). When this was pointed out to the person who told the story he was quite perplexed, but after an extensive consultation of the old men it was established that things had really happened in this way. It would also have been allright if Sawuri had been wounded, as long as blood had flowed.

Among the Numfoor, setting out on a raiding party in order to avenge people who had been killed was called *munsu mbrob*, literally "to requite the war", which should be clearly distinguished from the *rak* (p. 198), where the spilling of blood was not one of the points at issue. If during *munsu mbrob* blood was spilt, this was sufficient in itself, even if it has occured on the side of the raiding party itself. Then one said *rik isiper kwar*, the blood has already become one. If the raiders, who had started out to avenge spilt blood, caught a person, this person was not considered as *womin,* a slave, but as a *mansren* (Waropen: *sera*), a free man, who was retained as a hostage until peace was concluded. He was *mansren wekainuk mbrob,* the man who sits out the war.

Among the Waropen this rule was not kept so systematically. Once blood had been shed, the affair was considered as being temporarily finished, but people still waited for an opportunity to spill the blood of the opponents, unless in the mean time peace had been concluded and an arrangement had been reached. However, that it was not actual retaliation which was at issue is proved again by the fact that the raiding party which set out to revenge a life on the other party, caught or killed any person who happened to come within reach. Then it was said that such a person had "crossed in front of the bow of the prahu", (*woiari gharegha*), which was considered as a challenge. On the other hand, informants at Weinami stated that people at Weinami were likewise responsible and likewise had to contribute to the ransom, if raiders from Kai, for instance, who had started out to take revenge for a raid

---

[154] Feuilletau de Bruyn reports that among the Biak formerly even slaves who had been caught in a raid during the period of light mourning were sacrificed on the grave of the deceased, but for this information no certainty could be given; see his *Schouten- en Padaido-eilanden,* p. 116.

from Weinami, happened to catch a man from e.g. the Moor Islands. Revenge was not directly aimed at the person of the offender, but at every person of whom the ancestors decreed that he might be caught, mostly a person from the same village or clan or region as the offender. The following story relates how after having drunk a potion of the leaves of the kedondong-tree people observed in the ashes of the hearth the footprint of a certain person who in this way was said to have been indicated by the ancestors as a certain prey. This was not the actual culprit (Texts 193).

"The men from Nubuai here went into the hinterland (backwoods) and passed by Woinui. When they had arrived there, they looked around for some time and went back, but the next day they returned in earnest. The whole village went there. And when they had tied the prahus at the mouth of the river, they went to the village by land, in order to surround them in the village. When the village had been surrounded, Ghabi climbed into the houses of the people of Woinui. But Munamberi had posted himself in his house and killed Ghabi with an arrow on the staircase, and he fell down and remained lying there. Thereupon the men from Nubuai wanted to flee, leaving Ghabi behind, but Miteri and Raiwowi roused themselves and fetched Ghabi, because they said to themselves: "If we run away from him, the people in the large village there will scoff at us for having grown afraid and leaving him there".

"And then they went to fetch him, Ghabi, whilst Raiwowi held him back, by the legs. The raiders from here kept the men of Woinui at a distance with arrows. They took him along and went on board and reached the river-mouth with the prahus and rowed him to Nuwuri. When they had taken him there, his sisters lamented over him and hung him in the house (to be mummified). Whilst they watched over him, people went to catch fish and to beat sago and when the women who had laid out the corpse had been paid, they danced the funerary song.

"After having danced that funerary song they tried to learn the will of the ancestors concerning him by drinking kedondong-leaves and then they saw the footprints of Soruki from Woinui in the ashes of the hearth. They were jubilant and raised the war-cry. Next day they took out their prahus and then the same raiding party went out again and on their way they found men from Woinui and also Soruki, whom they caught and beat to death. When they had killed him, they took his head and they cut off his privy parts. The people from Nuwuri

danced around with his privy parts. The body they loaded separately
into a prahu and with this trunk they rowed singing to Manieghasi.
And when the body had lain there until the next day, Ghabi's sisters
went to fetch fire-wood and placed it on the shell-mound (underneath
the house). Thereupon they kicked his (i.e. Ghabi's) house to pieces
and fetched ribs of sagopalm-leaves. Next they took the body down
and put it on top of that shell-mound. The fire was lit and when it
blazed up, they brought out the body and it burned completely in the
fire".

This system of blood for blood has of course a certain tendency to
develop eventually into a vendetta of an ever increasing extent, so that
in the course of time no traffic would have been possible at all due
to the danger of raids threatening everywhere. However, for this purpose there existed an arrangement which functioned like a safety-valve,
viz. the conclusion of a peace. On this occasion the account of the past
period was drawn up, as it were, and the debts among the parties
were settled. However, the conclusion of peace did not imply that one
of the parties owed itself beaten and recognised its defeat. Nor did
anyone expect that from then on such raids would no longer occur
between the two parties; it was simply that the people wanted for once
to wipe out all the ancient claims and counter-claims.

When two parties wished to arrive at an agreement, village A sent
to village B, often through the intermediary of others, a knotted cord
with the request to come to A after the number of nights indicated
by the knots, in order to conclude the peace. First goods were exchanged
to satisfy the mutual claims. Then a bamboo was taken, which was
wrapped in string and from the two ends of which stones were suspended
by strings, like from a balanced scale. The *sera* of both parties each took
hold of one end of the bamboo with the left hand, and then they "threw
lime" (*so rosa*). Then they counted "one, two, three" and at this signal
the two *sera* had to sever the bamboo simultaneously at one single blow.
If one blow was insufficient, this implied that matters had not been
fully cleared up. Thereupon both parties went to the *seraruma* of A to
eat sago from one and the same bowl. At a certain day party A then
went to village B to eat sago there with the people from B, also out of
one bowl. Usually the agreement was sealed by giving a woman in marriage to the other party. A party which comes to conclude peace or to
ransom slaves is called *pareriaira*, from *pareri*, to release, to come loose,
and *aira*, group, party, band. One therefore speaks of *Otigha kiri-pare-
riaira*, a group from Wonti which wants to come loose, to conclude

peace, and *raipareriaira,* the group of people who come to ransom me, when I have fallen into the hands of others as a slave.

"Throwing lime" also belonged to the conclusion of the peace; this custom is also known from other localities on New Guinea. In spite of the verb *so,* to throw, however, in the Waropen area the lime was not thrown, but it was rubbed on the body of the other party. With lime a dot was painted at the pit of the stomach of all members of the party with which peace was to be concluded; it was said that internal inflammations occur in the ears, the nose or the lungs if this custom is not strictly maintained.

Obtaining a skull is, at least in modern times, not the sole object of a raid, and not even the most important object. Fame attributes to the well-known Maseiori from Nubuai sixty-three names of people whom he had vanquished; among these there are fourteen *sema* which he is said to have actually killed. Of course, so the explanation went, he made the others his slaves, but he did not kill them or take their head. As is shown by the Controller's diary for 1913, mentioned earlier, the killed enemies were sometimes simply left without cutting off their heads, and this also became apparent from the stories told by old Waropen men.[155] This must also be partly due to the evident fear of and revulsion against the intentional killing and beheading of a human being. Even experienced warriors had to work up their courage by drinking before they had become sufficiently *nikako,* hard, cruel, and it was often attempted to spare at least people one knew. Old participants in raids stated that fear of "shame" was one of the main reasons and that they only became really frenzied when blood had flown; for in the raid economic motives are also active, and this to a considerable extent.

This is due to the rule that a *ghomino* may be ransomed by his relatives. And because not only the severed heads were placed in front of the *inggoro* in the *seraruma,* but also the slaves who had been captured, whilst it was the *sera,* moreover, who profited most from the ransom received, it follows logically that the *sera* preferred a living *ghomino* to a severed head. From a matter-of-fact point of view, a severed head is always a loss, because spilt blood is only removed by blood and also because at the conclusion of the peace a head necessitates the payment of wergeld to the other party, whilst for an em-

---

[155] The possession of skulls as such was evidently not so highly prized, because it must have been only too easy to rob the skulls of the deceased persons who had been laid to rest at the river-mouth, downstream from the village.

prisoned *ghomino* one always receives purchase-money. To a certain extent there exists therefore a conflict of interests between the *waribo* who joins because he wants to become a great and redoubtable warrior, and the *sera* who summons his clan and who mans and equips and provisions prahus, because he wishes to collect goods. It is for this reason that initiation-festivals and funerary feasts so easily lead to raids, because a Waropen feast is not a pure demonstration of joy, but also (and often particularly) a purely matter-of-fact exchange of goods.

In view of the ransom, the *sera* has the greatest interest in avoiding manslaughter. This is only to be expected from his companions and even so not so much because of blood-lust, but because he feels himself wronged by his *sera*. As has been said before, the *sera* was under the obligation to offer a feast to his *waribo* on the completion of a successful raid. Now, if somebody believed that at this feast he had not been rewarded according to his proved courage, he became ashamed, and this "shame" rendered him so violent that on the occasion of the next raid he only thought of killing. Dumoi, the old *sera* from Apeinawo, expressed this as follows: "Then (i.e. on the occasion of the slave-feast) the *sera* gave them all manner of things to eat and tobacco to chew. And if one of them obtained too little, he became angry. And when they caught anybody, he killed him, because he was so angry, because when he was given of those things and they took them from under his nose and ate them, and they then captured a slave, he killed him", (viz. if a slave was caught, he was killed by the person who had felt himself slighted at an earlier slave-feast).[156] The following story also shows that a *sera* only proceeded to kill a person in the last extremity (Texts 191):

"Awurai and Maghami once went to Sumberbaba and they summoned Panduamin, and the people from Panduamin took out their prahus, rowed back and summoned the men of Manawiri.[157] And their three villages brought two Strangers along with their guns. They rowed inland near Cape Rombausa (at the mouth of the Nubuai-river). And having rowed upstream during the night, they stopped in Fafarui(-creek) and the others stopped in Rafande(-creek), where they remained until morning.

"In the morning, when it was still dark, they hoisted the flags on

---

[156] *Maika kiriseragha we kiriaigha kikana we sabaku kikanggi. Maika karisa nogha maika iurako. Maika kiraghakieka iurabeka munio. Maika we kiriaigha kikanggi wuarieka ragha ghominghaika munina.*

[157] Three villages on Japen Island.

the prahus, and when they had made ready, well, then they rowed along the cape, upstream.

"But Ghafai looked down and saw the raiding party. He summoned those of Apeinawo and warned his wife: "Get up! A raiding party down there!"

"The women fled. Whilst they fled, the men lined up to shoot arrows at the raiders. First the raiding party entered Manieghasi, but there the palissade kept them back and they did not succeed in climbing into the house. Then the raiders came here, over to the other side.[158] And whilst the men of Nubuai shot arrows at them, the Strangers fired with their guns. But their arrows missed and they climbed into the houses and took the net which was named Siraworu and threw it into the prahu.

"But Ekamai [159] kept them away from his house by means of arrow-shots, together with his *waribo* Kewowi. Here the whole village kept them away from the houses with their arrows. And the men of Nubuai here shot arrows at the raiders so that they became afraid and fled, leaving Araghi and Anikakoi behind. The men of Nubuai pursued them; Ekamai pursued Anikakoi, whilst Siki and Moki pursued Araghi. Araghi was killed.

"The women beat Anikakoi around the head with their red aprons. Whilst they beat Anikakoi around the head, he tried to pull off their aprons.[160] Ekamai hurriedly took Anikakoi with him and kept him in Manieghasi, otherwise the people of the village would have killed him.

"When they had killed Araghi, they cut off his hands and his feet and hung them in the forest near Rafande Creek, but the head they split into two and one half came to be suspended in Erari and the other in Pedei.

"Ekamai kept Anikakoi imprisoned in Erari. Once Ekamai took him along to Waren where they were staying in order to beat sago. And when their stay for beating sago had come to an end, they felt hungry for fish. Therefore they fetched the net in order to go and fish with it. Maidawofi, Ekamai, Dumoi and the slave Anikakoi cast the net at Waren.

"Then that slave noticed a bow on the outrigger. They were sitting back to back. Anikakoi took that bow and aimed at the centre of

---

[158] I.e. towards Apeinawo, where the man who told the story was living.
[159] The chief of the Apeinawo clan.
[160] See p. 208.

Ekamai's back and he shot him with a *safawa*-arrow. Thereupon the slave fled.

"But they jumped after him and Ekamai himself caught him. When they had caught him, they took him to the prahu and bound him with ropes. When they had finished doing so, they beat him on the head with the sharp edge of the oar and on the body with the flat.

"With this thing straight in his body they rowed Ekamai upstream. And when they had rowed up, Aighora opened Ekamai's back and pulled out the *safawa*.

"Thereupon they took Anikakoi up into the house and tied him standing to the central pillar. Siki was so angry about his *sera* Ekamai, that he wounded him with a bush-knife. He had to remain standing there for four days without getting any sago.

"Then they rowed Ekamai and the slave to Nubuai and they took Ekamai into his house. There they packed hot ashes into leaves and made him grow hot. And his wounds were healed and also those of the slave. And when they were better, men of Waren came to make an offer for Anikakoi and after some bids they ransomed him".

One of the main motives for catching slaves was therefore the ransom, especially for the *sera*. And that the ransom was not low, in Papuan currence, is proved by a notice from the former missionary J. Metz Gzn. concerning Roon Island, where in those golden days of the hunt for birds of paradise the payment for each prisoner consisted of: "Forty pieces of blue cotton, sold at the store for five guilders, twenty plates, four kilograms of white beads, four silver armlets, and shell-armlets. All in all per person a value of hfl. 230.—".[161] It is easily understood that the *sera* prefers a living slave to a severed head.[162]

In the relation between the attackers and the attacked in these raids there exists an element of reciprocity, which also determines the limits to which raiding can go. The issue is not that a certain tribe tries to acquire the supremacy over the others, but it is a fight of one village

---

[161] J. Metz Gzn., "Op Roon begint de Victorie" (Victory starts on Roon Island), in *Tijdschrift voor Zendingswetenschap, Mededeelingen* (Journal of Missiology, Communications), LXXXIV, p. 62.

[162] The story of the battle between David and the giant Goliath, and the exasperation of king Saul because of the praise of the women for David, greatly appealed to the imagination of the Waropen, because they immediately transposed my story to the relation they knew, viz. that of *sera* and *waribo*. What a loss for the *sera* when his *waribo* simply cut off the head of that big giant, and that the *sera* was moreover compelled to praise and to reward his *waribo*, because the latter had been so *nikako*, brave!

against the other, and actually of one clan against the other. Clan A
has obliged clan B to make a payment, because clan A has caught a
ghomino. Conversely, clan B will also compel clan A to assume the
same kind of obligation. But, especially when the clans are regularly
in touch with each other, as is also the case for the villages in the
southern part of Japen Island and in the Kai district, the clans A and
B will also maintain other relations: there are trade connections, there
are marriage relationships, there exist *kamuki*-relations between the
clan-chiefs, there are common interests directed against a third village
C, etc. In this mutual intercourse certain restrictions will therefore have
to be observed.

The first restriction is that one cannot simply kill each other recklessly
on all occasions. Anciently already such explosions of pure blood-lust
were therefore probably kept in check and countered, especially by
the more important members of the clan, who realised that every
manslaughter would be followed by attempts on one's own clan.

Restraint had likewise to be observed in the choice of the *ghomino*.
One might have expected that the raiders would try to capture especially
the prominent members of the other party, their *sera*, because in that
case one might hope for a large ransom. But this was not done, particularly when attacking near-by villages — with which one was therefore in constant touch — because then one might count on serious
counter-measures. As a result, the *sera* were relatively safe and they
could also move practically unprotected without any great danger. If
reckless or ignorant companions brought in a *ghomino* of rank and
station, the latter was sent back to his village with presents (*kiwei
nitera*).

Even towards the *ghomino* the reciprocal nature of the obligations
between neighbouring villages was recognised, for there existed the
curious custom of offering presents to the *ghomino* who were ransomed.
Old Dumoi formulated this as follows: "The people said: Take these
goods along to cancel the fact that we have caught you" (*kikafa*:
"*Mimboiwa aini maisa dagha muo*").[163] Only when after the receipt of

---

[163] A similar custom is reported by H. Zahn concerning the Jabim, as quoted
    in R. Neuhauss, *Deutsch Neu-Guinea*, III, p. 317. Upon the completion of
    hostilities there the vanquished go to the victors, from whom they receive
    presents. The *laimuki*, the bravest of the victors, paints a bar across their
    forehead with lime (used for chewing areca), in order to prevent their exposal
    to the wilfulness of the gods, as it is said that the spirits loosen the teeth of
    the defeated and that they have an unfavourable influence on the increase
    of the dogs and pigs so that they die.

the ransom the slaves had been given a present of e.g. ten "pieces", the *ghomino* was finished with the raiding party which had captured him and both parties could again enter into relations with each other. This was not called a present, but a payment to the slaves (*kikapoiki*).[164]

This shows again that the raiding activities were not simply fights with aliens, but rather a ritual game in which the parties maintain certain contractual barter-relations. If the issue were solely the highest possible ransom, the restitution of the expiatory payment would be out of the question. The *ghomino* has fallen into the hands of the opponents; he has cost his fellow-clansmen as much as they believe him to be worth;[165] however, the payment which he has received makes him not return completely empty-handed; he has some "initial capital" and so he is able again to try his luck in the raids.[166] For, although no clear notion was observed among the Waropen that a person might fall into the power of others as a *ghomino*, due to guilt towards the ancestors, the *ghomino* was in the general esteem a person of lower status and of less value. A slave who was not ransomed was considered simply dead; no more enquiries were made for him. Shipwrecked persons likewise became *ghomino*. If the ransomed slave came back to his village, a mollusc was broken and he was given some of it to eat; this probably indicated that he had been taken up again into the community of the clan.

I presume that in the raiding activities this economic motive was tending to become increasingly important. Initially, hunting for a head was perhaps the final test the initiandus had to perform under the guidance of the adult men. After the presentation of a head to the ancestors of the clan the initiandus could be taken up again into society. The head is therefore also the sacrifice to the ancestors, by means of which the initiandus was freed from religious isolation. Perhaps the head originally replaced the killed initiandus. With the development of differences in

---

[164] The word *apora* which may be translated in general as "to pay" therefore possesses the additional meaning of "retribution". "expiation" (see p. 98). The concepts of "buying" and "paying" in this sphere are to be considered differently from our society.

[165] It also often occurs that he is considered to be worth less than the ransom demanded and so he simply remains a *ghomino*.

[166] Sometimes even a kind of friendly relation seems to grow up between the *ghomino* and the clan-chief who captured him. Once when I stopped over at Risei with a prahu from Ambai, one of the rowers, a clan-chief from Ambai, was very kindly received in one of the houses. He said that this was due to the fact that formerly he had once captured a slave from this house.

status and the accompanying disintegration of the ritual of the clan and of initiation, the chiefs, whose position rises above the clan-organisation, might try to increase their influence — which was gradually growing due to the system of the obligatory exchange of gifts on the occasion of the feasts — by means of the ransom for the *ghomino*.

Perhaps the regulated raids might have developed in the future into a battle of gifts, like this was described concerning the presentation of the dowry and of the countergifts among marriage-partners. Already at this time there existed a potlatch relationship between the clan-chiefs, at least when acquiring a name, when one party challenged the other with a stake consisting of an honorific title and presents on one side and a slave on the other. Real potlatch customs were likewise known among the Waropen. In this way people could challenge each other with their stocks of food, A challenging B that B will never be able to collect so much sago that A would be unable to buy this by means of silver armlets. If B does not want to become "ashamed", B has to accept the challenge. The winner acquires the goods that have been staked by the opposing party. Or A buys a large prahu and challenges B to buy an exactly similar prahu, at the risk of losing a certain quantity of goods (*kisionamakudaruko,* they taunt each other, they make sarcastic remarks about each other).[167]

STORIES ABOUT RAIDS

By way of illustration I include here some stories concerning raiding, told by old men who had participated in these raids. (Texts 187).

"I shall tell you about a raid we undertook. Suaiwini, Maseiori's wife, had died and then he called together the men of Mambui and those of Apeinawo and those of Pedei and those of Kai and those of Nuwuri, to go raiding. 'Collect palm-wine to-morrow', he said, 'in order to take out the prahus. Let us go with these, so that I may take off my mourning cap, you going in those (prahus) of ours'.

"We remained lying in Epagha Creek and the next day we rowed to the Wapoga river where we beat sago. And when we had beaten sago, we roasted part of it and we put away part of it unprepared. For the purpose of the raid we lined up at Foarama Creek and then we

---

[167] Two men often hold a diatribe against each other concerning the number of feathers each is allowed to wear in his *manggotio,* warrior's comb (*kisionamakuraba maniwuro*). In this way they therefore taunt each other about the number of *ghomino* they have captured.

established ourselves at Mambor. But when we had passed the night there without finding anything, we rowed to the Moor Islands. There Kawori said: 'Go and catch the people from Makimi'.

"Kawori was related by marriage to Maseiori, because his wife's younger sister was married to Kawori.

"Then we rowed away from there and fetched the other part of the sago we had stored near to the Wapoga. Thereupon we sailed again to Epagha Creek.

"The people of Makimi came sailing from Nau Island and then we saw their sails and we lined up against them in Epagha Creek. They came sailing up to this place in the evening and then we rowed out from there. Maseiori asked them: 'Who are you?'

"We are people from Makimi', he said, 'I am Embauri together with Biriki'.

"And when it became apparent that he was dealing with strangers (lit. 'when he had questioned them unsuccessfully'), he ordered the raiders: 'Come on, you prahus over there! Let us catch those people from Makimi! Cohabit with your wives! [168] Come on, you prahus over there! Let us catch them!'

"Maseiori shot his arrows at them and they shot at us. We shouted our war-cry at them and they fled without stopping. Again we shouted our war-cry at them. We caught one prahu and one other fled away. We had shot an arrow into the bodies of a boy and his father so that they died. Near the Binataboa river a pit was dug for the dead, in which they were placed.

"Two others we took along also in the prahu called Ikirowi. The men of Apeinawo took Berei and Moroghai in Ekamai's prahu from Erari, the prahu being called Eiwui. The men of Kai took along one person, and those of Nuwuri one, but those of Mambui did not.

"We spent the night near the Binataboa river. And the next day we rowed home singing. We fetched triumphal branches at the Nugari river and entered our water here rowing. We set up our triumphal branches and rowed to the other shore. The men of Nuwuri threw theirs (viz. their triumphal branches) and then we rowed singing to Sawaki, with our triumphal branches. Sorei came with two slaves, those of the *ruma* Maia with two, and the crew of the prahu Moabawa with one, i.e. only with Biriki.

"Whilst we were keeping them, their people came here to make a

---

[168] A term of gross abuse.

bid for them. Otobori — him they ransomed first, and the girl later.
Berei from Apeinawo was ransomed and still another from Maia, and
then still another from Kai, and those from Nuwuri, and the man from
Tanatirewo, Manggepamai, and then Biriki and Gagai, and finally·
Ambeinuri. All were ransomed.

"When those from Makimi came here, the people were afraid and
remained inside the village, whilst those from Makimi stayed at sea,
downstream. We went and sat in our prahus and rowed down with
them and then those from Makimi mentioned the name of some person.
Then we sold him. And then they made a bid on another. All those
with whom we came sailing along, they ransomed. And when they had
ransomed all of them, they rowed away".

An attack on a whole village (Texts 187):

"Tabefi and Rarumi were living at Ambai, but they had married
women from over here. They came here and they called us out, whereupon we took out our prahus and rowed after Tabefi and Rarumi.
They rowed out ahead and they lined up near Uwi Island, where
they lit huge fires. And when we had seen these, we went and lay
alongside. Our bows, bush-knives and lances of iron and bamboo we had
taken along, and also our medicine. We went to tap quantities of palm-
wine and we drank palm-wine.

"Maseiori said: 'You just stay and wait; then we shall go alone with
our raiding party' (viz. the party from Maseiori's clan).

"Maierisagha commanded Aiafari to take his prahu outside. And
when father took the auguries, the result was that we were to have
slaves (lit.: 'he augured our slaves'). And when he had obtained this
outcome, he said: 'We may perhaps catch our slaves, if we go'.

"Thereupon we rowed out and lined up near Uwi Island, where
we consulted the ancestors. And also the swaying of the boat indicated
that we were to obtain slaves.

"We remained lying close-by, and when we came rowing in, in the
morning, the houses became visible. And in the morning, after having
slept, we appeared and surrounded them in their houses. The rain
kept them back in their houses, so that we could catch them. Our
ancestors retained them in their houses by means of the rain. We
blockaded their houses. Because of the rain their prahus were water-
logged, so that they could not sit in them. Then they shouted: 'Hey!
A raiding party down there!'

"But only one single person whose prahu was dry, took to flight, but

the others, whose prahus were full of water, we all caught. We climbed
into the houses and took away all their goods: pieces of cloth, bands of
shell, old beads, nets and bows. The prahus were hewn in pieces and
their houses knocked asunder. Singing, we rowed straight here. We
rowed up-river and thereupon we fetched triumphal branches. We
took leaves and adorned ourselves. We took lime and painted ourselves
and put on the raiding-paint around the nose. We raised our triumphal
branches and thereupon we danced in the prahu".

A tale about a raid which ended badly (Texts 187):
"The men of Makimi sent a summons and they came here, at Cape
Usa. The next day we prepared our prahus for the voyage and rowed
towards them, outside. We rowed in the redwood-prahu of the house
Tanatirewo. We had manned two prahus. We rowed to the back-
country and passed the night near Cape Usaba.

"Then we took the auguries concerning slaves by heating the *sase*-leaf
and the result was that we were to obtain slaves, but that we also were
to incur wounds. Near the people of Sasora village we entered, rowing
into the river, and during the night we rowed upstream, until we saw
their houses.

"And the next day we immediately surrounded them in their houses.
We brought up alongside with our prahus and climbed into their
houses, but they put up a strong resistance inside their houses and shot
arrows at us. The men from Makimi shot at them with their rifles. When
they jumped (out of the houses), the men from Makimi caught one.
But because they shot at us with arrows and because we became afraid,
we climbed into our prahus and fled towards the sea.

"But the (captured) boy jumped into the water and ran away. Then
the people from Ghoisano climbed into their prahus and raced for us,
whereupon a man from Makimi jumped into the water and ran ashore.
The men from Ghoisano pursued him and killed Kodioi".

TRADING VOYAGES

On the one hand the trading voyages promoted raiding. This was
because the fully loaded trading prahus on their long and lonesome
course practically invited attack, or also because there existed no single
reason to prevent an unsuccessful trading expedition from becoming
a successful raid if the circumstances were favourable. I am inclined

to find the connection between raiding and trading there where the raid, which had developed from the head-hunting expedition of the initiation ritual, changes over into the less sacred catching of slaves, an undertaking in which the growing class of the *sera* is especially interested. When finally the profit-motive has become predominant, it is hard to determine whether the clan-chief intends to trade or to hunt slaves.[169]

In the old days, arms were always worn, even when one was simply sailing through one's own village. On the other hand, commercial intercourse placed raiding under certain natural restrictions, because the inhabitants of the different villages still stood constantly in need of each other. According to my informants therefore a certain truce was maintained in the native markets held on Japen Island, and likewise in those of Serui which are still held every other Saturday.

Trading extended over the whole of Geelvink Bay, the inhabitants of the Kai district being of course more interested in the southern part of Japen Island, and more especially in the south-eastern part, whilst those of Napan also looked towards the Wandamen, Kai, Numfoor and Roon. In principle, the organisation of a trading voyage does not differ from that of a raid, the raid having only a wider scope and being explicitly aimed at capturing *ghomino*, whilst a trading expedition with its more peaceful aims was sometimes also accompanied by women and children. Trading is therefore usually as much the concern of the *da* as raiding.

When undertaking trading voyages, the *sera* — and other "great men" likewise — were again in a far more favourable position, because they

---

[169] The inhabitants of the Torres Islands made "hunting or so-called trading trips" to those natives with whom they exchanged garden-produce, according to W. N. Beaver, *Unexplored New Guinea*, p. 102. J. van Baal reports of the Marind-Anim in his *Godsdienst en samenleving in Nederlandsch-Zuid-Nieuw-Guinea* (Religion and society in Netherlands Southern New Guinea), p. 192: "The relations between the tribes were far from being exclusively inimical; sometimes there was trade between them, and they were even taken as allies in war. In fact, there existed a regular ceremonial for concluding peace. However, we are insufficiently informed concerning this side of the head-hunting expeditions to be able to draw any conclusions. It is only certain, that these head-hunting expeditions were simultaneously trading journeys as well as marauding expeditions, because in preference to buying the desired goods these were robbed in the conquered villages". — The data I have collected do not prove convincingly that trade might be mentioned in connection with raiding and war. However, the same arrangement has been chosen by W. H. R. Rivers in his paper "Trade, Warfare and Slavery", in *Psychology and Ethnology*, p. 291.

possessed sufficient riches enabling them to buy one or more large prahus.[170]

Although other people were not prohibited from possessing prahus, they could not easily collect the purchase-price. They might perhaps carve a prahu themselves, but for this purpose still some co-operation is necessary. And inevitably a certain amount of capital has to be invested in a prahu, because there have to be people who are willing to provide the daily necessities of the prahu-builders during the rather long period which is needed for building a large prahu. And because especially during feasts an accumulation of riches took place, the *sera* who was able to give many feasts had the greatest chances to have funds at his disposal for buying prahus. The *sera* acquired fresh profit by entrusting the management of large prahus to his meritorious *mano-bawa*, prominent men, who were obliged to render him certain counter-services. For this reason it is not inexplicable that a Waropen *serabawa* established a connection between giving feasts and undertaking trading voyages in large prahus.

However, also in this case the *sera* cannot assert personal property-rights on the large prahus, as the members of his family-branch also have a say in the matter. When a prahu falls to pieces, each one of the fellow-owners may have a piece of it to use as a seat in his private prahu. In case the *sera* were to make arrangements for the large prahus in too independent a fashion, his brothers would be quick to resent this. That is the reason why at present in the Waropen area it is occasionally difficult to hire a prahu, because so many people have to be consulted.[171]

Persons intending to make a trading voyage may hire a prahu. For a long voyage the rent is about three "barang". The hirer is responsible for all damage incurred during the voyage, whilst he has to find oars, sails, etc. himself. Just as the *sera* is responsible for victualling his raiding party, the person who undertakes a trading voyage has to take care of the victualling of the crew he invites. The crew is not paid, but they are allowed to bring a limited quantity of trade goods on board at their own risk. The organiser of the voyage, besides transporting goods at his own risk, may also take goods for friends and relatives, against a certain commission. This type of giving goods in commission is called

---

[170] At Weinami the price for a good-sized prahu was given as 60 to 100 "barang", i.e. as much as a good dowry.

[171] The traveller who wants to hire a prahu therefore has to pay not only the rowers, but also the prahu, for which one pays the same price as for a rower. In our days this was twenty-five Dutch cents a head for a day's rowing.

ana aigha we naio, to despatch goods as being deposited.[172] To trade on one's own account is called *we nuai.*

Delivery often takes place on credit, sometimes for longer periods, and occasionally also outside the village. On these occasions one often uses the *keri,* the counting sticks, of which both the buyer and the seller retain a number which is equal to the amount due. No interest is taken, even in those cases where a simple loan is concerned. However, according to an Ambumi informant the price would increase, if the seller had to wait a long time for payment. When somebody denies a loan-debt, an occurrence which my informants stated to be rather frequent, the only means usually left to the lender is brute force, unless he be able to compell his debtor to accept an ordeal, viz. by hot water. For that matter, making debts is also considered as a moral evil, which the *inggoro* may visit upon the whole clan. When a member of the clan is publicly pressed for payment by his creditor, the clan-chief considers this as being also a reflection on his good name, with the result that often the whole clan will assist in settling the claim.[173]

On the other hand, however, dunning a person for debts is a serious violation of good morals. In such a case the dun is accused of greed (*nihaiha,* N.), which is taken as a serious insult by the latter.

In the Kai district, goods were obtained from various regions: from the tribes of the interior (Kaipuri, Sorawi) birds of paradise, horti-cultural produce and — especially from Sorawi — decorated mats (*egharo*) and rain-hoods (*saiwu*); from Kurudu forest-areca and dried fish; from Japen (Ambai Islands, Serui and Ansus) native pottery, cases of pandanus-leaf, and store-goods; from the Schouten and Padaido Islands smith's work (arrow-heads, spear-points, bush-knives, harpoons); from the Moor Islands coconuts, prahus and drums. The main exports from Kai were sago and basketry-work. At Weinami many prahus and

---

[172] The same expression was also used for goods which relatives, who had remained in the village, sent to members of their family who had gone to work on an estate. For every consignment there was an indication of the counter-present which was expected, the latter nearly always being things the value of which was completely unrelated to the small value of the goods which were being sent "as being deposited".

[173] Making debts is not a purely private matter, as is proved by a Numfoor custom, according to which the creditor goes and takes a quantity of goods away from a third party, with the latter's knowledge; in this way the debt becomes publicly known. See J. L. van Hasselt, *Gedenkboek van een vijf-en-twintigjarig zendelingsleven op Nieuw-Guinea* (Memories of 25 years as a missionary on New Guinea), p. 54, and W. K. H. Feuilletau de Bruyn, *Schouten- en Padaido-eilanden,* p. 60.

drums were sold to Wandamen Bay against ancient pottery, whilst from Japen also pottery and store-goods were obtained.

The trade-route from Weinami to Japen ran along the East coast of Geelvink Bay via Nau Island to the Ambai Islands or Serui. The first day's voyage usually was from Weinami to the Binataboa river, the second day to Waren, where people stayed for a few days, crossing finally to Japen, where the whole South coast was visited. There was no fixed time for travelling, because in Geelvink Bay the seasons do not show great changes. However, according to the old men the beginning of the East monsoon, approximately in April, was considered to be the most favourable time for travelling from Kai to the Moor Islands. And in fact in the months April, May and June a good number of prahus were seen, sailing towards these regions for trading purposes.

———

CHAPTER SIX

THE SACRED AND THE PROFANE

THE CONCEPT OF SACREDNESS

The opposition between the ideas sacred and profane as developed by the French sociologist Emile Durkheim, an opposition I have attempted to apply in order to understand Waropen religion, has brought us in contact up to now with rites which concern the group rather than the individual. The ritual of the *saira* and of the *munaba* clearly belongs to the "système solidaire de croyances et de pratiques relatives à des choses sacrées, c'est à dire séparées, interdites, croyances et pratiques qui unissent en une même communauté morale, appelée Eglise, tous ceux qui y adhèrent".[174] Even if occasionally in the *saira* hardly anybody but the initiandus and his mother's brother are concerned, both parties still represent the group and even the whole of the community, because in a mythical sense brother and sister occupy the central place in the mythical view of the world.

The *saira* we have described as a positive ritual (as the Waropen called it: intended for life), connected with the clan. However, unlike the old-fashioned initiation ritual, the *saira* is not in the first place the admission of the uninitiated child into the world of religion. The negative element of seclusion, although clearly present, is no longer as prominent as the positive attempt to make the initiandus continually more sacred by means of a prolonged series of rites. The successive lifting of different taboos on food is likewise rather intended in a positive sense of increasing the sacred powers of the initiandus, than in the negative sense of fear of the contact with sacred things.

The view still prevails that in any case the final rite of the *saira*, the perforation of the nasal septum, is compulsory for every individual, and this is the reason that this rite is even performed in effigy on dead children. However, besides this obligation to undergo initiation for all members of the clan, we have noticed a very evident tendency to

---

[174] Emile Durkheim, *Les formes élémentaires de la vie religieuse*, p. 65.

reserve the *saira* for the descendants of the senior branches, i.e. for the social group of the *sera*. In this connection I spoke about the disintegration of the ritual which tends increasingly to become dissociated from the clan and to bring about the heightened sacralisation of the higher class.

Also, the *munaba*-ritual, although clearly intended for death, is rather more positive than negative in nature. In this case the element of gaiety and joy of living is of course not so prominent, but for the rest the *munaba*-ritual is also not completely dominated by taboos or by the fear of sacred matters. In this ritual the difference in status is likewise beginning to work. The fact that for old *serabawa* a separate ritual is performed, a ritual which is considered as exceptionally sacred, is not only to be explained by the stronger emotions caused by the decease of prominent warriors, but above all by the circumstance that the dead man belongs to a higher class. And although this ritual for the *serabawa* is accompanied by a taboo which in Waropen eyes is very strong and which is imposed on the whole of the village, it is again also quite positive. One of the elements of the Serakokoi ritual is again a *saira*-episode. In the next chapter I shall explain in more detail that the initiation of women is perhaps connected with this.

It is likewise curious to observe that the myths are more clearly connected with the *munaba* than with the *saira*. The most poetical, i.e. the most positively sacred versions of the myths, are transmitted in the form of *muna*, funerary songs. In other words, it is especially at the *munaba* that the myths are recited.

The songs of the *saira*-ritual also occasionally treat mythical themes, but usually not in the ordered version of the *muna*. The men's song during the *saira* is rather more in the nature of a religious song used to add lustre to the ceremony, whilst the *muna* in and by itself is the text of the religious service and not an accompaniment [175]; the song of the dead is itself the *owa*, the dance.

The very fact that the myths are most closely related with the ritual for death proves that no direct influence on nature is expected of the myths. The myths are not danced in a ritual for life, in order to support and to develop nature by means of this dance. They are sung during the funeral ceremonies, when the community has suffered a loss which it attempts to remove by a more intimate contact with the world of

---

[175] On p. 161, note 122, I proposed to find the reason for the difference between *saira* and *munaba*, in the origin of the *saira* from the *dama*, the young men's house.

the powerful ancestors. The funerary ritual is partly intended directly
to increase the prestige of the family-branch concerned.

For this reason little attention is paid to a dramatization of the
myths. Their language is so obscure that the dancers themselves cannot
have a clear idea of the deeds of the ancestors as sung in the myths.
To a considerable extent the *munaba* becomes a feast to glorify a certain
family-branch, but its sacred nature has by no means become lost; in
the Serakokoi ritual it is even quite pronounced. However, there exists
a certain tendency towards shallowness, a tendency which pervades the
whole of Waropen religion. This shallowness is the result of the disintegration of the ritual.

The opposition between sacred and profane shows only little tension
in Waropen religion. In general, the ritual of both the *saira* and the
munaba is positive, but the Waropen are far from being either priests
or prophets. Their religion is very far removed from e.g. Vedic religion
where man, by means of his ritual, commands the gods and where it
is even said of the sacrifice that it creates the gods.[176]

In the land of the Waropen the human spirit never fathomed the
depths of mysticism, nor climbed the heights of religious contemplation.
Waropen thinking never ventured the brilliant leap into the other
world. They are ensconced in the quiet, every-day life of the village,
safe behind the walls of a religion with few onerous taboos or severe
rules of avoidance.

The Waropen is as little devout in ritual as in daily life. The intensity
of the emotions shown is a bad guide for determining the border between
sacred and profane; occasionally ceremonies are performed without any
outward show of solemnity and as it were in passing. A Waropen may
make astonishingly cynical remarks. One of my best informants, and
a man who for the rest did not take the obligations of the pagan
religion lightly, made merry of the pretended miraculous powers of the
ancestors who were said to have descended from heaven to open up
the creeks in the tidal forest single-handed.

This does not mean to say that in Waropen religion the sacred world
does not impose laws on its subjects. Once ritual has established contact
with the sacred world, this cannot be arbitrarily interrupted. A rite
imposes obligations and it has to be fulfilled up to the end. When a
village abandons the pagan religion, a number of concluding ceremonies

---

[176] Hermann Oldenberg, *Die Religion des Veda*, p. 313 ff.; Emile Durkheim,
Les formes élémentaires de la vie religieuse, p. 51.

are often still held, whilst various feasts are celebrated in order to put an end to the past. Only then one may safely proceed to the new.

The concept of sacredness is expressed in Waropen by the word *mi*. *Mi* is in the first place the story of the olden days (*tina garo*), i.e. the myth, mainly the myth in its poetic, sung or danced version. The myth is also called *owa*, dance, because the sung myth is danced at the religious festival. *Mi* are likewise the figures who appear in the myth; in this way one speaks of a *mibino*, a woman from the mythical world.

But, what is holy always possesses a double aspect; it is not only an all-pervading and sanctifying power, but it is also a danger which is delimited by means of taboo-regulations. Durkheim speaks of *le sacré faste* and *le sacré néfaste*. This same double aspect the word *mi* posses-ses likewise, i.e. in the word *amina* which refers to the awe in which sacred things are held, and to the distance which man has to maintain. *Amina* means "to be taboo as regards certain things".[177]

When a person wants to protest his innocence or to swear to the truth of something by some object, e.g. a lamp or a rifle or a bamboo, etc., he says: "*Aminara we padamara ineni*", "taboo for me by this lamp". When after a violent quarrel one wants to break off all intercourse with another person, one swears: "*Aminagha afa asitiraruko wewomo*", "taboo, we shall not know each other again". After this oath the two people will avoid each other, unless the conjuration is broken by rubbing each other with a certain leaf.

Furthermore, all sacred things are dangerous in principle. Therefore a person's soul, *roséa*, is *amina*. The same holds true for those places in the forest where dangerous powers are to be feared, powers who strike the innocent visitor with evil sickness (*ghoghoido*). A person's totem-animal (*ghori*) is likewise *amina* for him; he is not allowed to eat it.

In modern times, and especially in Christian circles, the word *popono*, pure, smooth, clean, is used as a synonym of *mi*; it is the direct translation of the Malay word "suchi" in the sense of "holy". It also denotes the two aspects of what is holy. At Napan one says: "*Poponibea*

---

[177] The element *a* in the word *amina* is not quite clear. *A* might be the personal prefix for the exclusive first person plural. If this were true, forms with the other personal prefixes (*kimina, mimina*, etc.) could be expected, but according to my informants these forms are impossible. However, at Napan I noted down *yaminaha*, I am taboo for this. A second possibility is that *a* is the formative prefix treated in my *Grammatica van het Waropensch* (Waropen Grammar), § 11.

nengga reraha auo", "what is holy will eventually expose you to super-natural influences".

A third term, but with the stress on the dangerous aspect of what is holy, is the Napan word *hai*. In this way the above mentioned *ghoghoido*, the dangerous places in the forest, are called in Napan *nahuhaiwea*, places where it is *hai*. When somebody is about to do something danger-ous, he is told: "*Hai maha auo*", "that is *hai* for you". Of mourners who after the decease of a relative have to dwell in seclusion, i.e. who are in a condition of taboo, it is said that they *haido,* dwell in the *hai.*[178] Because of their contact with the dangerous aspect of the sacred world, the mourners themselves have become "contageous", as Durkheim calls it. They therefore have to be segregated.

A curious train of thought is shown by the use of the word *rera* (*reraha,* N.), which again refers to the dangerous side of what is sacred. In normal usage, *reraka* means "to take apart", "to comb", e.g. *reraka worai,* to comb the hair. In combinations the word is often used with the sense of "apart", "asunder", "away", e.g. *sorera,* to scatter, from *so,* to throw. But in the connection meant here, *rera* may be translated as "exposed to supernatural influences". For instance, somebody has no objections against telling a myth in the less sacred prose-version, but he does object against reciting the poetic version. "*Ronausari rerara. Mangga reraikangga ipero*", "if I were to tell this I would expose myself to supernatural influences. Then we would even die because of this". Morals are also built on a sacred base: "*Auofaro reraha auo*", "if you cheat, you expose yourself to supernatural influences", one says to sinners.

The logical opposite of the loss of sacred powers due to taking apart, is the concentration of power by binding together. This idea lies at the bottom of the "binding up" of affected limbs by wrapping these in tightly drawn rattan. When a person feels ill, he winds a strip around his body. Especially for little children with their small powers wrapping the belly is believed to be highly favourable.[179]

However, binding may also be used to activate the dangerous powers of what is sacred; practically the whole technique of black magic is

---

[178] In the language of the Waropen this is called *seado.* The same element *sea* is to be found, I believe, in the word *roséa,* soul, where *ro* means "interior". In the Kai language *sea* means "enclosed", "blocked off". The variation in meaning from "enclosed" to "holy" occurs quite often. It is not certain whether in Kai *sea* also already means "holy", and neither, implicitly, whether *sea* like *hai* has to be taken *in malam partem.*

[179] Could this perhaps be connected with the custom that mourners tie bands around their arms, shoulders, head, and hips?

based on this idea. The blocking and closing of openings is equally dangerous.

In Waropen religion there are also numerous acts which do not embrace the whole group, but which are the concern of only a few individuals. In all these cases the starting-point is provided by the generally accepted and recognised religious ideas and in so far these ideas have a social foundation, like the ideas which lie at the base of the "total" ceremonies. These acts are the many individual rites which are more or less secret. But once one has been initiated into these secrets, one addresses in a perfectly legitimate way the same sacred world as that with which the rites of the group are concerned. A number of these individual rites differ only because of their a-social character, although they belong to the sacred domain when judged according to their form and execution taken by themselves.

In concluding this section I want to remark that the Waropen language hardly knows any linguistic taboos. The myths have a completely separate language of their own, but in everyday speech the sacred prohibitions have no influence. One example of the sparing use of linguistic taboos is the avoidance of the word *moiwa,* iguana, by the person who wishes to catch this animal. When one goes collecting *roio,* a kind of mollusc, one has to say that one hopes to find *ado,* fish. A person collecting snails, *ka,* speaks about *fiku.* This is called *ona weritio,* to talk obliquely or metaphorically. Also in other cases figurative expressions are used; when going on a raid, one says: "*Bo sa tira diaroi*", "We are rowing upstream to look at our sago-forests". But the use of such figurative ways of speech is not invested with any influence on the result of the undertaking.

AIWO

As in many other religions, also in the religion of the Waropen a large number of practices are found which aim at the use of impersonal powers. Every individual may use a certain favourable combination of objects (leaves, stones, bones, etc), by means of which he obtains a certain power. Their use differs from the public rites, in that one does not need to work in co-operation with others. Everybody who desires to do so, may buy or acquire the desired knowledge and so exercise a certain power.

This use of impersonal forces is often called magic, but one objection against this appellation is that in that case many things in religion should also be called magic, because religion is far from always dealing with personal powers. The application of the term magic becomes quite dangerous, when consequently the magical practices are grouped separately and are not evaluated at the same level as the evidently religious usages. In this way, judgements of value are applied in the field of sacred things in a way which is not confirmed by the facts. In religion personal and impersonal forces are not sharply differentiated entities.

In this way the *mamano,* a pungent herb, is used by young girls in order to win the hearts of men, by mothers in order to alleviate the stomach-ache of their children, by the person who concludes a marriage for rubbing the sleeping-mat of the young couple. If this last mentioned case is also called magic, then nearly every ritual has to be called magical, because in the end cutting the hair, putting on the leg-rings, perforating the nose, etc., are all intended to increase the sacred powers of man.

The means by which the impersonal powers are brought into action consist of an *aiwo,* usually a decoction of certain leaves, or a collection of objects to which power is attributed, e.g. the tooth of a deceased ancestor, the closing piece of a shell, a white porcelain-shell, a.o. The *aiwo* always has something supernatural about it, however little of this sacred character may be apparent. An *aiwo* against sickness e.g. means something different to a Waropen from what medicine means to a Westerner, if only for the reason that to the Waropen illness means something different. There also exist several activities where people fully rely on their own powers and believe an *aiwo* to be unnecessary. So I was assured that an *aiwo* is not absolutely necessary when making a prahu or building a house. On the contrary, nobody will succeed in making a good glass ear-pendant (*dimbo*) without a suitable *aiwo.* Composing or understanding the sung versions of the myths is impossible if one has not taken the appropriate *aiwo.* And this, I was told, was the reason why people were unable to get me a translation of these poems.

Furthermore, there are numerous cases when one may use an *aiwo* in order to obtain powers which are greater than usual. When faced with a calm at sea, one may have recourse to a *ghamaiwo,* a wind-medicine, two match-sized sticks tied together in the form of a cross, which one lets jump on the waves behind the prahu. In order to promote the growth of the beans, the garden is sprayed with a decoction of *kawaruirana,* bean leaves. In order to shoot well, the bow is smeared with an *aiwo,* or small beads and snippets of cloth are hung on it, by way of an *aiwo.* Starting from the general principle that one obtains power by locking in or by tying together, people often put an *aiwo* in a closed bottle. In this way one often finds, a corked bottle with a *ghamaiwo* for favourable wind in the travelling-prahus. When a man wants to charm a young woman, he encloses a certain *binaiwo,* lit. woman-potion, in a bottle, whilst he pronounces the name of his beloved.

Aiwo are also frequently used to protect the right of ownership to coconut-trees and areca-palms. These *aiwo* are hidden in the ground near the tree and they act on everybody who comes too close; in order to prevent unsuspecting passers-by from being struck by the power of the *aiwo,* a bamboo is placed or tied against the tree by way of warning. Such a warning is called *dora,* lit. defence, *aheta,* N.

An *aiwo* may also be what we would ordinarily call medicine.[180] The trees *kagheri* and *siriwino* were often mentioned as producing good medicine. In one of the tales it is said that somebody stole from the snakes an *aiwo* to cure a fever (Texts 167). The *sema* (suangi) likewise use leaves (*semarana*) in their anti-social activities, bij means of which the people they have killed are temporarily called back to life.

It is rather questionable whether any medicinal properties may be ascribed to the leaves which the Waropen use. Against coughing, e.g. a mixture of lime and lemon-juice was recommended, to be applied to the hollow at the base of the throat and to the pit of the stomach. For that matter, an *aiwo* is not intended as a medicine against specific diseases, but as a remedy with general favourable effects. It is therefore not necessary to be ill in order to use an *aiwo,* and so the Waropen did not understand at all that we only wanted to give our medicine to people who were ill, and even not to all sick persons indiscriminately. They were very fond of eucalyptus-oil, which they hung around their neck in a small bottle, or used to prepare love-potions, or to protect small children against illness.

In order to make rain, the Kai people went to places where fresh water was found (*ropeio*). There the leaves of the *bora*-tree [181] are beaten against the tree-trunk, whereupon some earth from this place

---

[180] A modern term for *aiwo* is *oba,* from Malay obat, medicine.

[181] A tattooing-pattern is believed to be shown by this tree, as mentioned on p. 28; the leaves are used as wrappings for hot ashes, sometimes placed on painful parts of the body, like in the story told on p. 223.

was packed in *komburana*, a kind of leaf, and taken to the village. At Weinami rain was made by beating specific leaves against a stone.

It is not only the things to be used that are important when applying a remedy, but also the right way of using it. Once upon a time some young people, who had the task of fetching fresh water, wanted to alleviate their duties by making rain and catching the rain-water. However, in their unexperienced zeal they had taken so much earth, that the rain could not be stopped, to the great displeasure of those people who had just planted beans in their gardens. When handing out medicine we gave the people some water to drink. This water was also expected to possess a special effect, although they knew quite well that it had simply been taken from the rain-barrel; sick people carefully took the water home in a small bottle.

Magic formulas are little used in general with *aiwo*, although there are some combinations of words which possess a dangerous effect. If one says to a fisherman: "*fo saro anggoro*", "pig cassowary-bird crocodile", he will not catch anything. Sneezing is generally dangerous, and if a person sneezes and one says *tupatirere*, this has an unfavourable influence on the sneeze and no fish will be caught that day. In the stories there is mention of formulas which enable people to open and to close caves, in the style of "open, Sesame". The Waropen word for pronouncing a magic formula is *afasa*.

Specialists in the application of *aiwo* are unknown among the Waropen; certain persons have at most the name of possessing a great many or especially powerful *aiwo* and for this reason they are somewhat respected. When an *aiwo* does not produce the desired effect, one assumes that a mistake has been made, or that an obstruction has been created. In this way the strongest *aiwo* for the plentiful catching of fish will fail, if in the village an incestuous relationship has been established, because in that case the ancestors refuse to provide fish.

It is curious that some *aiwo* do not act on impersonal sacred powers, but represent a personal supernatural power.[182] For *aiwo* are also the innumerable roughly carved ancestral images which the men hang around their neck, especially when they go raiding; it is mostly a short

---

[182] The same is reported by F. J. F. van Hasselt concerning the Numfoor *or*: "The *or*, as explained to me, in all respects greatly resembled a protecting spirit. On the one hand the *or* is conceived as impersonal, but it may be as it were materialised. Attached to spears, bows and other weapons a tiny parcel is to be found, containing small leaves, pieces of bark, or root. And this is supposed to be *or*", in *Tijdschrift voor Zendingswetenschap, Mededeelingen* (Journal of Missiology, Communications), LXXIV, p. 236.

stick, of which only one end has been carved like an ancestral image. In many cases it is even nothing more than a simple chip decorated with a small bead. In my opinion these small sticks, with or without an ancestral image, represent nothing else but the central pillar of the young men's house or the central pole of the dwelling, on which the skulls of the vanquished are hung.[183] This pole is the ancestor, now considered as a person, now felt as an unseen power, as we shall see in the next chapter. At Napan these ancestral images worn around the neck were called *hesoaiwo*.

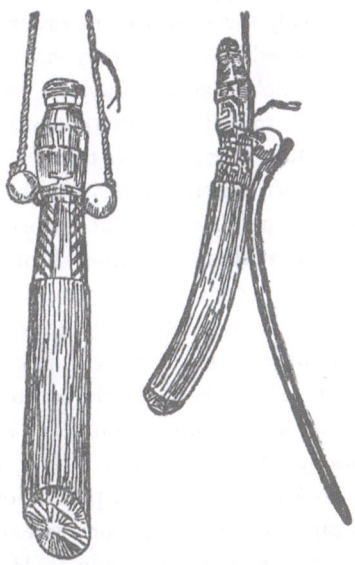

Aiwo, amulets worn round the neck on a string.

The relation with the ancestors is often so intimate that these *aiwo* are given the name of an ancestor. In this way the objects belonging to the Pedei clan at Nubuai were called Rerai, and those of the Sawaki clan Donggori. When fishing, an *aiwo* was used which seemed to represent a fish-like creature.[184] Still, the fisherman does not only

---

[183] Perhaps all ancestral images in Geelvink Bay are originally meant as carved pillars; this explains the curiously elongated form of the images. If the maker had been able to design the image freely from a piece of wood, different shapes might have been the result.

[184] See the illustration in F. S. A. de Clercq and J. D. E. Schmeltz, *Ethnographische beschrijving van de West- en Noordkust van Nederlandsch Nieuw-Guinea* (Ethnographical description of the West coast and North coast of Netherlands New Guinea), plate **XXXVIII**, fig. 14.

rely on his *aiwo*, he also invokes the ancestors to help him: "*Inggoiaee, ripangga rafo adogha eno*", "Ancestors, let me pull in some fish when I throw the net". Nevertheless, the *aiwo* carved like ancestral images are real "medicine", means to influence impersonal powers. In this way an old man who wore an *aiwo* carved like an ancestral image around his neck used this object as medicine against tooth-ache, by continually biting on it.

OATHS. DIVINATION

A person may also consciously expose himself to the dangerous influence of the sacred world, for instance, in order to demonstrate that one is not guilty of some misdemeanour; one may show one's innocence by breaking a bamboo on one's head and saying: "*Aminara we anasani*", "By this bamboo I expose myself to supernatural influences". This is called *awoura,* lit. to mention preventively, to curse, to conjure.

The heaviest oath is sworn when a person wishes to clear himself of the suspicion of being a *sema* (suangi). The suspect has to take a stone from a pot of hot water without burning himself, and he swears: "*Ramumbe semangga rasara urani —, rasarangga kaparao*", "If I have committed 'suangi'-murders, may I burn myself when putting my hand into this pot". For the suspect this is a very risky procedure and the unfortunate person will only take this desperate step under strong pressure of his family. The last victim remembered at Kai actually burnt his arm and was consequently considered guilty, to be attacked and murdered later when working in the sago-plantation.

Now it is curious to observe that both when swearing oaths and when taking the auguries, an appeal is often made to Heaven and Earth.[185]

In Geelvink Bay they function as supreme beings, especially Heaven. Therefore, also when swearing and auguring one appeals to impersonal powers, but this impersonal power has the tendency to become personal. The question arises, whether swearing and auguring are not partly meant as an *aiwo*, by means of which one obtains this power, i.e. as if these acts would not only enable a person to know what is favourable or true under the circumstances, but also that swearing and auguring would exert a favourable influence on these very influences.

Taking auguries (*we sora*) is done in different ways. At Kai the simplest method was augury by means of an areca-nut. The nut is split, one blows on it and one says what one wants to know. Then the specialist finds the augury in the lines of the split nut. E.g. one wishes

---

[185] W. K. H. Feuilletau de Bruyn, *Schouten- en Padaido-eilanden,* p. 67.

to know whether a prahu will soon come; a dot in the design of the cut surface represents the prahu, another the village, and from the distance which separates the two dots the distance which separates the prahu from the village is calculated.

The result of a raid is usually divined in another fashion, viz. by heating a previously moistened strip of rolled *sase*-leaf. The specialist reads the result from the way in which the heated leaf curls. In doing so one says: "*Utora sora roragha ado*", "Divining. Divining. Heaven. Depth"; this is called *utora*, from *una,* to fry, and *sora,* to cast the auguries.

Another method of divination which is also used on the occasion of a raid is the rocking of the prahu, a means which is likewise used in order to establish the cause of death (cf. p. 173). In this case one does not appeal to the supernatural powers in general, but to the deceased ancestors personally.

For again another method of divination one takes a bamboo, weighted with *korombowi*, white porcelain-shells, and this is thrown into the water. The specialist reads the result from the direction in which the bamboo drifts, or from the way in which it sinks or floats.[186] This method is applied on important occasions, e.g. for raids, when looking for a suitable place to build a village, etc.; this is called *adumanggusara.* The same term is used when a person has unintentionally committed a very serious offence and then throws himself into the sea as a willing prey for the animals of the deep. In this way the brother who has slept by mistake on the thigh of his sister says: "*Radumanggusara, rara ndau, rasunatau. Sokabuigha kimunarao, kaigha kimunarao*", "I throw myself into the sea. I go to the sea and jump into the sea. Let the sharks and swordfish devour me" (Texts 85).

Still another method of divination is used by the *ghasaiwin* in order to foretell the course of a disease, for in a plateful of water she is able to see whether the patient will die. The auguries are also taken by counting the number of spans needed to measure an arm, and at Weinami (when establishing the cause of death) by the trembling of the hands. One may also put the hair of the deceased person in a carrying-bag and then blow smoke against it, keeping the bag hanging free from two fingers; if the bag starts swinging this is considered as an affirmative answer to the question, e.g. whether one will catch fish.

---

[186] See also J. L. van Hasselt, *Gedenkboek van een vijf-en-twintigjarig zendelings- leven op Nieuw-Guinea* (Memories of 25 years as a missionary on New Guinea), p. 56.

GHASAIWIN

Not everybody is daring enough to venture forth into the realm of sacred things. Among the Waropen there exists a special group of persons who may enter into the supernatural field with impunity, and these are the *ghasaiwin*.

Ghasaiwin are old women who, after a specialist training, are able without danger to enter into contact with sacred matters and who make it their occupation to mediate in this contact on behalf of others. Usually these *ghasaiwin* are women who already have acquired a certain authority due to their rich experience; often they are well acquainted with the mythology and so they frequently appear to "weep the funeral song", although this is not a task reserved for them *ex officio*. Every clan has a *ghasaiwin* of its own, who mainly also acts only on behalf of members of her own clan, and who when she appears in any case, addresses herself to a certain series of ancestors. In many cases the *ghasaiwin* even maintains relations only with certain local representatives of the mythical beings. A person, therefore, who calls in the help of a *ghasaiwin*, e.g. in case of sickness, can only cover a small part of the sacred realm. The effect of an *aiwo* is much more general and it may be obtained without difficulty, but the effect is usually weaker.

The person best fitted for *ghasaiwin* one believes to be an old *munaba*, an old woman of high status, who also is *binapa*, i.e. whose oldest child has died. One *ghasaiwin* transmits her knowledge to the other, but nobody could or would tell me what was the actual training. It is certain, in any case, that the disciple has to dwell in isolation for some time; according to my informants she has to refrain from certain things during this period, i.a. from eating red and black kinds of fish. The only time I heard something about this training, it concerned the widow of a *manobawa* who had died shortly before. It was said that this woman was able to arrive at some contact with the world of the ancestors through her deceased husband, who manifested himself in her by means of convulsive trembling, i.e. in the manner also customary among the other tribes on Geelvink Bay.[187] In this case, however, the training was not completed, because the oldest son at an untimely moment kicked the little room to pieces, in which she was passing her

---

[187] See the interesting articles entitled "Levend Heidendom" (Living Paganism) by F. C. Kamma, in *Tijdschrift voor Zendingswetenschap, Mededeelingen* (Journal of Missiology, Communications), vol. LXXXIII, pp. 187, 289, 387.

period of seclusion. Probably the consideration that a good *ghasaiwin* has to be *binapa*, i.e. a woman whose oldest child has died, had something to do with his action.

The *ghasaiwin* of Pedei told me about her training as follows: "The former *ghasaiwin* insisted that I should become *ghasaiwin* when my boy had died. She slept together with me and then we went together (in our sleep) to become acquainted with the water-monsters and the *auo* (a kind of demons). My predecessor instructed me as follows: 'You have to go like this. If the *auo* have taken away somebody, you see him and you have to kill him. And if the *dareo* passes and you look in this fashion, you see him coming. Then you take a knife and I shall hold your hand and then we shall be able to stab him together'. My predecessor gave me the magic stones. When the villages first arose here, all adults were *ghasaiwin*, but now there are only a few of the adults who are *ghasaiwin*, but formerly there were many. When these (former) *ghasaiwin* were adults, they died and now there are only a few".[188]

The *ghasaiwin* therefore relates that in mythical primeval times mankind could have relations with the sacred world without any danger, because at that time everybody possessed the powers of the *ghasaiwin*. The Waropen imagine the time of the ancestors to have been a golden age without sickness, death or misery; from their point of view, the modern world has deviated from this old ideal and it has decayed. In this way we have also to understand that the sacred world is identical with the world of the ancestors, i.e. with the world of the myth which speaks about these ancestors. The *ghasaiwin* enters into contact with this sacred, i.e. mythical, world of the ancestors.

Her work is partly preventive, viz. to prevent that the *dare* (see p. 216) harm the young children. But her main task is still to bring back those small children whose soul has wandered off too far, a danger to which children seem to be constantly exposed before their first *saira*. It is mainly the *auo* and the ancestors who keep back the souls of the children. In the descriptions I was given, no clear distinction was made

---

[188] *Ghasaiwinarenggaghaika raiwarimagha feika wewa raika rawe ghasaiwini. Ienakiwara asi anggede atira ananggigha, atira auigha. Ghasaiwinarengga ionababe iona ineni iafa: "Ara unaineni. Auo raiwakiengga asirio. Amunio. Dareo ra mangga aghanimgha onainegha asiri wa ra maini. Fimbo awu naiwiro mato ramonggado awanggéagha asikoamieni". Ghasaiwinarengga we inaiwipino. Nuni boako renggabawaigha kiwe ghasaiwin besiaka, nemani no bawaigha kiwe ghasaiwino. Maika inokini kiwe ghasaiwini. Renggawegha ari fabo. Ghasaiwinggigha kiwe bawaika kipero. Maika no nambora kini wea.*

between the *auo* and the ancestors, although they should be kept separate.

In the following I reproduce the *ghasaiwin*'s own words concerning the method used to fetch back the soul of a child: "When I am asleep, my soul wanders away. In my dream I arrive at the house of the *auo*. This house is not far. It is large house. A banyan tree is this house, in the forest, with rooms on both sides, and the stairs are a rattan, which is like this heavy rope hanging here. The people who have been taken away by the ancestors are living in that house. The *auo* ask me: 'Where are you going?', and then I say: 'They have sent me to look for their children. The people have ordered me: 'Go there with some goods to ransom them (the children) with those'. In the morning I return and then I report to the mother of the sick (child). Thereupon they give me various objects and shell-bands. Two days later they (the *auo*) bring him along again, and he is better and then I see him. But when I now treat this sick person and he does not grow better and I have no success, then he dies perhaps. My treatment might perhaps remain unsuccessful and then he dies. When the soul is brought to me, they hand it over to me. Then I am given soul-water in a small bamboo tube, in order that I should treat him (the sick person), and then he recovers".[189]

In other words, the *ghasaiwin* relates that during the night her soul undertakes the dangerous journey to the house of the *auo*. The idea seems to be that in one's dream one sees what the soul experiences on its wanderings during one's sleep. To dream is called "to sleep in the wrong way" (*enapabo*). This may quite well have some bearing on the fact that my informants maintained that they dreamed only very rarely, probably because one does not like to admit that during this "wrong sleep" the soul moves about restlessly. And when one dreams, I was told, one dreams about good things, about travelling and fishing. Sometimes one dreams that one dies or that somebody else dies, but such

---

[189] *Renakangga raroséani rao. Renapabora rara ma auigha kiriruma. Kirirumagha karaba ewomo. Kirirumagha bao. Raghangha we ruma anana regha. Ne risigha we aradogha, ne risigha we aradogha. Epamanigha we erera ona sesani saraini. Nungguigha kikoaina ruma inggoigha kiraiwi. Augha kikuanara kikafa: "Auari ara ghoe?" Fimbo rafa: "Kitawa wa kirikuigha rarawaiki". Maika nungguigha kikona kikafa: "Ara awu a ghare iato araiki mangga aghowukio". Saitopéana rara ghéa ma fimbo ronausara ghoiwarugha inaiki. Fimbo ke a ghare, ke sapari. Rana woruo fimbo kiraiwi niro. Fimbo rasiki. Ghoiwaruni raweiengga afa niro iwomo, rawepaira, io kabo fero. Mangga rawe paiki feinina. Kiraiwa roséangga roséani kiwea ma rao. Fimbo kiwe roséamasini nana kaifiadogha mato raweiengga niro.*

dreams need not be given a prophetic meaning. However, only the *ghasaiwin* would be able to see the deceased relatives, as they pass in their small prahus, just as it is she who is able to dream about mythical beings, like the initiation-demon. Ordinary men occasionally have disagreeable or shameful dreams, about which it is better not to think or to speak. According to my informants people not infrequently had very embarrassing dreams about illicit intercourse with their wife's sister.

The *ghasaiwin*, however, tells about her dream-experiences without any reticence. The world into which she enters is inhabited by monsters and it is more beautiful than the ordinary world. One *ghasaiwin* told me that in her dreams she often saw beautifully worked mats; perhaps she wanted to indicate that the patterns of these mats had been adopted from the sacred and not from the profane world. In the same way a man who wants to decorate a prahu, first has to dream of transferring the patterns, if his work is to be eventually successful.

The *ghasaiwin* therefore starts out to trace the souls of sick children. Sick children often cry and then they often have, it is supposed, some internal inflammation (*waweru*), e.g. in the throat, for which reason the soul does not want to stay in the body. If the *ghasaiwin* is unsuccessful in fetching back the soul, the inflammation will soon increase and the child will die in the end. Old women often maintained that with their fingers they could clearly feel the holes caused by *waweru* back in the throat of sick children. In case the *ghasaiwin* succeeds in ransoming the soul with the presents which the parents give her (and which, for the rest, the *ghasaiwin* keeps for her own use), she receives *roséamasino*, soul-water, in a small bamboo tube. Water plays a role in various rites in Waropen religion, where the association heaven - rain forms the starting-point. The patient has to be rubbed with this water. At Ambumi the soul is fetched in a carrying-bag and then one makes it enter into the mouth of the patient. Proof that the patient will recover lies in the fact that he mentions his name.[190]

In many cases it is not enough that the *ghasaiwin* simply traces the soul of the sick child and returns it. Then some kind of a doll is previously made of a gourd, knotted in a fine piece of cloth, so that the two loose tips represent the arms, everything being decorated with many beads and a beaded dancing-apron, and placed inside the covering

---

[190] The relation between the name and the person is also shown in the application of various *aiwo* (p. 240).

of a case plaited of pandanus-leaf.[191] This doll is pushed five times from left to right around the sick baby which sits on its mother's lap, and thereupon it is placed underneath the stand, on which the inhabitants of the house put their boxes with beads, armlets, old porcelain, etc. This was called *kipaba warimagha wa owuki ri iomofikio,* one binds on behalf of the child with the intention to ransom it from the ancestors. Later I was told that this doll represents a *ghomino,* whose soul is presented to the *auo* in stead of that of the sick child. Here two *ghasaiwin* are necessary, one who has to take the soul of the doll to the *auo,* and one to fetch back the soul of the sick child.[192]

Once this doll had been made, the *ghasaiwin* stated that she would need four nights to enable her to obtain an answer concerning the soul of the sick child during her dreamed peregrinations. This period is long enough for an experienced woman to arrive at a prognosis of the disease with quite some certainty. When after four nights hardly any change had occurred in the situation, the *ghasaiwin* stated that her search had not yet been successful. The mother in her dismay then moved into another house, just as one often moves to another house or even to another village in order to bring about improvement during an illness. The mother was not even held back by the fact that the restrictions of the segregation inside the house had not yet been lifted by the ceremony of taking the child outside. But even after this move the child remained ill.

By means of divination with an areca-nut it now appeared certain to some people that the fault lay with the father, who was said to have had his hair cut at an unsuitable moment, which would have given the child an internal inflammation, with the result that its soul had run away. This rather haphazard pronouncement deeply shocked the young father; he became visibly stupefied and had recourse to an excessive use of palm-wine. In order to demonstrate that in any case he had no evil intentions, he undid everything which he had tied up: a football which he had helped to pump up, atap which he had tied, etc., for by

---

[191] The illustration in F. S. A. de Clercq and J. D. E. Schmeltz, *Ethnographische beschrijving van de West- en Noordkust van Nederlandsch Nieuw-Guinea* (Ethnographical description of the West coast and North coast of Netherlands New Guinea), plate XXXV, fig. 13, may perhaps refer to something similar.

[192] I refer to the remark on p. 215. Here the *ghomino* is evidently intended as a sacrifice to the *auo* or to the *inggoro.* It is perhaps for this reason that two *ghasaiwin* are deemed necessary, viz. one as the sacrificer and the other as the priestess. Concerning these dolls see also A. C. Kruyt, *Het animisme in den Indischen Archipel* (Animism in the Malay Archipelago), p. 93 ff., esp. p. 98.

binding or blocking one exercises power, just like the *ghasaiwin* acquires power by tying the doll into a piece of cloth. At this dramatic moment it became clear to us that administering a very simple remedy would be sufficient to cure the child completely within a very short time. For the *ghasaiwin*, however, even this short period was long enough to declare that before long she would be successful in fetching back the missing soul.

Once the *ghasaiwin* has brought back the *roséamasino*, the soul-water, from her journey to the sacred world, this is used, at least at Kai, to rub the patient vigorously. Whilst she is rubbing him, the woman utters stifled groans (*doasaro*): *eeeiaki eeeiake*, like one does when straining every effort. In the Kai dialect this is called *kipapidera kugha na masingha*, to liberate the child by means of water. *Fapidera* is probably a combination of *fapi*, to bung up, to knock a plug into something, and *rera*, asunder, i.e. to break the charm which has been created by blocking an opening. In the Napan dialect this is called *wasaiwina irawadera waitéa*, the *ghasaiwin* catches the child out of it.

It is often assumed that the disease has been caused by evil magic. Somebody has blocked a hole with leaves (*fapi*) and in this way he has obtained power over the sick person. Now the *ghasaiwin* has to learn from the ancestors — where the soul is dwelling temporarily — where this closed hole is to be found. If she succeeds, she pulls away the plug, which she puts with certain leaves into a bamboo tube filled with water and with this water she rubs the child.

Stroking a person with wet hands in this way is a method the *ghasaiwin* also applies in general, without first having caught the soul, i.a. when children have a fever. The patient is rubbed with water, stroking from the head towards the feet; with every stroking movement the *ghasaiwin* shakes off her hands behind her, as if she had rubbed something off the child and threw it away behind her.

The *ghasaiwin* may also render assistance in cases when people have incurred damage due to contact with the sacred world, for it is supposed that various internal pains are caused by carelessly approaching a corpse too closely, due to which splinters of wood and small sticks penetrate into the body. The work of the *ghasaiwin* is to remove these splinters from the body of the sufferer. For this purpose she uses round blackish stones (*ara*) which she often takes over from her predecessor and of which it is assumed that they have been brought above ground by a snake. *Ara* is likewise the name of the spherical closing-pieces of shells, put into the pandanus-cases in order to increase one's riches. Mysterious forces are attributed to these stones of the *ghasaiwin* and they are therefore handled with great care; they are usually called with the special name of *inaiwino* or *inaiwipino,* in the Napan dialect *inderi* or *inderipino.* I was told that once upon a time such a stone had been inserted by the *ghasaiwin* into the head of her patient, and to have reappeared again after five days.

With the stone she rubs the painful spot where the splinter is supposed to be, and in actual fact she suddenly shows a small piece of wood in the hand which holds the stone. The patient declared that the pain had now been taken away and he rewarded the *ghasaiwin* for her labours with a celluloid arm-ring, i.e. a rather small reward. It is difficult to verify what people really think of this treatment, because even those persons who submit to it and who therefore evidently believe in it and find it efficacious will maintain at one moment that the splinter actually came out of their body, to state laughingly one moment later, that the *ghasaiwin* of course takes the splinter in her hand beforehand. For the rest, it was not proved that the *ghasaiwin* possesses a better insight in diseases than other people. She is simply called, because, like other old women, she has more experience than the younger women.

Quite often also the *ghasaiwin* is powerless. The woman in question told me further: "When the *sema* (suangi) have touched him, our treatment is of no avail and the patient dies. But if the *auo* have taken him along, they improve after treatment".[193] Moreover, the *ghasaiwin* may always pretend that the ancestors did not want to tell her which hole was plugged during the magic practices against the patient. In other cases she simply says that her soul lost its way and that she therefore was unable to find the house of the *auo* and the *inggoro.*

Other actions undertaken by the *ghasaiwin* follow here in her own description: "When somebody has intercourse with a woman outside the house, the *anano* (the water-monsters) snatch them away. In my sleep I then see the *anano* who have taken these two away. Just as if they were human beings they sit among the *anano* here, and later they become *auo.* The fish here asks me: 'Where are you going?' Then I say: 'I have been summoned and now I am looking for the young people'. And if they would be heavy, the child dies, and if he is light, then I take them away.

---

[193] *Sema muniengga be paiki ma fero. Auini kiwukiengga bekiengga niro.*

And after I have taken him away, he recovers after treatment. Of these water-animals two are sitting on the woman and two on the man. If the man and the woman (again) become human beings, the *anano* fetch their children so that these die. But when the mothers die, the children remain alive. And when the children die, the father and the mother remain alive.

"If somebody has had a wrong dream, he summons the *ghasaiwin* and says: 'I dreamt that I was sitting in the water outside'. And then I say: 'You have certainly lain with somebody in the front part of the house, that you talk like this. You just wait, and then I shall go and sleep. And then I see him outside in the water'. I say: 'That is the way things are with him, that he says this'. When I sleep, I see that they are sitting with him as if they were making love with him".[194]

This information concerns the prohibition which obtains in the whole of the Kai area, viz. to have intercourse with a woman outside the house or in the front part of the house. The people believe that the mythical animals, which are living in the water in the village, in such a case drag the souls of the transgressors towards them and sit with them like a boy who is making love with a girl. Now it is the duty of the *ghasaiwin* to venture into the domain of the *anano* in order to try to bring back the soul. If the souls are too heavy, this is a sign that eventually the child will die. And if the souls are light, this signifies that the mother will die. It is therefore only the souls of children and young people which the *ghasaiwin* is able to fetch.

Finally, one may appeal to the *ghasaiwin* in case one wants information concerning expected travellers. Then her soul goes out at night on the sea to look for those who are expected, giving her answer the next day. Probably she will clothe her pronouncement in obscure oracular language, because most of the *ghasaiwin* are by no means lacking in cunning.

---

[194] *Eno io bino na sedoangga anani wuio. Rena fimbo rasira ananggigha kiwukisio. Babona kiwe nunggu oai di anangha sini. Fimbo kiwe auo. Adoni iuanara iafa: "Auari ara ghoe?" Maika rafa: "Kitawaika rawarawa warimaigha". Mangga nimaiwa kuani feigha. Nirofa iani rawuieni. Buiengga beiengga niro. Nimaiwa kuani feini. Anadoangga bingha na enduo, mangha na enduo. Mangha kisi bingha kiwe nungguangga ananggigha kiwu rikuiwake kipero. Inaiki peiangga kuigha oaro. Kuigha kipeiangga imai kisi inai kikoaro.*
Eno ienapaboi kumaia ionéa ghasaiwini iafa: "Renapabora wa raminana ghaidorauo". Maika rafa: "Kabo amina ri eno na rumarenggaika aghona onamaiaini. Na fimbo rena fimbo rasiri na ghaidorauo". Rafa: "Nionainenieka iawoaini". Renakangga rasira soana bina imaia ioaiwi.

DISEASE

Although the medical lore of the Waropen is mainly of a sacred nature, one also recognise normal and everyday causes for bodily harm. These are mostly wounds and injuries, received in a fight or due to the clumsy use of sharp objects. The people possess some skill in treating these wounds. In this way they are sometimes successful in excising an arrow from a wound. Once a man had seriously hurt himself in the belly with a knife, due to an unfortunate fall, so that his intestines protruded. The intestines were pushed back and the opening was closed with a shell, whereupon the patient fortunately recovered. For internal pains or closed abscesses one sometimes applies packs of hot ashes, packed in leaves. Occasionally the attempt is made to open swellings by perforating them with a pointed piece of bamboo.

However, in most cases the treatment of wounds is restricted to plugging the opening, whereupon the damaged member is wound with rattan. In this case the people evidently bear in mind the influence of blocking and binding.[195] Sometimes they succeed in safely closing a fresh wound with a small plug of pounded masoi-bark (*manggasa*) or with some leaf, but mostly the use of dirty leaves, old scraps of cloth or even filthy pieces of newspaper causes an infection, whilst without this treatment there would have been a fair chance of the wound healing without more ado. People with sore eyes sometimes make an eye-shade of cassowary-feathers which they tie over the forehead.

However, the idea that at least adults should not die has taken root so deeply, that even in cases of wounds resulting in death one thinks of supernatural influences. Then it is said e.g. that the wound or the accident was caused by a *sema* (suangi), who in this way wants to divert the attention from his evil doings by making it appear that the patient has died because of the wound, whilst in actual fact the cause of death is the voracity of the *sema* who has previously eaten away the patient from the inside.

Of course, the Waropen only have a vague idea of illness, so that they only rarely come with distinct complaints. Internal complaints they reduce to internal inflammations (*waweru*), mostly caused by super-natural causes, e.g. because of the omission of applying lime on the conclusion of peace, or because the patient has come too closely in the vicinity of places that are dangerous from a religious point of view; it is said that "the inflammation has dived into it" (*waweru suna*). This

---

[195] See also R. F. Fortune, *Sorcerers of Dobu,* p. 143.

causes holes in the interior, and often the old women will push their fingers deep down the throat of sick babies to feel for these holes. Due to these inflammations one swells (*pino*), or one will have a fever (*wisi*). This is the reason why a Waropen when ill does not answer the questions concerning the specific nature of his complaint, for he is simply full of holes inside and now he only wants a strong *aiwo* in order to obtain fresh powers. According to the views of the Waropen, Western medicine only looks at the exterior symptoms of a disease, whilst it neglects the actual cause which is to be found in the harmful contact with sacred things. Once the people were convinced that the illness had been caused by *sema,* neither the patient nor his surroundings expected anything of the medicine given to him.

There was e.g. a woman who suffered perhaps of tuberculosis of the lungs, but who repeatedly explained in great detail how she had drunk by mistake of water in which an ancestor had been sitting, with the result that now she had a snake in her belly which devoured her from the inside. It is said that this happens quite often; at death the snake crawls out of the patient's anal opening or the vagina. Several people even maintained that they had personally seen this snake, but they had been so frightened by its appearance that the animal had been able to disappear in the water underneath the house before they had an opportunity to look at it. The woman in question wanted medicine from us, preferably strong laxatives, which would cause the snake to leave her body.

When somebody has been bitten by a snake he runs the risk that his belly will start to swell.[196] This will cause him to die, unless the poison is removed from the patient and his soul is brought back. In order to achieve this, the patient is laid on a stand, under which a thickly smoking fire of wet mangrove-leaves is lit; the wounded member is tightly wound round with a piece of rattan. Due to the suffocating smoke the patient soon starts to groan and to writhe and this is considered as a sign that the soul is returning. Then water is squirted into

---

[196] Swelling of the belly is usually considered as the result of the infringement of a taboo; perhaps it is believed that only sacred snakes bite people when these trespass on their domain. In one case of snake-bite the swelling of the belly — which I was unable to find, in spite of the protestations of the bystanders — was accompanied by violent shocks of the extremities and a spasmodic stiffening. I wonder whether these symptoms are purely physiological results of snake-bite, or whether the whole condition is the result of the trance into which all those people fall who are possessed by sacred beings. Similar phenomena are quite common in Geelvink Bay (and also elsewhere among the Papuas).

the face of the unfortunate patient. Due to the biting smoke mucus comes forth from his mouth and nose after some time and this is considered to be the venom the snake had introduced; his oral cavity and his nose are cleaned with pieces of cloth as far as possible, then the sufferer is laid on the ground and if he answers correctly when asked for his name, the treatment has been successful. Otherwise it has to be continued.

In case of unconsciousness one often attempts to recall the soul by tormenting the body, i.a. by beating it with cloths soaked in hot water or, in serious cases, by burning it with pieces of glowing coal.

Of the insane it is said that the sea-monsters have played tricks on them (*adoanggi okofari*) and that the resulting fright is within them (*niawaweio*). In general the people are very patient with the insane.

It was said that formerly the population had suffered greatly from an epidemic disease (*nipura,* probably smallpox), but since long years this had not recurred. On the other hand, in recent times they suffered more from "strangers' fevers" (*wisiamberi*), probably influenza.[197]

Diseases are very often attributed to too close a contact with tabooed places, like dangerous spots in the forest or on the water which are *hai,* as it is called at Napan (*nahuhaiwea*): in the Kai dialect *ghoghoido,* i.e. where there are snake-like creatures (*ghoro,* snake), in Napan *audo,* where the *auo* dwell. Such places are numerous in every village, e.g. stones thrown down there by a mythical creature, or human beings changed into stone, like a brother and a sister who had committed incest. These may be also places where dangerous monsters dwell, like a large crab or a cassowary-bird, or a squid. Sometimes these are whole islands, like Nusariwe Island near Napan, or localities where mythical beings dwelt or where they left foot-prints or planted trees, etc. At Napan a pool was pointed out to me, where the sacred giant crab Imbarawai was said to dwell; this animal causes diseases and it shows itself only to descendents of the mythical hero Kuriserai, i.e. members of the clan of Waratanoi. Near the cape of the island of Nusariwe dwells the yellow giant turtle Iriwoni which every morning is served with salt water by dolphins; it shows itself when a death occurs in the Waratanoi clan.

The great contagiousness of sacred things is also shown by the fact that contact with the taboo places is especially dangerous to children, so that mothers with small children have to avoid rowing too closely

---

[197] See further the article on health-conditions by Dr H. de Rook (W. C. Klein, *Nieuw Guinee,* III, p. 835).

52. Kuriserai and his pig Tarumeni

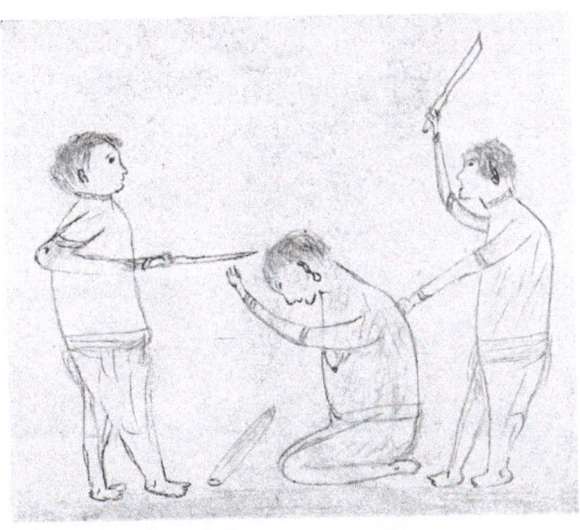

53. The *sema* Notaperei and Notarerei killing Motibai,
Kuriserai's wife

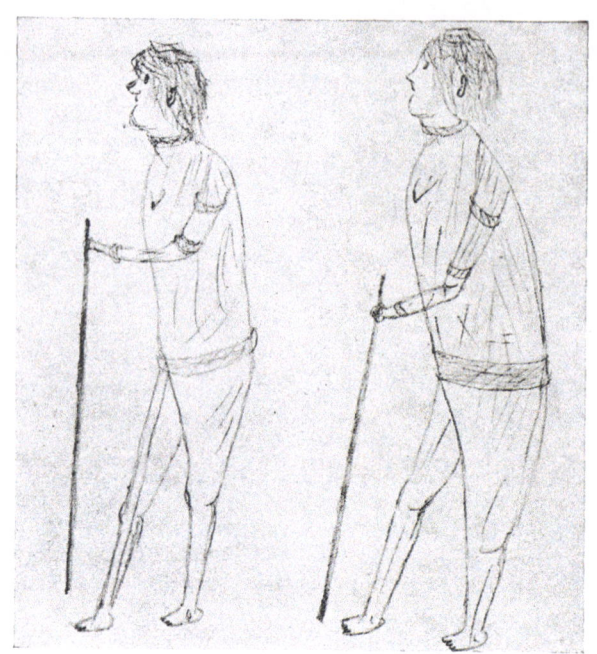

54. The *sema* Notaperei and Notarerei

55. The *sema* Nafaidoserai and his wife Bituraki

past these *ghoghoido*. If one is afraid of having still passed such a place too closely, old people at Napan advise to fetch a bambooful of fresh water and to let this evaporate by letting it drip on a heated knife, the steam touching the endangered child. Fear for the *ghoghoido* is often the cause that people stubbornly refuse to enter into the forest, e.g. for collecting dammar-resin. Tropical ulcers on the legs and framboesia (*afafuro*) are also often attributed to carelessness of the patient or of his relatives.

Disease may also be caused by the infringement of other, especially sexual, taboos, and particularly by incest. It was rumoured of a woman, whose nose was affected by some disease, that she had had intercourse with a relative with whom she stood in a forbidden relationship (probably a brother). The man in question had then jokingly pulled out her nose-ornament with the result that later this part of her face became affected. And because such an infringement of a taboo is also contagious, the infection was passed on to her daughter, whose finger came to suffer from a persistent serious infection. The woman attributed her sufferings to supernatural causes and therefore she hardly cooperated with the treatment, which in her opinion could never remove the actual cause. The villagers considered it especially praiseworthy of this woman that she did not bear a grudge against her mother who after all was the cause of her suffering.

EVIL MAGIC

It is also possible to use the sacred powers in order to harm others. It is fully permissible to present offerings to one's ancestors with the request to kill one's enemies; in such a case one will say: "*Siko Ghafa inggoigha, mindiwiwira ineni. Mimbogha mimuna Saghakukigha*", "Ancestors Siki and Ghafai (the names of two deceased ancestors), here is mush for you. Go and kill the people of Sawaki". However, this is not condemned as evil magic.

True evil magic is the plugging of a hole with leaves, while pronouncing the name of the victim. This is called *wea* (*kiwe ape wea*, evil magic is performed against us), or *fapi*, in which we find *fi*, closed, plugged, and *fako*, to bind, or *fama*, to knock. As became clear in the description of the work of the *ghasaiwin* and of the ideas concerning illness, special importance is attributed to binding and untying. The more general meaning of *fapi* is "to bewitch"; as we saw, this word combined with *rera* means "to free from a spell".

17

Every action which entails the closing of a hole or tying something up is potentially dangerous, even if it is done without evil intent. And because it is especially small children who are susceptible to this kind of bewitchment, the father has to be very careful, as he is so closely related to the child and as he so often thinks of it. The people readily conceded that it would be absurd to suspect the father of practising evil magic against his own child, but they still maintained that it was he who had to be suspected first of all in this respect. From a sacral point of view the husband is dangerous to the child during the critical first period of its life, and the taboos the husband has to observe on the occasion of the birth of his child are connected with this.

The most frequent motive for practising evil magic was said to be the jealousy of a neglected lover or mistress. At Weinami the following examples were given. The scorned lover ties a knot in the aerial roots of the banyan-tree, pronouncing the name of the woman; this will cause her children to die. He may also tear strips of bark of the aerial roots of the pandanus-palm, cut up these strips and push these into the aerial root; this will cause the victim's breasts to wither away. Again another method is to chew ginger and to spit this at the tree called *wonimbo,* and then to shake the tree until its fruit drops off; the children of the woman are bound to fall like the fruit of this tree.

The people from the interior are considered as being especially dangerous, as it is said that they are able to assume another form at will. During trading expeditions one has to be very careful not to excite their jealousy by exhibiting too many riches or precious objects. They are also considered capable of poisoning (*audo*) people by means of mysterious leaves. The Waropen do not consider it necessary to have at their disposal the victim's hair or nail-parings or excrement, etc. Neither do they believe, like the Numfoor do, that it is possible to introduce some dangerous medicine into the body of the victim by magic, e.g. by throwing the medicine against the body. Magical practises whereby a pointed piece of bone is pointed at the victim, or an image of the victim is destroyed are likewise unknown, whilst for magical practices in general charms are unusual.

The most dreaded form of evil magic is the "suangi" (*sema*) which is feared throughout the Moluccas. Here the *sema* are true specialists whose sole object is to kill others. It is said of them that they have an unsatiable desire after human flesh (*sema anggino ano nunggudaio*). They are able to approach their victim unobserved, quickly making a cut in his body and then proceeding to devour him inside. Then they touch the wound with their leaves (*semarana*), whereupon the wound is closed and becomes invisible. The victim is completely unconscious of these happenings [198] and from then on he is completely at the mercy of the *sema*, who compel him, for instance, to dance for them or to have sexual intercourse with them. In the long run the unfortunate person cannot continue to live, his body being devoured from inside.

Often the *sema* tries to divert people's attention by seemingly reviving his victim, after having emptied him inside, and ordering him to drown after some time, or to fall out of a tree. This is the reason that even in these cases the Waropen do not think in the first place of a fatal accident, but rather try to establish which *sema* could have caused this death. Another result of this attitude is that the Waropen set little store by rendering first aid, as they argue that of course nobody can have an accident if he had not been first killed by the *sema*, being alive in appearance only when the accident occurred. So what sense would there be in treating somebody who is actually dead? This I was told quite clearly when in a case of drowning I remarked that according to Western ideas one should have applied artificial respiration for quite some time, instead of being satisfied with swinging the child by the legs just for a moment.

In the light of these views we have to understand the statement that some nobles are not only able to revive the dead, but that they actually have revived deceased persons. This refers to those cases where according to the assertions of the *sera* a person has fallen into the clutches of a *sema* who has already harmed him to such an extent that the patient has grown unconscious (*fero*, dead, unconscious). At the last moment the *sera* then proves able to determine the nature of the disease and the person of the *sema*, and to restore the victim by giving him a suitable *aiwo*. For the rest, however, one is practically powerless against a *sema*; only in some cases one may be slightly forewarned, because there are people who pretend that they are able to foresee the coming of a *sema*.

It is said that one becomes a *sema* after having learned a number of secrets from another *sema* and after having drunk a certain medicine. A *sema* is believed to get his training from his father or his mother, and for this reason children of a *sema* are likewise soon suspected. Neither age, status, nor sex are safeguards against this horrible suspicion. In the *seraruma* Pedei there lived a man whose father was murdered, being

---

[198] To the Waropen a surgeon's work therefore possesses a frightening similarity with the activities of the *sema*.

accused of being a *sema;* when suspicion was likewise directed against this man, his wife took his part, but without avail, as he was murdered in the same way. And when his wife continued to defend his memory, she became the victim of her loyalty. This horrifying tragedy was concluded with the murder of the son of the unfortunate couple.

When the first incriminating rumours start, the family usually violently contradicts the accusations, but in the long run, when the rumours continue to circulate, the unfortunates are also deserted by the members of their own *ruma.* They may submit themselves, usually under strong pressure of their relatives, to the perilous experiment of the hot water ordeal, but usually they still have to expect the revenge of the raiding party at a time when they are working alone in the tidal forest or in the sago-gardens, or when they have been left alone in a room by all the members of their *ruma* without having noticed it. Probably there are also people who are personally convinced that they possess the powers of a *sema,* at least I was assured that they often blackmail others with hidden threats of using their power as a *sema.* Curiously enough, mythology attributes to the *sema* what is usually said of the *auo,* e.g. that they desire sexual intercourse with men, or that they possess certain treasures, like the miraculous chicken which is able to make money.

The most unbelievable stories about *sema* are readily accepted. People are convinced, for instance, that the *sema* possesses some *semamani,* small red or yellow birds, which are found inside a bamboo-joint or which escape from his mouth when their master dies. These birds assist him in his grisly calling and e.g. simply replace the head on his trunk when this has been struck off. At Mambui an old woman had been accused of being a *sema* and had been seriously wounded; she was taken to the missionary hospital at Serui, where she recovered. The physician, dr. Höweler, wishing to fall in with the ideas of the people, stated that upon investigation he had found that the woman was completely normal and not a *sema,* but the villagers shook their heads pityingly; one man stated even to have seen that the woman had replaced her head on her trunk when the raiding party had cut it off. And moreover, her recovery by itself was sufficient proof, as a normal person would never have recovered from such wounds. Some persons assured me that they had seen with their own eyes that when persons have been killed as *sema,* the rectum protrudes from the body, whilst under their armpits they have long, black flight-feathers like those of a cassowary-bird.

Consequently, no pity is shown to a *sema;* it seems as if the *sema* are no longer considered as human beings and the *sera* took a pride in ridding their clan of this pest. For, although also other causes of death are recognised, in actual practice every death is still attributed to the activities of the *sema*. This is the reason that in the eyes of many insecurity has increased now that the government has forbidden to kill the *sema*, because at present the latter are enabled freely to continue their nefarious activities under the protection of the authorities.

A suspicious sound during the night is sufficient to make people grow rigid with fear; I have seen adult men weep because they had been startled out of their sleep with the idea that something had touched them. The frightened imagination sees the *sema* rowing through the village in the dark of night, his tongue hanging from his mouth, and observing the houses of his victims with fiery eyes. Shivering with fear they hear him softly climbing up into the house and crawling stealthily along the eternally creaking slats of the Waropen house; they hear his hands searching for the door, touching the walls which creak mysteriously because of the gnawing of the termites. And if the Waropen were not such astonishing sleepers, they would never close an eye in their houses which are merely tied together and where the night-wind creates all kinds of creaking, squeaking and shuffling sounds, whilst underneath the house the tidal current adds its share to the symphony of mysterious noises. It is therefore hazardous to go wandering through the house on a sleepless night. Parents explicitly warn their children not to go about through the village all alone, looking for amorous adventures. In the village suspicion is very swift to arise. And, people will say, it is exactly a trick of the *sema* to roam about at night and to give a random answer to the question what they want, e.g. saying that they are looking for some tobacco.

By way of illustration I here include the story of a person who pretended to have caught a *sema* (Texts 131):

"Ghominarada had gone to the other side and was sitting with his sister at Pedei. In the middle of the night he left the house, went up at Koisi's and felt for the door along the wall. Koisi was working on his net and he heard how Ghominarada went to the door. Thereupon he sat down at the side in front of the doorway. Ghominarada cut the swivel-string of the door, lifted out the door and took it inside. Thereupon he went in with it. But Koisi seized hold of him.

"Ghominarada made no sound when he was caught. Then he said to Koisi: 'Koisi, what are you doing?'. That was in order to mislead him.

" 'Why are you coming to sit here?' said Koisi. 'Why are you sitting here in the middle of the night?'

"Then he held him with one hand and stretched out the other hand to·take the axe. Nobody else heard anything of this. Thereupon Koisi cut through his backbone and his neck with his axe. And when he had completely severed the neck, he threw his head on the ground (underneath the house) and the trunk he also threw down at the same spot. The head was lying apart from the trunk.

"And whilst Koisi was looking at that *sema* in the space underneath the house, the phantom birds put him together again. They put him together by putting his head on his trunk again. Because he was lying down, he wanted to get up, but he fell down again. When he stood up again, he fell down. Then he got up and remained standing. After having stood there for a long time, he went away to his sister at Pedei.

"And being there, he took his prahu early in the morning and went to their family on the other side. And then he climbed up into their hous and took the sero (fish-trap) to set it. And just when he was about to set it, he fell and remained lying. When the women saw him, they wept over him. Weeping, they carried him here and laid him on the ground.

"And since the moment he had died, Koisi waited for a long time before he said to his wife and children: 'Last night we were sleeping, without you noticing me, but it was I who gave him a blow when he climbed up. I have killed him. You have noticed nothing that last night I killed a *sema*'.

"Then his boy said: 'How were you able to kill him?'

" 'He was with our neighbours next to us', he said, 'and then I killed him. He was already a dead man, but you do not remember anything about that. It is for him that they are making such a clamour just now, about his death. He came from another clan and then I killed him, as he was sitting here".

SACRED AND PROFANE ANIMALS

Among the sacred animals the Waropen venerate and fear, the sea-eagle (*manduko*; *ngganggani*, N.) occupies an important place. It dwells below the village, close to the sea. Every village worships a bird of its own; at Nubuai it is called Rambairoi, at Waren Ghaisarei, and at Weinami Mambewari. On Japen Island and among the Numfoor the sea-eagle is likewise a sacred animal. All the mythical birds are local representatives of the heavenly bird Manieghasi, whilst according to some Manieghasi is the *sera* and the local birds are the *waribo*.

The curious point about this bird is that at sunset it changes its appearance, becoming a sea-animal (*rina*) but as a mythical sea-animal it is not associated with one specific creature of the sea. Occasionally it appears as one kind of fish, and again as another, but mostly the *rina* is conceived as a large marine animal, a shark or a saw-fish. The large fish in which according to the biblical tale the prophet Jonah dwelt for three days was immediately called a *rina*. However, a relationship is often established between the *manduko* and the winged ray (*ofu; manofu*, N.), or the hammer-headed shark (*sawai*), the peculiar structure of these animals being presumably the basis for this association. The mythical connection extends even further, because both the winged ray and the hammer-headed shark are constellations.

The constellation Sawai in particular has an important role, for from the position of the Hammer-headed Shark one calculates the right period for undertaking longer voyages and for planting beans. At the beginning of the calm season (April—May) Sawai raises his head above the water in the east, in the morning. At the beginning of the western monsoon Sawai dives head-on into the sea in the morning, and with his tail he stirs up the waves. According to this view Sawai stands in the west in November in the morning, i.e. at the wrong place, and then the time is unfavourable. The change of the seasons and the submergence of Sawai is therefore a repetition on a larger scale of the daily transformation of the heavenly bird.

Perhaps we may enlarge this mythological complex even further. In note 120 on page 160 I referred to the representation of a winged moon, used as a float when fishing for turtle. According to the ideas of the Waropen, the *manduko* assists the men in looking for food at sea; this is of course in agreement with actual fact, because wherever one sees a sea-eagle hovering over the waves one may expect to find fish. Moreover, there exists a connection between Manieghasi and the young men's house, because according to the myth, his mother was thrown into the sea locked up in a *dama*.[199] So there probably exists a relation between the heavenly bird, the moon, the young men's house and the catching of turtles.

---

[199] De Clercq, *op. cit.* p. 179, when describing this object, notes: "It was said that it meant obtaining much cheap food, and that certainly a struggle were to follow, if it were taken away. Such disks are e x c l u s i v e l y suspended over the *rum sĕram*".

The daily metamorphosis of the *manduko-rina* characterises the moment of sunrise as a favourable time for the proper rites of initiation and marriage. In my opinion, the observation of the dawn, the climbing of the bamboo which leeds into the loft, the torch-dance, etc., are connected with this metamorphosis and they likewise aim at realising the change from night into day. An element of struggle and pain is not alien to the *saira*, if we think of the *Inurisaira* and the perforation of the nasal septum, whilst the shooting at the coconut-shell at the wedding also indicated a struggle. A struggle is also implied in the figure of the *manduko-rina* itself, because the sea-eagle lives on the fish, the form of which he assumes.

The *manduko* is most dangerous in his hidden fish-form, because nobody can know whether he is dealing with an ordinary fish or with a *rina*. If one were to shoot at a *rina*, the error becomes apparent soon enough, because then the *rina* resumes its bird-form. Several stories bear witness to the revenge a *rina* took on some fishermen who shot arrows at it (Texts 160). The person who commits such an error does best to hurry to the *manduko*'s dwelling to notify him of the mistake. At Nubuai there is a man who maintains that this has already happened to him twice.

Another two-fold creature which dwells upstream from the village is the bisexual initiation-snake, to be discussed in greater detail in the next chapter. I connect the position of the *manduko-rina* below the village and that of the snake in the forest above the village with the division into above or upstream and below or downstream, a division which governs the Waropen view of the world.

Besides the sea-eagle which one is not permitted to kill, let alone to eat, there are several other animals which certain people are not allowed to eat, although they are permitted to kill them; in view of the "contagiousness" of the concept of taboo, the transgression of these prohibitions would make itself felt on the children. If one were not to keep the required distance from the sea-eagle, this animal would clutch the guilty man's children in his claws (*manduka sama kuigha*), which would at least cause large wounds all over their body. These animals which one must not eat are called *ghori*.[200]

In the dirge, to which a *sera* from the Kai clan at Nubuai is entitled, ond which I quoted on p. 190, a number of animals are enumerated which must not be eaten during the period of mourning. These are:

---

[200] On Japen Island *wori* or *hori* are the dangerous beings in the taboo localities.

fo	pig*	*ghana*	tree-kangaroo*
saro	cassowary-bird*	*ghaia*	bat*
sisa	small cangaroo*	*sinei*	kind of bird
mafo	jungle fowl*	*kokofi*	ditto
ghama	wreathed hornbill*	*kui*	„
manggauno	white pigeon*	*sawa*	„
opeiruno	grey pigeon	*pesaruna*	„
mi	crowned pigeon*	*aifamande*	„
tire	honey sucker*	*toimbo*	„
aumani	phantom bird	*konande*	„
manduko	sea-eagle*	*kapora*	„
agha	green cockatoo	*koi*	„
mandara	white cockatoo	*tusi*	„
ghegheri	lory	*samai*	aquatic bird,
bafu	owl	*susai*	ditto
		nutuka	„

The animals marked with an asterisk are those of whom the informants stated that these were *ghori* for them; we may safely assume that the other animals mentioned in this dirge also have to be counted among the *ghori*. If somebody were to eat his *ghori*, he would faint (*mamura*) and die. The prohibition to eat one's *ghori* is based on the feeling of the relationship between the person and the animal. Sometimes this relationship implies the belief that one is the descendant of a certain animal, like the turtle is the ancestor of the Pedei clan; sometimes also the belief that the ancestors have been transformed into animals, be it as a punishment of these ancestors for certain deeds, be it that they themselves assumed animal form in order to escape the evil intentions of others. That in this case we are undoubtedly dealing with totemistic conceptions is proved by the statement of one of the old men who said: "*Fofoki renggawe kiwe kora. Maika nemani amina kora, amina saro*", "Anciently our grandfathers were marsupials and therefore we are now taboo for marsupials and cassowary-birds".[201]

In contrast to the Numfoor, I was unable to find here among the Waropen any connection between totemism and exogamy. The members

---

[201] G. A. Wilken, *Verspreide Geschriften* (Collected Works), IV, p. 114, establishes a connection between totemism and the different representations of animals on ancestral statuettes and utensils in Geelvink Bay. — Concerning the metamorphosis of human beings into animals (and into the banana) see Texts 137, 138, 139, 140. — The article by P. Wirz, "Die totemistischen und sozialen Systeme in holländisch Neuguinea" in *TBG*, LXXI, p. 30 ff., gives an exaggerated impression of the importance of totemism in Geelvink Bay.

of a *ruma* are far from all possessing the same totem. I was not successful either in discovering from whom a person inherits his totem, as the people quite simply did not wish to talk about this subject. The only thing they were able or willing to say was that the totem may be inherited from the relative whose name one comes to bear. In particular, the person who is given the name of his mother's brother would not be allowed to eat his mother's totem animal, and in general one should be very careful of one's mother's father's totem. At Napan it was said that a man inherits his father's totem and a woman her mother's, and also that people of the same totem feel more or less related. But to my mind no certainty is provided by this assertion which may well have been constantly repeated because the informants had found out that this information interested me highly. Among the Numfoor the totem is inherited patrilineally; among the Waropen the not strictly unilateral computation of the membership of the *ruma* perhaps plays a role. The totem does not seem to function either as a protective demon, because it is only of the sea-eagle that it is said that it indicates the places for catching fish when the men are fishing.

However, it is not the whole animal genus as such which is considered as sacred. A cassowary-bird, e.g., is not treated with special reverence because it is a *ghori,* not even by the person whose own *ghori* it is. Only those cassowaries are sacred which show in some way or other that they are not ordinary animals, just like only one certain fish, into which the *manduko* transforms itself, is a sacred animal. The facts have to show first that one is dealing with an extraordinary animal, and this is demonstrated, for instance, by the animal not letting itself be caught, or even attacking human beings.[202]

Due to this conviction the crocodile is therefore considered as a harmless animal and so it is attacked without fear. If the animal were to inflict a wound upon the man, this would prove, on the other hand, that this crocodile was not an ordinary specimen, but a sacred representative. The hunters had already met a sacred being in the form of a cassowary, because this had been proved by the fact that the bird had not only been able to escape, but had even attacked them. On the contrary they said that a sacred being in the shape of a tortoise had

---

[202] Among the Australian aborigines of the Kimberley district the "conception totem" is distinguished from the other totems also by the fact that not every representative of the genus considered as totem is believed to be sacred; this is exclusively the individual animal which has been "found" in a sacral manner. See P. M. Kaberry, *Aboriginal woman,* p. 195.

never been seen yet, because never a turtle had been found which knew how to escape from experienced fishermen, and even less a turtle which made dangerous attacks.

Beside these animals which are directly sacred, there are still others which different people do not like to eat. In this way there are people who do not like to eat ground-shark, and others who show the same aversion to sago-larvae, but this is not directly due to these animals being sacred, but often to personal preference. There exist also numerous cases where certain persons are temporarily forbidden to eat various animals or plants. So a man with open wounds on his legs was advised against eating sweet fruit and leaves, whilst he was recommended to eat all kinds of tart and bitter ones. Although this renders these animals, fruit, leaves, etc. sacred in a lesser degree, they still are so in a different manner.

CHAPTER SEVEN

MYTHOLOGY

THE MYTH

The word *tina* means both "story" and "proceeding", "the manner of doing things". It therefore immediately establishes a connection between manners and customs and tradition, or, in the sphere of sacred matters, between rite and myth. A myth is generally indicated by *tina garo*, ancient story, or also in short *mio*, which term not only signifies "myth", but also simultaneously "sacred" and "mythical"; a *mibino*, for instance, is a woman from the myth, a mythical or sacral woman.

A myth is likewise *owa*, dance, because it is danced as a *muna* in the funerary ritual, or celebrated and sung as a *rano* in the ritual of life. And it is particularly in this danced or sung form that the *tina bawa*, the great story, or the *tina garo*, the ancient story, is most highly sacred. The poetical version is at the same time the most dangerous from a sacral point of view. Occasionally a woman may object to telling the prose-version of a mythical tale, because she is afraid that this might endanger her child or a relative who is away from his home, but as a rule everybody is quite ready to tell the stories in prose, both by day and at night.

This does not mean to say that the people are especially hesitant about reciting the mythical songs unconnected with the rites, but there still are certain limits. Whilst nobody objected to my noting down the prose-version of the Serakokoi story, only one single woman was willing to recite the poetic version of this tale; the others considered it too dangerous, too *mio*, too sacred.

The mythical songs may be divided into men's songs (*rano*), and women's songs, either *ratara*, i.e. songs sung by women on the occasion of birth and marriage, or *muna*, songs sung by women (and men) at the funerary feasts. *Rano, ratara* and *muna* each possess a different melody, but as far as I know they deal with the same mythical material.[203]

---

[203] At the request of the Phonogramm Archiv of the Museum of Ethnography in Berlin the different songs were recorded on wax cylinders, to be studied there by the musicologist, Dr. Marius Schneider.

Among the *rano* a distinction is made between:

soitirano, sung when rowing around a bridal couple or a new prahu;

ghomindano, slaving-songs, sung during raiding expeditions, both during the raid, and during the subsequent feast at home;

amairano, morning-songs, sung during initiation feasts;

damadorano, songs in the young men's house, sung at initiation feasts;

nuarano, trading songs, mostly sung in the prahu;

ramasasiri, artificial songs, usually in a foreign tongue, sung during prahu voyages;

ratisara, love-songs, sung in the prahu and at home.

The last three kinds of songs are also sung during prahu voyages by the Christians. At Napan I was assured that the long *muna,* as it is known at Kai, did not occur there; at Napan the *ratisara* was also sung when a person had died, whilst there the Kai funerary ritual was likewise unknown.

The *rano* are sung more or less as catches (see p. 338). With regard to the construction of the songs I may add that in the case of the *rano* there is mention of an *euo,* foot, basis, which probably will have to be taken as the first part of the first strophe which is started by one group.[204] In addition to the *euo* or base one might expect an *uri* or top, but questioning on this point did not lead to a result.

The language of the mythical poetry differs greatly from ordinary speech, so much so that hardly anybody was able to provide any explanation as regards the meaning of these chants. With their endless repetitions, their artificial sound-associations and their curious combinations of words they remained practically a closed book to me.[205] In some cases the meaning is also incomprehensible for the Waropen because the song was in a foreign language, e.g. the language of Ansus on Japen, in the language of Windessi, or even in the language of one of the tribes of the interior of the Vogelkop Peninsula. Usually such songs are introduced by a person who has made a distant journey, or who has heard the song sung by another person. In that case the alien words are

---

[204] The Numfoor likewise distinguish between *rwuri,* head, *randak,* beginning, or *widom,* top, over against *fuwar,* trunk; see F. J. F. van Hasselt, *Noemfoorsch woordenboek* (Numfoor dictionary). See also F. W. Hartweg, "Das Lied von Manseren Mangundi", in *Zeitschrift für Eingeborenen-Sprachen,* XXIII, p. 47. The inhabitants of the Trobriand Islands distinguish in their sacred formulas between 1) *u'ula,* foot of a tree, trunk, basis, origin; 2) *tapwana,* body; 3) *dogina,* crown, end. See B. Malinowski, *Argonauts of the Western Pacific,* p. 433.

[205] Poetic distortions are e.g. *rumboi* for *ruma,* house; *auambinui* for *auini,* female *auo; aiwui* for *awa,* high tide.

mostly adapted to the exigencies of the local sound-system and associated to known terms from the local language, with the result that the whole becomes a rather meaningless jingle.

However, although for the Waropen personally most of the songs are untranslatable and incomprehensible, it would be incorrect to conclude that they sing completely senseless songs. For in the first place, the song forms part of the ritual and as such it is connected with a certain world of ideas. And secondly, the people understand something of the contents in so far that everybody knows approximately what is the essential point. For the mythical subjects are also known in a prose-version, which is transmitted in the evening at the fire-side by the old to the young, and so a name, a word, a vague indication, reveal to the dancer much of what remains hidden behind the incomprehensible language.

The sacred language covers the figures and the intentions with a veil, but to the Waropen themselves these veils are not completely opaque. When he is asked, however, to transpose the language of the songs into clear and plain prose, he is landed in considerable difficulties; he cannot and dare not take the veils away.

For him it is therefore much easier to reconstruct the mythical material in a prose-version, but even so, for many there remain such a number of obscurities that anything but a fluent story is the result. It was especially old women who often still relapsed into the poetical style. Some examples of the hybrid products which were arrived at in this way have been included in the texts (173, 174, 175).[206] For this very reason the prose-versions have no standard-text, as every narrator has to reconstruct his version from memory and from his imagination, and so there are no two people who tell exactly the same tale.

This lack of clarity in the reproduction of the mythical material finds its cause not in the low degree of development of the literary powers of the Waropen, or in the confusion of their ways of thought, but it results from the typical character of the myth which is intentionally cast and transmitted in mythical, mysterious and poetic language.

According to Waropen views it is therefore impossible either to understand these chants or to compare them, unless one has first drunk the required medicine. When noting down this poetry, the ethnographer

---

[206] During my research I also noted down a number of songs, but these have not been published, because neither the text nor the translation is clear. In this chapter the figures between brackets refer to numbers of the collection of texts. [See p. 15 note 21. Ed.].

is faced with considerable difficulties. The poems do not aim at presenting clear language and so one has no objection to changes and mistakes, etc., which are unavoidable when taking down dictation, and one even admires such changes as being very good and poetical.

The mythical language did not prove to be an older stage of Waropen, although it is of course quite natural that in poetry old-fashioned terms occur and that these are sometimes corrupted. Clarity and simplicity are studiously avoided, because the obscure and unusual form of the language by itself is an active element in the ritual, just as the sacred formula supports the sacral act. When the mythical beings appear in the myth, they likewise use an obscure and incomprehensible language. And the cave where the *sema* dwell can be opened by a mythical "open Sesame" by every person who knows the formula (133, 134, 135). The chant has the same object as the ritual, viz. to activate the contact with the sacred world, and for this very reason the poetic language has to be different from that of every day. The most active categories of this language are the repetitions of sounds and the symbolism of the sounds, which have overgrown the syntactic and the morphologic rules of the language.

It is also difficult to determine in how far the myths use metaphor and enigmatic language as a means of expression; this point can only be cleared up by a long and intimate contact with all sections of the population. The Protestant missionary F. J. F. van Hasselt became increasingly convinced in later years that the Papuas are fond of associating all kinds of ideas, especially playful and sexual ideas, with numerous seemingly innocent words. By way of an example he mentioned the obvious sexual association which the school-children of both sexes proved to connect with the word "sea-shell" which he had applied jokingly to the children.

Although Waropen does not possess specific linguistic taboos, like the Numfoor *sor,* it occasionally uses metaphor and enigmatic terms. The triton-shell is the image of the men's house; in the myth women are called "small birds" or "fluttering leaves"; Kuriserai describes the killing of two *sema* as "beating out those two fires". The primary incest is meant when it is told how the brother lays his legs in the lap of his sister (85). But if one wishes to proceed further along this course, it soon becomes mere guesswork.

For instance, is it permissible to recognise the mother-son incest in the fact that the son nails his mother's red apron to the pole which forms the stairs, especially when it is said in this connection that the son has to kill his mother, whereupon the new community is generated by her body? Is there an element of sexual symbolism in a woman's name like Wawitaburapi (Wandamen: "Woman Triton-shell"), or in the white porcelain-shell which is so frequently used for decoration by women and girls, and which the mythical beings desire to possess (128)? Or in the circumstance that the initiandus, when finding the heavenly woman, shoots arrows into her ear-pendant or her hair-tube? Does the banana of which Mimiandeisei is so fond (119) possess sexual significance, particularly when we know that the male semen is likened to banana-juice (4)? Or is this true for the cleaving of the coconut over the heads of the newly wed couple, and of the shooting into the coconut-bark by the bridegroom? I shall not venture to answer these questions, but I certainly consider the existence of this trait in the myth to be possible and even probable.

That the myths aim at veiling rather than at revealing is also shown by the curious use of names in mythology. When in normal language a proper name is mentioned, one definite person is meant; when in a conversation a certain mister Smith is mentioned, the speaker and the listener necessarily have to mean the same person.[207]

However, a superficial acquaintance with Waropen mythology is sufficient to convince the reader that in the mythical sphere names are used in a very different manner. One and the same mythical name is used by one narrator to indicate a sail, by another for a mythical ancestor, and by a third for a totem animal, whilst in another connection the same sail, the same ancestor and the same totem animal may equally well be given completely different names, which in other stories may again refer to something else. In two obviously identical stories the hero has completely different names and he is descended from wholly different parents, whilst later the same couples are mentioned in again another connection.

In this way, different narrators said that the mother of Werani (87) had the names Nuai, Rakoi, or Raimi. The names Nuai and Rakoi we then meet again in the longer name Nuawirakoi, as being the name of the snake Siroei (24, 25, 26), also called Waisimaimuni or Wokui. But this is still far from proving the truth of the obvious assumption that Siroei would be identical with Werani. Even more striking, Siroei's mother is also called Motibai, whilst in another story Motibai is again the daughter of Nuawirakoi.

---

[207] For the function of proper names see O. Jespersen, *The philosophy of grammar*, p. 64 ff. For the relation between name and object, see J. Vendryes, *Le langage*, p. 216.

It may be said in general that it is never clear that figure A in one myth is identical with figure A in another myth, and whether it would therefore be permissible to construct the image of figure A from the data provided by the two myths. Or to mention another possibility, it is not clear whether figure A in one myth is identical with figure B in another, of whom exactly the same things are told. When one asks the Waropen whether the two figures A of the first instance, or the figures A and B of the second are identical, they mostly reply in the negative. It is as if somebody at one time states that David was the man who killed the giant Goliath, and at another that he was the brass snake which Moses erected in the desert, and again a third time that David-Goliath was the name of the son of Adam and Eve.

From all this one might be induced to conclude that Waropen mythology has become greatly disordered, with the result that the people themselves are no longer able to find their way in this confusion, but this conclusion does not seem to be warranted. The Waropen themselves do not consider this system of nomenclature confusing and they accept it as a matter of course, even when their attention has been expressly drawn to it.

Therefore, no other conclusion remains but this, that this phenomenon agrees with the mythology, i.e. in the mythical language nomenclature possesses a function which differs from that in normal speech. In mythology it does not serve so much to specify, as rather to qualify. A name is simply a sacred ornament, a sacral epithet, indicating that the being with this appellation belongs to the mythical world.

For the same reason the people are fond of giving names to objects close by, to the house, to the front part and the back of the house, to the prahu, to its stem and its stern, to the net, the drum, the bow, the lance, and in short to everything which at the same time possesses or might possess a sacred aspect. This is also the cause that there is hardly ever any unanimity between the informants concerning the sacred names. Just as the obscure language of the myths is an exponent of the religious value of the rites, the sacred name is an essential part of the sacred being, a formula which expresses the sacredness of the object thus named. If one knows such a name, one obtains power over the named object as a result.

It follows from this that the mythological name is not quite arbitrary. In the first place, the name reveals that the being or object is mythological. The names in mythology are often very flamboyant and made up of different parts; they are frequently double names, like

Ghondumi Sisari, Rumboi Doramana, Ghorupi Warimadamai, etc. Mythical couples mostly bear names which are related in sound, like Romaniserai and his brother Faidaniserai, or Aka Weroroi and Fema Weroroi, two *sema*. And secondly, a relationship between the names is likewise indicative of a certain degree of relationship in essence, as it is quite clear e.g., in the initiation-myths. Although identical names do not prove identity, they still indicate the possibility of association in the sacred world.

The figures of the myths do not move on a stage but rather in a shadow-play, where the shadows continually coincide more or less; although they may be distinguished according to their individual form, it is hard to differentiate them according to their essence. In this way the artistic form and the literary rendering of Waropen mythology have become bogged down in vagueness. Its heroes do not possess a character of their own or an individual personality, like in Indian epic poetry where the noble Arjuna is depicted quite differently from the figures of the righteous Yuddhiṣṭhira or the imposing Bhīma. This of course causes great difficulties for an ethnological analysis.

It is only a few figures who are beginning to assume form and shape in this fog. The fierce Aimeri, the hidden father and the divine trickster Uri possess more or less a form of their own, but for the rest the mythical figures differ as much or as little as the stiff and unreal ancestral images of Geelvink Bay.

The analogy with a shadow-play might be pursued also in another respect, i.e. as regards the lack of an historical perspective. The search for any chronological sequence in the mythical events is bound to end in disappointment; in any case, the myth does not provide sacred history in our sense of the words. According to one myth, the female sex originally lived in the forest and the males on the sea-shore; guided by the pig the women were then said to have found their way to the men (1, 2, etc.). But according to another myth it is the man who discovers the road to the women's village and who there finds women for the first time (82). And a third myth makes society come forth from the death of the mother (52, 53).

To the Waropen the chronological, causal relation of the sacred happenings is of little importance. The scene is laid in a timeless world, which extends undivided over the past, the present and the future. The myth localises the sacred event in a dim past, but the sacral world is nevertheless very near. People dying to day also become *inggoro*, ancestors, and they are incorporated in their world. After their death they are invoked together with the ancestors of a distant past, to render assistance in the daily work, in fishing, in sailing, in building, in raiding, etc. The ancestors continue to have a lively interest for the world of man. They become angry when their descendants do not maintain their institutions, when they treat carelessly what had been dear to them, when they shame their family with their unpaid debts, when they are careless in their contacts with what is sacred and so endanger society, e.g. by transgressing the sexual taboos. In their anger they then deny people their fish and overturn the prahus of unsuspecting sailors. Then they punish man with sickness, in his person, or in his nearest relatives, especially in his small children. This does not mean to say that the ancestors are the righteous and inexorable judges of good and evil, for the Waropen does not believe in this kind of punishment for sins.

The ancestors as people of the past are always venerable old persons. And because myth speaks about the past, the mythical personages are likewise mostly old. In many cases the mythical sphere is already determined by the statement that the story deals with "an old man" or "an old woman", even if the subject proves to be an initiandus. What has happened in the historical past is simply projected into the myth. The well-known ship-master from Tidore, Amos, who in the last decades of the nineteenth century often visited northern New Guinea, was in Waropen eyes brought from Tidore to Nubuai by no less a personage than the divine culture-hero; at Nubuai he bestows the honorific title of *sanadi* on a clan-chief on behalf of the sultan of Tidore (48).

In the myth, time draws no sharp dividing lines between to-day and yesterday. Arming mythical figures with rifles and guns is to the Waropen no quaint anachronism at all. In one of the newer stories it is said that Sièn (Wandamen: "God") descends from heaven and helps the poor, hungry boy (the initiandus) to a sewing-machine and the king's daughter.

The mythical world is not limited as regards space either. The central portion of the house belongs to it when the row of dancers winds around the central pole. Still, this sacred world is infinitely far away. A person who, therefore, makes a long journey in the myth, lasting months or even years, is on his way to that world. On the other hand, the man who has found the woman, makes a long journey to return to his home village. By his visit to this other world the traveller has become so sacred that the cooking-pots in which the people want to prepare his food, fly into pieces.

The mythical world is at the same time the jungle, or the mountain

Ghamusupedei in the Kai area, or Mount Inggorosai on the upper course of the Woisimi river, or also the sky above, from which the mythical ancestors descend. The person who is flung into the sky by a resilient palm-tree (4), arrives in the mythical world just as well as the flying heavenly woman who returns on her wings to the Ghamusupedei mountain (72), or the *ghasaiwin* in her dream. The question for the definitive localisation of the mythical world always caused great embarrassment and violent discussions among the informants.

The mythical world is therefore simultaneously the far distant realm of the old people from the past, and the present world of youthful dancers, men and women, who continue the myth in the ritual. They are two worlds, at once far apart and very near to each other. This curious relationship has not been artificially constructed, but it has grown from the actual relation of song and dance.

Anyone desirous of interpreting the myth has to realise full well the difference in attitude of the ethnologist and of the believer. The ethnologist tries to find in the myth an explanation of the primitive religion. He attempts e.g. by means of sober analysis and careful comparison to construct the figure of the initiation-demon and to establish a functional connection between initiation and myth. He searches for the figure of the Supreme Being, of divine tricksters, etc. To the Waropen, on the contrary, the myth is not the complicated formulation of a certain dogma, but the meaningful sacred dramatisation and justification of the ritual in force at the moment. He is not interested in a theological explanation or in an unprejudiced dissertation on the religious event. He neither considers nor analyses the myth; he believes in it. The myth is not a search for eternal values; it only states how certain religious or ritual acts or a certain conviction is connected with the sacred world. Truth or untruth, the possibility or impossibility of the myth itself are not at issue, as long as the religious reality, which forms the starting-point of the myth, continues to exist.

Only when the myth becomes fixed, or when the ritual suddenly changes, the criterion of truth may become important to the Waropen; it is likewise only at this moment that religious doubts concerning the truth or the possibility of the myth may be awakened. As long as the practice of initiation exists, where the initiandus is segregated and brought into contact with the world of the ancestors via a ritual death, to be enabled thereupon to marry a woman as a full member of the tribe, so long is there sense and true content in the myth which relates the trials which the young man has to undergo in the other world in order to obtain the heavenly woman. But when the practice of initiation comes to lapse, whilst at the same time the myth is fixed in a literary form, then the question concerning its credibility and the truth itself is bound to arise in the long run. Real theological interest in the myth is born when the question regarding the truth of the myth arises.

Now, as I explained above, the religion of the Waropen was in a curious state. The ritual of initiation was losing its character of a "total" ceremony, changing more and more into a potlatch ritual. The ensuing disintegration caused the disappearance of the sacred men's house. The myths are not acted in a dramatic form; they are only sung. The obscure mythical language became an independent active part of the ritual. What is important is not which myth is sung, but the fact that it is sung. One does not dance what is being sung, but simply whilst one is singing. The connection between myth and ritual in Waropen religion is no longer very intimate, because the myth is still mainly centred on the former initiation ritual, whilst the modern ritual is mostly aimed at an increase in status of the noble first-born.

The participants in a ritual only feel themselves to be officiants in a ritual drama to a very slight extent. Although the myth is also *owa,* dance, the dance is increasingly becoming the essential part, whilst the song is simply the accompaniment of the dance. The row of dancers forms the *ghoisaira,* together they form the snake, but the consciousness that the writhing row actually represents the snake is no longer alive. The snake is still present in effigy, represented by an unsightly sago-cake; masks and decorative paint are unknown.[208]

Only in the Serakokoi ritual we witnessed the appearance of the mythical figures of the divine trickster and the heavenly bird. Nevertheless, the idea that a myth may be enacted is not so alien to the inhabitants of Geelvink Bay. On page 135 I quoted a notice by the missionary Van Balen, describing a ceremony which is undoubtedly nothing but a dramatisation of the myth known in the Waropen area as the tale of Roponggai and his battle with Kirisi Aimeri. And when in 1939 the Papuas of Manokwari wanted to add lustre to the festivities in honour of the birth of H.R.H. Princess Irene, they gave a performance of this same myth, and of that of the hidden father which is likewise well-known in the Waropen region. On this occasion they had also taken great trouble in impersonating snake-like initiation-monsters.

---

[208] In the village of Rowdi, near Manokwari, a certain family used a mask at weddings, a custom they said to have borrowed from Ansus on Japen, from where the ancestors of this family were said to have come.

The narrowness of its dramatisation, the obscurity of its language and the vagueness of its figures make the Waropen myth a difficult object for an ethnological interpretation. The complete material of the myths on which this interpretation is based will be published separately, together with a translation, so that the road to a different grouping and interpretation of Waropen mythology is open to everybody.*

THE SUPREME BEING

The figures in Waropen mythology possess the power of infinite divisibility; practically every being may be split in a duality, either of brothers, or of a married couple, or of parent and child, their unity being expressed again in the relation existing between their names. This mythical arrangement in couples also corresponds with the contrast between the clans which are joined together in couples, related like the "head" and the "tail" of the triton-shell, or also as the elder and the younger brother. Due to the heavy obligations which the raids put upon the women, the division of the original clan takes place (176, 177). Just like every being may be split into two new beings, every double creature may be combined again into one.

This cycle of constantly repeated division and unification of the mythical figures ultimately also embraces the representation of the primeval being. The problem of a first cause, of a really creative original being, has hardly interested the Waropen, because they are not interested in the historical perspective or in scientific problems. The problem of a *creatio ex nihilo,* of the prime origin of matter never occurred to them. What interests them is the origin of a fantastic curve in the coastline, the haphazard division between marsh and forest, between hills and valleys; the origin of an extraordinary whirepool in the seemingly quiet current of the river; the origin of large rocks in the channel; the origin of numerous smiling islets in the middle of the ocean, etc.

To them the creation is at most a certain ordering and division of things. The origin or the creation of the sago-palm by itself is not explained by the myth, but the curious distribution of the sago-area is (149). The nipah-palm comes simply rolling downhill (152). A sense of orderly creation has led to the development of a mythical figure: the

---

* See p. 15 note 21. Ed.

"builder of the hills". This hill-builder (*iyatana boirawa kua*, N.) was the hidden father Manserendaki, according to my informants.

In every village the people tell numerous tales which at one point or another mention this ordering and creative activity, but in many cases these are truly local stories about a stone, a dangerous taboo place in the forest, etc. A number of these stories have been collected in the Texts under the title "etiological tales". This does not mean to say that other stories are not etiological. In this way the distribution of the islands in Geelvink Bay is attributed to the action of the divine trickster or of the hidden father. Many of these tales only provide what might be called a mythical local topography.

At the point where the understanding of the organising work of this kind begins to lose itself in something colossal, we find the vague figure of a Supreme Being. In principle the creative work of this Supreme Being is not different from that of the other mythological figures who have instituted or organised or regulated something. Every divine figure is able to assume unthinkably huge dimensions, as large, e.g., as the divine trickster, whose single eyelash or fingernail is capable of overturning a prahu (47), or as large as the snake Roponggai, whose yawning mouth is as wide as the central portion of the house (38). What is larger than the largest object imaginable is bound to become extremely vague in the end.

At the limit of the imagination we find Heaven. This concept shows most life in the Napan area, where this Supreme Being also has a name of its own, viz. Naninggi. At Kai heaven is also occasionally the dwelling-place of the supernatural powers, but there the idea of a separate Supreme Being is still less clear than in the Napan region; perhaps there influences from the sphere of the Waropen and the Numfoor have made themselves also felt. Naninggi is unmistakably the same word as Numfoor: Nanggi. According to Kern, Nanggi is Austronesian *langit*, heaven, the sky.[209]

Heaven is associated with rain. When it rains, the Kai expression is *dora muna ghero*, the sky is falling down. The Biak people call on Nanggi when making rain.[210] Throughout Geelvink Bay we find the invocation of heaven and earth (*rora* and *kakofa* on Japen) when swearing heavy oaths or an oath of purgation. When swearing a very heavy oath at Napan one testifies : "*maha doradoruna hua maha anawaini, tingguo*", "By heaven above and by the earth here, it is

[209] H. Kern, *Verspreide Geschriften* (Collected Writings), VI, p. 63.
[210] W. K. H. Feuilletau de Bruyn, *Schouten- en Padaido-eilanden*, p. 102.

true".[211] This oath-formula shows again that in the end even the
Supreme Being itself is a double figure of heaven and earth. Heaven is
also the ultimate guardian and preserver of morals; when somebody
perseveres in stealing or lying, he is warned with the words: *"Naninggi
iso auo* (N.)", "Heaven will cast you away".

The way in which the Supreme Being is visualized hesitates between
a person and a supernatural power (mana). At Weinami several informants stated that heaven is a power which concentrates itself, so to
say, in several places, e.g. in the *dama,* the young men's house, and in
the *masa,* the first central pillar in the central portion of the house
which represents the ancestor. It is invoked before starting out on a
raid: *"Naninggi, asira andimasani! Anggidana andimaiwani amunggi"*!,
"Heaven, behold our central pillar here! Strike our enemies when we are
on their tracks"!

Among the Numfoor the relation between the personified Lord of
Heaven and the impersonal power is particularly curious; they are called
Nanggi and Nanek respectively. According to Van Hasselt "Nanggi is
also occasionally called Nanek, and *nanek* is the Biak word for rain.
Rain is therefore conceived of as being very closely related to Nanggi.
The prayers addressed to Nanggi are very often for rain on the dry
fields".[212] Nanggi and Nanek are not sharply distinguished, because
Nanek is also addressed as Nanggi.[213] On the other hand, one speaks of
"somebody's Nanggi" as being stronger than the "Nanggi" of another
person.[214] Numfoor informants told me that, after prayer one sees
Nanek descend like a stroke of lightning on a newly built young men's
house (*rum sĕram*).

Although belief in a supernatural power is not very pronounced in
Geelvink Bay, and although the available data do not allow of an
extensive adstruction, there is no doubt in my mind that the relation
between the Supreme Being and mana here may be compared with
that pertaining between Brahmā and Brahman in ancient India.

---

[211] It is probably to this that the divining formula refers, mentioned on p. 244.
When taking auguries with the ancestral images, one goes by the movements
they are said to make. Perhaps we may think here of occult phenomena, like
our table-tilting.

[212] F. J. F. van Hasselt, "Iets over de Roem Seram en over Nanggi" (Notes on the
rum sĕram and Nanggi), in *TBG,* LX, p. 113. See also F. C. Kamma, "Levend
Heidendom" (Living Paganism, in *Tijdschrift voor Zendingswetenschap,
Mededeelingen* (Journal of Missiology, Communications), LXXXIII, p. 313.

[213] F. C. Kamma, *loc. cit.,* p. 309.

[214] F. J. F. van Hasselt, *loc. cit.,* p. 113.

In the shape of other mythical figures the vague form of this Supreme Being comes closer to mankind, until at last it works and lives amongst men as one of the culture heroes. One of these vague representatives of the Supreme Being I believe to be recognisable in the helping old man or woman, who fulfil an important role in the background in several myths. For we are told that originally the primeval women dwelt in the forest under the protection of an old woman, whilst the men lived on the coast under the care of an old man who had found an abode in the garret of the central portion of the house, i.e. in a high place, in the space to which the marriage ladder leads and where the skulls of the deceased persons are hung during the funerary ceremonies (see p. 134, note 103). Within the framework of Waropen mythology it is also of importance that this aged helper often has no name, whilst in all other cases the multitude of names practically covers the essence of the mythical personages.

This helpful old person renders assistance by smoothing the way which leads the two sexes to each other, betraying the presence of the woman to the man or presenting the woman with a man (1, 82). She shows the man how to catch the heavenly woman by stealing her dress and hiding it (75). Sometimes a Mother Carey motif has been elaborated in this myth. The old woman is e.g. the mother of the turtle (this animal often is a clan-ancestor, e.g. of the Pedei and the Nuwuri clans at Nubuai (12). She requests the women to gather molluscs for her turtle-son, but the elder sisters refuse to do so. Only the youngest sister is willing to render this service to her and as a reward she is allowed to marry the tortoise, who proves to be a man (113). In another story the old woman produces a man out of her cooking-pot for the obliging youngest sister (201). The old man who lives amongst the primeval men not only betrays the presence of the women to them, but he also teaches them to eat the pig, and he guards their treasures (2).

This helpful old man or woman is often pictured as an ancestral image (*inggoro* or *nurawa*). This ancestral image, named Koghei, was preserved by the family of the primeval women. Therefore it was stolen by Faidaniserai (56) or by Kuriserai (133) — perhaps the divine trickster — and brought to mankind, who ever since then also consult their ancestor when setting out on a raid. The people of Weinami believe Koghei and Warari to be the ancestors of the oldest and the youngest clan of their village; during divination they gave information concerning the result to be expected from a raid (14).

Now this Koghei is undoubtedly identical with the mythical woman

Kokowei, of whom we are told that she had no breasts and went around like a man, refusing to perform woman's work. This Kokowei was a warlike woman who went out raiding all alone (182, 183). In other words, Kokowei is a bisexual creature.

As already explained on p. 205 and p. 209 at Weinami the ritual of the raid was concentrated on the figure of Kowei who showed his people the way and to whom also the captured skulls were presented. The representative of the Supreme Being, the bisexual figure of Kowei (= Kokowei) is therefore directly associated with the ritual of the raid. He is the ancestor, who was venerated in the first central pillar of the house on which the skulls of the conquered enemies were hung. And we may probably retrace the name of his Napan partner Warari to that of the Kai initiation-demon, Gharaisimai. Anciently, the raid was probably the most important part of the initiation ritual of the *dama*. The initiation-demon was likewise bisexual.

When we observe the duality of the Supreme Being, we need not wonder when we hear also from other myths that the man who wants to marry the heavenly woman, finds her under the care of two old women who are grown together at the back. It is his main duty to split this double creature again by separating them with the knife (65, 104). In one myth it is told that "grandfather" Waiseri, who comes to fetch the ancestral statue of Kowei, was descended from a lory (a bird) and a snake (135).

THE DIVINE TRICKSTER

The concept of the Supreme Being is more or less the result of a spiritual short-circuit. The idea of constantly larger double creatures, who constantly embrace more and more, finally comes to assume the vague shape of an endless, far distant, all-enveloping heaven, which still proves to have a mythical counterpart in earth itself. This Supreme Being is so vague that finally it can hardly be personified any longer as an independent being. When the distance between the two partners of the bisexual creature is decreased, the picture becomes constantly sharper. The vague figure of the helpful old man or old woman in the end becomes the rather sharply outlined figure of the god of the raid, whose image was venerated at Weinami where it had coincided with the ancestor of the clan. Although it would not be wholly correct to state that the Kowei of Weinami is fully identical with the Supreme Being, it would not be justified either to deny their essential unity.

Whilst on the one hand the final unity of the double creature is personified in the Supreme Being, on the other the ambiguity is most clearly expressed in the figure of the divine trickster. Just as the figure of the Supreme Being may be reduced and accentuated to become the ancestral image Koghei, the figure of the divine trickster may also conversely be enlarged, until the two eventually are hardly to be distinguished. These — and most other mythical figures, which after all are all ambiguous to a certain extent, — cannot be drawn with sharp and immutable outlines.

For these reasons it is easily understood that the figure of the divine trickster Uri, which is well-known and popular in the whole of Geelvink Bay, is occasionally considered by the Ambonese Christian gurus as being the main god of the Papuas. In the imagination he lives as an enormous creature; on Japen he is said to be so large that the deepest sea only comes to his knees, and at Napan, that one of the hairs of his head or one of his finger-nails would be too heavy a burden for a prahu (47). He is not only equally unthinkably large as the Supreme Being, but he is its personified ambivalence.

Usually he is a younger and an elder brother, and according to the Weinami people moreover a snake, which may also be transformed into a bird. Uri is so sacred that he is only to be referred to by numerous names, but preferably not by the name Uri (= behind), because then, the people of Nau Island say, he might grow angry. As a couple of brothers, the oldest is often called Noi, Mamberanoi, Fai, Faisai, Faisai Kukuri, and the youngest Uri, Kurupaisai, Kuri, Paisai, Noi. The identity of the names already proves the identity of their being. The Numfoor call him Uri and Faisei.

Uri is the figure of the trickster in Waropen mythology, fully in agreement with the orthodox image designed by De Josselin de Jong, "a combination of the ridiculous and pedestrian and the cunning and malicious".[215] Because of his comical ideas and his amusing antics he is as popular as Tyll Ulenspiegel in our part of the world (73), but by means of these he even drives his older brother out of the country of the Papuas, for he made him believe that he could use his mother's

---

[215] J. P. B. de Josselin de Jong, "De oorsprong van den goddelijken bedrieger", *Mededeelingen van de Koninklijke Akademie van Wetenschappen, afdeeling Letterkunde* (The origin of the divine trickster, in Proceedings of the Royal Academy, section Literature), LXXXVI, ser. B, no. 1, p. 3. In his course at Leiden University Professor de Josselin de Jong stressed the connection between the supreme being, mana, ritual, the divine trickster and classification, which formed my starting-point in the present section.

skin as a drumhead (47). When the older brother at last ran angrily away from the younger, on Mount Inggorosai (Ancestor's mountain) on the upper course of the Woisimi on Wandamen Bay, he also took the inspiration and the capacity for higher civilisation away with him. He travelled in a north-westerly direction to the country of the *Ande,* the strangers, i.e. to the land of the Europeans and the people of Ternate, Ambon, etc., whose father he became and whom he gave the clothes he had taken with him from the country of the Papuas (47). From Tidore he brought captain Amos, who bestowed the honorific title of *sanadi* on a clan-chief at Nubuai (49).

The younger brother remained behind and became the ancestor of the Papuas. Finally, however, he had to pay for his roguish tricks by becoming in his turn the victim of a crude prank, when the people told him that they were able to draw his liver out of his anus in order to eat it. But this practical joke is, of course, not attributed to the Waropen, but to the uncivilised inhabitants of the interior (47, 48).

In his death-struggle the giant kicked loose several small islands which came floating down from the back country where he died (Kurudu Island), and which became attached here and there. Nusariwe Island (*nusa raiwa,* floating island) near Weinami was hauled in by Amaponi and Romaniserai, the ancestors of this village, whilst it carried a number of mythical inhabitants and animals, and fastened at its present place (8, 9, 10, 70).

It is only to be expected that this double figure will not remain completely outside the opposition of man and woman. In the Serakokoi ritual, Mimiandeisei (the relinquished wife of Serakokoi) is i.a. also called Kukuraighei and Inuriwanderi. This last name is likewise used to indicate the mother or the wife of the divine initiandus. According to the informants the names Kukuraighei and Inuriwanderi actually refer to the divine trickster. The curious mock-battle which is performed in the *Inurisaira* ritual is also concerned with Uri, but unfortunately it was not made clear to which episode in the myth the performance referred. Does it perhaps imply that, like Uri, the honoured dead man has only fallen due to trickery and that in actual fact he is still alive?

The closeness of the relations of Uri and the female sex is likewise shown by the statement of the informants that he is always running after women. The Numfoor conception of Uri, according to Van Hasselt, is as follows. "Uri is a somewhat mythical personality. The stories concerning him are half *kakōfein* (fable) and half *kakajīk* (legend). He is described as an enormous giant. But he is especially popular because of his lies. In this respect he rather shows the type of the Papua.

"I was told that, when Mangundi, the creator of the Papuas, left them because they did not trust him any more, he appointed Uri in his place. Uri lacked the miraculous powers of Mangundi, but in order to acquire influence among the Papuas, he taught them the feasts and gave them ornaments.

"Besides for his lies, he is also known for his passion for women. In his description an abnormally large penis is therefore never omitted. Wherever he went, he asked for a daughter of the country as his wife. When he obtained one, he shot with a small arrow made from the rib of a sago-palm leaf; this arrow, which was still fresh and green, developed into a sago-forest. When he was not given a wife, he still shot a similar arrow, but this arrow was desiccated and no sago-forest came into being".[216]

His deceitful character is the direct result of his intermediary position between heaven and earth, which made him a truly ambiguous being. In his nature there is a struggle between the elder and the younger brother, one being more inclined to good and the other to evil. This is the struggle between the joined clans who live in a never ending rivalry as the "head" and the "tail" of the triton-shell, whilst they are also related as brothers. It is only through deceit that the one clan is able to best the other, just like Uri was only able to overcome his older brother by means of trickery, to be finally vanquished himself again by cunning.

To the believing pagan he is a figure who excites both ridicule and fear. The worship of this coarse divine jester also reflects the insignificance of man. Ordinary life, with its unexpected ills, its tragic deaths, its dangers and hardships provides daily proof of the fickleness of the supreme powers who make game of man and his desires.

AIMERI AND HIS BATTLE WITH THE SNAKE ROPONGGAI

Another culture hero is Aimeri, or, as another double creature, Kirisi Aimeri. Like Uri, this twofold being is partly good and partly evil, but the dangerous aspect predominates. That Kirisi again should be a snake is therefore quite what might be expected. For, when Kirisi and

---

216 F. J. F. van Hasselt, "Nufoorsche fabelen en vertellingen" (Numfoor fables and stories), in *BTLV*, LXI, p. 581.

Aimeri were once out hunting together, they found a couple of eggs, which later proved to be the eggs of a snake. By drinking the eggs by mistake, Kirisi was changed into a snake, and a snake which could spit money at that. Just like the older brother of the divine trickster and several other mythical figures, Kirisi left by ship for the Country of the Strangers (36).[217]

One of Kirisi Aimeri's most important deeds is his battle with the snake Roponggai and the consequent introduction of the *munaba* ritual. Roponggai is an enormous snake, as large as an iron-wood tree and with a mouth larger then the central portion of a house (33). It lies in wait in the back-country on the upper course of the Woisimi river on Wandamen Bay, in order to devour the people, who at their wit's end finally scattered and went to live at Numfoor. This is the explanation for the similarity to be discovered between Numfoor and Waropen customs.[218]

Only two people remained behind in the village, viz. Kirisi and his mother, or Kirisi and an old woman (is the old woman identical with the Supreme Being?). Both took their position in their house and in the end they succeeded in killing the snake by means of a stratagem by throwing hot water and glowing stones into its mouth when it opened it in order to devour them. Thereupon Roponggai was transformed into a prahu, in which Kirisi and his mother travelled to Numfoor to spread the news of their victory (33, 34, 38).

On the other hand, Aimeri is depicted as a brute, despised by all, a creature which does violence to everybody and which everywhere causes annoyance by its pugnacity. In his prahu Ghondumi (Ghorumi is also the name of a mythical sea-creature) or Ghondumi Sisari he visits all the villages on the coasts of Geelvink Bay, especially those on the East coast, and on the South coast of Japen Island. He left the Woisimi with

---

[217] The significance and the function of the snake in a primitive religion is discussed by G. W. Locher, *The serpent in Kwakiutl religion;* there the snake is also depicted as the possessor of riches. That the snake in Waropen lore possesses the same characteristic is connected with the potlatch-character of the feasts. The organiser of the feast grows actually rich. The people who attend to the snake acquire riches due to this, and during the *saira* they sing in its honour the *amairano*, the morning-song (51). The snakes also possess medicine against diseases (167). Near Manokwari a stone was shown, under which a snake was said to live which guarded a large treasure; a wooden image of this snake had the shape of a naga.

[218] J. A. van Balen, "Windèsische verhalen" (Windessi tales), in *BTLV*, LXX, p. 504.

the fire, but he was unable to carry it into the back-country of Waropen Kai, as it had already been extinguished half-way, near Weinami. During the remainder of his voyages he was therefore obliged to devour his fish raw (37). He is so fond of palm-wine that he tapped a creek near Waren completely empty (40).

Near Waren he has also opened up the river, in order to catch Aighei Dorai (the heavenly woman) and Aighei Ghafai (the moon woman) who were living there in the interior near Mount Ghamusupedei. According to some, Dorai is his mother's brother's daughter, i.e. his lawful potential spouse, according to others he committed incest with her (42), or again, Aighei Dorai offered herself to Aimeri, when the latter preferred Umai Siriwini (star Siriwini), the woman who went drifting down the sea in a gourd (42; see also p. 304). Even his own daughter was not safe from his amorous desires (36, 44).

It is quite in agreement with these incestuous relations that according to other myths the two women retired together with their father, whereupon Aimeri blind with rage destroyed their house and pulled the poles out of the ground. In his excitement he threw his rifle into a creek which still bears the name of this incident (40, 43). Traces of his hunt for women at Waren are still to be found in the form of two stones near Marierotu village on Japen Island which remained attached to his giant prahu when he rowed up the Waren river (41). In another stone the people recognise the wine-gourd (41) and the comb which he flung away (45, 40).[219]

He is also closely connected with the Auo, who are his fellow-raiders (40). Another myth relates that the snake which violated the wife of the divine initiandus changed in the fire into an Auo called Aimeri; the initiandus changed into a bird of paradise (99).

His victory over the snake Roponggai is brought into direct relation with the institution of the funerary ritual. A question about the origin of the *munaba*-ritual was simply answered by means of the tale of his battle with the snake Roponggai (34). Aimeri came rowing with his first dead person from the Woisimi, trying everywhere to find a house where he would be given the opportunity to celebrate the ritual for this corpse, but he was refused to do so in all villages. Only at Kai he found people of the house Manieghasi (at Nubuai?) ready to admit him. This is the explanation for the different funerary ritual at Kai and on Japen

---

[219] The people assured me that this comb had also originally been a warrior's comb. See p. 151.

Island where (differently from the Napan area) the custom existed of mummifying the dead (37, 38).[220]

It is not quite so easy to decide who Aimeri is exactly, and which place he should occupy in the pantheon. As I shall explain below, there exists a certain relation between the snake Aimeri and the snake Roponggai. The heavenstorming and warlike Auo is in any case a figure of death who wants to conclude the dangerous incestuous marriage. His constantly travelling about in his prahu perhaps finds an explanation in the fact that he is one with his prahu Ghondumi (Ghorumi is also the name of a mythical porpoise), just like Roponggai also changed into a prahu. Although he fights and conquers death, he still personifies death itself, just like Roponggai will prove to be both the dangerous initiation-demon and the initiandus. In Waropen mythology life and death are as ambiguous as the divine trickster. It is striking that Aimeri proved likewise unable to carry the fire further than half-way into the Waropen region. Roponggai is the initiation-demon in his frigtening aspect, who is overcome by the initiandus Aimeri with the assistance of the old woman. But in the end Aimeri is again one with Roponggai. Both are snakes. And Aimeri sails away again in the prahu into which Roponggai has been transformed.

Concerning the ritual of the raid, of course only oral (and therefore imperfect) information was obtainable, but more detailed data would undoubtedly have made it easier also to understand the figure of Aimeri, for he is also described as an indefatigable raider. Formerly, the ritual of the raid was an ordinary cosmic ritual, as is already proved by the place Koghei assumes in it.[221]

It seems quite probable that the struggle between Kirisi Aimeri and Roponggai is at the same time the battle between day and night, or between heaven and earth, the critical moment, at which the heavenly bird divests itself of its submarine form. The house where Kirisi dwells is heaven. Among the inhabitants of Biak Island it is said that the boy in his battle with the devouring snake dwells in a house with seven platforms, in which he retreats before the snake, each time to a higher platform. Similar houses with platforms are also erected when appealing to heaven on the occasion of a prolonged drought.[222] The hot water

---

[220] According to Text 39 the funerary ritual was instituted by Anirei who had also brought the two ancestral images Kowei and Warari to Napan.

[221] Concerning the relation between the raid and the funerary ritual see further p. 294.

[222] W. K. H. Feuilletau de Bruyn, *Schouten- en Padaido-eilanden*, p. 100.

is the water of heaven itself, used to beat back the attack on heaven, just like the moon washes away those men who want to take it down (140). In the ritual the bamboo is often used as a container for the water of heaven. The glowing stones (*wai*) which Aimeri throws into the mouth of Roponggai are perhaps the clouds, or "heavenly stone" (*dorawai*), as the Waropen call them. The battle of Aimeri and Roponggai is therefore also the battle of the dark underworld with the fiery hot sky. The fire of the day has to conquer the darkness of the night, and this is the reason why it is at this critical moment that the torch-dance is held during the *saira* (see p. 147) and that all essential ceremonies are performed.

Like a true intermediary figure between life and death, the fighter against the demon of darkness and of initiation, but simultaneously the pursuer of the heavenly women, Aimeri was unable to take the fire further than Weinami. He causes fright; he fights always and everywhere and therefore he is avoided by everybody just as Roponggai. Only after initiation the initiandus may return to normal society, no longer the lonely wanderer shunned by all and isolated from the community, but admitted as a fully qualified member of society.

It is also to be noted that during raiding parties nearly always the time of dawn is considered as being the most favourable moment for the attack. It seems highly improbable that the choice of this moment was left wholly to chance, although according to my informants any other moment could also have been chosen, but this hardly ever happened.

THE HEAVENLY BIRD MANIEGHASI AND THE SERAKOKOI MYTH

In comparison with the more frightening figures associated with darkness and death, the heavenly bird is less prominent in this mythology. The fact that Manieghasi actually is the bird of heaven is apparent from the direct answers of the informants, which state that at nightfall he assumes the form of a marine animal, e.g. of a fish, very often the winged ray, in order to rise again as a bird the following dawn, usually as a sea-eagle. Manieghasi is considered as being the chief of the clan of all sea-eagles, of which mostly one specimen is to be found somewhere in the vicinity of the village, near the sea. These are his companions. At Weinami I was told that during the night Manieghasi usually assumes the form of a hammer-headed shark (*sawai*) or a winged ray (*manofu*);

we find both again in the evening-sky as constellations. The constellation Sawai plays an especially important role in the regulation of the monsoons and the times for planting, and the determination of the suitable time for long voyages.

This does not mean to say that the Waropen show any special respect for those sea-eagles they find within their territory. At most they do not like to disturb the bird, because it always remains possible that it is a sacred being. But for the rest Manieghasi in his twofold form does not belong to the profane world, but to the sacred realm, where different rules pertain and where other possibilities exist.

What mythology has to say about Manieghasi is this. A man wants to do away with his wife because he suspects her of unfaithfulness, or because she has killed his first wife. Bound, or locked inside a *dama*, a young men's house, she is left at sea, but her brother has arranged matters in such a way that she is still able to reach the shore where she remains alone. She is in a deplorable state, having lost her loin-cloth, whilst the birds soil her. She is pregnant and she bears two children, a human being and a bird. The human child returns to its father. After having tried in vain to take it back, the bird has to return to the other world where the mother dwells without having realised its purpose. According to another version, the mother gives birth to one single child which is able to change itself into a bird.

The bird has to collect the fish on which they live by stealing it from the fishermen (102). The bird likewise catches slaves for its mother (103). The bird-man marries the youngest of three girls who have first ridiculed him because of his appearance; by her he has a child which is called after its father and its mother. However, the bird does not wish to stay among men; it is so sacred that the cooking-pots in which people want to prepare food for it are blown to pieces (104).

Because his mother has no fire, he goes to fetch it from two old women, who in view of their names (Andebami and Rofobimbami) again form a couple. He carries both in his claws to his mother, when it has become apparent that the two old women are his mother's mothers (102). In these two old women we may probably recognise again the heavenly women, the women from the moon, of whom another myth reports that they preserved the fire there and that it was taken to the earth from there by one of them (197).[223]

---

[223] In this connection mention should be made of the curious story which is found among the Kiwai and in Torres Straits. In these regions it is a creature which may be compared to the Waropen *auo* which succeeds in supplanting the

56. The prahu with the enclosure carrying triumphal branches

57. Inhabitants of Roon presenting the Roponggai myth

58. Schoolhouse at Napan

59. Elders and deacons at Napan

Now we see Manieghasi appear in the Serakokoi ritual as an old man
— i.e. as a mythical being — who steals food from the people (see
p. 296). This is one of the rare cases in this religion that episodes from
the myth are dramatized.

And so the question arises automatically which might be the connection between the myth of Manieghasi and that of Serakokoi. The
name of the woman who is thrown into the sea is also that of Mimiandeisei, Serakokoi's deserted wife. However, in another myth the same
name is again that of the wife of the divine initiandus (104) so that
we are not allowed to conclude right away from the agreement of the
names that Serakokoi's wife is the mother of the heavenly bird.

The myth of Serakokoi relates that Serakokoi goes sailing everywhere
with his wife Mimiandeisei (*mimi* might be a reduplication of *mi*,
sacred), because she is pregnant and wishes to eat bananas. Whilst
Mimiandeisei is bathing somewhere on the sea-shore, she is caught by
an Auo, who takes hold of her and who seals her eyes with resin and then
pretends to be Mimiandeisei herself, trying to imitate her pregnancy by
swallowing a cooking-pot.

Mimiandeisei remains behind alone. She gives birth to a child which
just like the human brother of the sea-eagle is found by its father when
shooting *soa*-fish. The father recognises the son and lets himself be
guided to his mother. Then follows a duel between the real wife and
the interloper, in which the Auo is overcome due to her stupidity, as she
accepts an imitation knife whilst the real wife holds a genuine
knife (119).

The only agreement to be discovered between the myth of Manieghasi and that of Serakokoi is the role which is filled in both by a
pregnant woman, who wanders about outside human society. Therefore
there has to be a reason why this woman is made to appear in the most
sacred *munaba*-ritual.

Although in this case also the material is anything but clear, I would
venture to suggest the following explanation: Serakokoi himself is the
noble deceased person for whose benefit the ritual is performed. We
are therefore told about the dead man that he loses his wife and is
temporarily united with an imitation woman, an Auo who pretends

wife in her husband's favour (i.e. like in the tale of Serakokoi). The woman
remains behind alone and gives birth to a sea-eagle, which carries fire and water
to its mother and which succeeds in the end in reuniting husband and wife;
thereupon the intruder is killed. *Reports of the Cambridge Anthropological
Expedition to Torres Straits*, V, p. 25; G. Landtman, *The folktales of the
Kiwai Papuans*, p. 225.

to be his wife, to be Mimiandeisei. This means to say that the deceased *sera* by his death loses his own wife, to be united with the dangerous, indetermined beings of darkness. Now the intention of the ritual is to liberate the *sera* from his pseudo-wife by performing the myth, and so to unite him again with his real wife, Mimiandeisei.

The separation and the reunion with his wife is in actual fact the division and reassembling of Serakokoi's being itself, for, like the myth says, Mimiandeisei herself is Serakokoi. And Serakokoi's other name is Saranaighei, Bird of paradise Aighei, i.e. he is as much an Aighei as his wife. And she again bears the name of the divine trickster, being called Kukuraighei, or Inuriwanderi.

Although now the deceased has temporarily fallen under the powers of the nether world, the child which is the very image of its father — and who in other words is again identical with the father — is born in the other world, the world outside the village, where Mimiandeisei wanders about. The child therefore has no name, a curious enough occurrence in this mythology. Through the intermediary of the child, husband and wife are again reunited.

Both the birth of the child which is identical with its father, and the reunion of the husband with the wife from whom he has been temporarily separated, intend to express the belief that, although the *sera* has temporarily come under the influence of the Auo, he will still continue his existence in the hereafter. In the hereafter Mimiandeisei therefore also continues the normal work of all women, i.e. the preparation of the sago. An attempt is made to effect that also in the world of the dead normal life is continued, and this is done by singing the Serakokoi-song on the occasion of the *sera*'s death.

Now as regards Manieghasi, who is also connected with the Sera-kokoi-episode in the ritual as we saw, we may continue to build on the interpretation suggested here and assume that the child which was born of the relinquished woman, i.e. the bird of heaven, again represents the deceased. But in this case it is the dead person before he is certain of his existence in the hereafter. Manieghasi appears at the beginning of the funerary ritual. The human form of Manieghasi (or of his human brother, for Manieghasi also is of course a twofold being) is one with the father. The father personally does not yet want to come along to the other world. The Waropen always imagine that the deceased person only leaves the land of the living with the greatest aversion.

The heavenly bird is probably not only the child, but also the husband of the relinquished woman. Therefore he sees her without her loin-cloth and brings slaves to her (103); formerly slaves belonged to the dowry.[224] It is said of the relinquished woman that, as has already been told of Manieghasi, she finds two old women who are grown together and that she has to separate them (104).

The heavenly bird goes to steal food for the relinquished woman from the mourning relatives. The people are not willing to part with it, because they do not yet want to consider the dear departed as a being of the other world. Manieghasi has to remain in the other world, but his father does not yet want to accompany him. Only when in the ritual Serakokoi has been united with Mimiandeisei the dead person has been united in the other world with his spouse. Then the food for the dead person is no longer stolen by the heavenly bird either, as it is offered to him by the surviving relatives who celebrate the *munaba*.

In this ritual Manieghasi is therefore the dead person who is about to take leave from mankind; perhaps also the name of the bird refers to this: *Manieghasi*, "the bird which takes apart".

The deceased person is in an intermediary position, being on the one hand in the power of the Auo of the underworld, for which reason the myth also calls him a snake, whilst on the other hand he has already been re-born as the heavenly bird. His younger brother (Serakokoi is of course a double creature) is called Numanimosi or Manimosi, in which we find perhaps *mani*, bird.[225]

The relinquished woman we find again in the myth of the snake Simundopendi (128, 129, 130). That it is permissible to connect this myth again with the *munaba* ritual is clearly evident from the fact that the woman concerned is a woman in mourning apparel.

Again man and wife are separated. This time an Auo has stolen the husband from his wife, so that the woman remains behind alone. Now she assumes mourning and puts on the mourning-dress made of tree-bark, which Waropen women still wear at present when mourning. This dress, however, has an extraordinary origin, being the bark from the central pole of Simundopendi's snake-house. When she puts her bark-dress in the sun to dry, it automatically comes to show the well-known figures which still adorn this type of dress and which are also identical with the tattooed patterns on the bodies of most women.

---

[224] It is a white slave who is brought by the bird; would this be an allusion to the white colour which the deceased person assumes when after mummification his skin is removed (p. 177)?

[225] The name Serakokoi might mean "Lord Chicken" (*koko*, chicken). In this mythology the chicken is a helping bird which brings riches (75, 77, 164).

Probably the woman in her mourning-dress made from the central pole of the house which is considered as a god, is nobody but Simundopendi, the snake itself.[226]

Now the snake Simundopendi comes to visit the woman and wants to unite with her. All her relatives flee. The love-making of the snake is extensively described, but in the end the woman is too clever for him and she makes good her escape. The monster pursues her, but it is vanquished by her brothers. According to others, the snake still tried to keep the woman back by creating all kinds of obstacles and an impenetrable darkness, but the woman bought the two birds which by their whistling call the dawn. The snake creates the darkness, the bird whistles the light.

According to a Windessi version of this tale, given by Van Balen, the snake, here called by the name of the divine trickster Inuri, changes into the breakers of the surf, wind, rain, thunder, lightning, a water-spout, a crocodile and a porpoise.[227]

Although the lonely woman in her mourning-dress wears the dress of the snake, she does not want to unite with it. Although at death there is a temporary relation with the nether world, one does not wish to perpetuate the power of the nether world. It is not the snake which creates darkness that has to be born, but the bird of heaven which fetches the fire from the heavenly women.

For the mention of the fetching of the fire can hardly refer to anything else but to the change from the dark night into the clear day. It is at the first dawn, as we saw, that according to Waropen ideas Manieghasi resumes his bird-form, putting off his underworld form.

Here we should remember another bringer of fire, viz. the snake Aimeri, who wanted to take the fire from the Lower Country on the Woisimi river to the Uplands of Waropen Kai, but who was unable to go further than half way (see p. 287). Aimeri also has his particular function in this connection, because it is to this pugnacious snake that the whole of the Waropen Kai funerary ritual is attributed. Assuming that Aimeri really represents the initiandus during the period that the latter is excluded from the life of the community, there must conse-quently have existed a relation between initiation and the funerary rites. This was actually the case, because formerly slaves had to be caught

---

[226] Would therefore tattooing be a snake-design and an indication of mourning?
[227] J. A. van Balen, "Windèsische verhalen" (Windessi tales), in *BTLV*, LXX, p. 447.

on the occasion of a funerary feast; a *sera* could not take off his mourning-apparel before this had been done.

Although the initiation ritual is a ritual for life, this is still life in a struggle with the dangerous snake. This creates a curious tension in the Waropen funerary rites. The feast for the dead is impossible without the ritual of men, because slaves are needed. On the other hand, it is impossible to do without the ritual help of women, because it is the women who take care of the corpse. In the funerary ritual there occurs a snake which plays the main role in the initiation ritual, as well as a bird which continues life after death. The women, associated with the bird of heaven, are therefore placed on the side of life in the funerary ritual, whilst the men, associated with the snake, are to be found on the side of death.

We find this curious connection further elaborated in the Serakokoi ritual, for there we observe the performance of the struggle which is called the *Inurisaira*, the men's ceremony of Inuri, the divine trickster. It is to be noted that this battle is not fought in the morning, as customary in other men's ceremonies, but in the evening, and also, that both men and women participate in it.

Basically, men's ceremonies and women's rites could hardly be combined in this religion, for the simple reason that at the men's ceremonies the women ought to be excluded. It is not very likely that the ritual for the dead, who were especially cared for by the women, was ever strictly taboo for the men, but conversely it is beyond doubt, that from of old the men did not perform their ritual inside the house.

The structural change in this society which entailed the transition from the secret ritual in the clan-sanctuary to the public performance in the open central portion of the house, also increased the importance of the women who had to lay out the dead. The men's feasts for the catching of slaves which belong to the funerary festival, take place in the same space where the women used to sing the funerary dirge. Once the ancient "total" clan-ceremonial has been linked to the *ruma*, the feast in which also the women make an appearance, i.e. the funerary feast, becomes the most inclusive. However, the exchange of presents on the occasion of funerary feasts is not common, so that for the time being there is little possibility that the funerary feasts will completely oust the initiation ceremonies.

It is evident that the shift of the religious centre from the *dama* to the *ruma* has also had as result that the women have become more involved in the ritual. In modern times we find them already in the *saira*-ritual (see p. 145). But it is not only the care for the dead, but also the process of birth which in this way is increasingly incorporated in the sacral structure. At present, *saira*-ceremonies are already performed for the still unborn child. In this way both birth and death, with which the women are most intimately concerned, are brought into contact with the ritual of initiation. As a result, the influence of the women in ritual increases; at least, they are no longer excluded from a ritual of men, performed in a distant, enclosed space. They now see all ritual enacted in their own home, a ritual in which they take an ever larger part.

AIGHEI

The Aighei is the primeval woman in Waropen mythology. She is Mimiandeisei, who continues life in death. She is also in general the female officiant in the mythical event which in spite of many changes of scenery still presents only few scenes.

She is the sister with whom Seraiani commits incest, whereupon he commits suicide for shame (85). She is likewise the woman whom Aimeri desires, or occasionally the sister who desires him; then she is called Aighei Ghafai (*ghafa*, moon) and Aighei Dorai (*dora*, heaven) and dwells with her father on the heavenly mountain Ghamusupedei. There Serawairusei with his long arm also dwells, who takes the possessions of men from their houses for his weeping daughters Aighei Fokei and Aighei Nanai (137, 138, 139). Mount Ghamusupedei is the mountain to which the heavenly woman flies back when she has been unjustly suspected by her husband. There heaven is also located (72).

The Aighei live there, in the other world, in heaven, in the village of women, where Seraiani, the initiandus, finds them, and where his younger brother has to experience the danger of sexual intercourse with these virgins, because he has to pay for it with his life (82). She also lives in the moon where she preserves the fire (197), and she is Aighei Wokakai (*wokako*, hot), Anirei's (i.e. Aimeri's) wife, who inaugurated the funerary ritual at Kai.

She is Aighei Indosi who is too clever for the snake Simundopendi and who introduces the mourning-garment of tree-bark for women (128, 129, 130). She is Aighei Rafenai who had intercourse with the deceased Seraghombarokui (and who celebrates the first *saira* for her child in heaven and who builds the first house?) (194, 195), or the daughter of Serasukarori (i.e. Serakokoi), the revived dead man (12). She is Aighei Ramasi, who together with Wisopi steals the leaves of life from the *sema* with which she is able to resuscitate the dead (132). She possesses the ancestral image Koghei which the younger brother has brought down to mankind (46, 82). She is the daughter of Gharopendi, the ancestor of Nubuai who descended from heaven (11, 12). She is, in general, the heavenly woman Aighei Inai (*inai,* mother) who is taken down to the earth by the man and who is ridiculed by the women there, whereupon she changes into a cucumber (82).

Among the Aighei there is also a constant rivalry between the elder and the younger sister. The elder sister is jealous of the younger because of the latter having a husband. When the younger sister marries the pig-man, the envious elder sister comes to kill him (114). The younger sister, Aighei Inai (*inai,* mother), Aighei Manai (*mana,* soft, sweet), Aighei Kuboma (*kuboma,* small child), Aighei Fasi (*fasi,* white) performs some task or other, usually at the request of "the old woman" or "grandmother", and as a reward the latter helps her to obtain a husband. It is e.g. only the younger sister who is willing to fetch food for the turtle-child of the grandmother (113). She is the only person who takes good care of the grandmother and who is able to drag ashore the floating house in which the primeval men live (201).

The grandmother (who also is an Aighei) is the mother of the turtle who is considered at Nubuai to be the totem-ancestor of the clans Pedei and Nuwuri (11, 12). She is likewise the mother of the *ghéa*-bird which regulates the seasons (see p. 340). One of the tales relates that once upon a time the *ghéa*-bird went on a journey together with the bird of paradise, some other birds and the cassowary. Because the cassowary due to his jealousy of the bird of paradise made the sailing of the boat very difficult, the others were finally able to tie him to the prahu by a ruse, whereupon the other birds flew up and left him. Due to the kindness of the turtle the cassowary is enabled to escape, but he requites the help he had received by catching the turtle in order to eat him. The frog is appointed gaoler, but he lets his prisoner escape. Thereupon the treacherous cassowary puts his heavy foot on the frog who has remained flat ever since. However, the cassowary also runs into a trap and loses his feathers. The *ghéa* and the bird of paradise have been able to spread, but the cassowary could not (161).

AUO

The natural rival of the Aighei is the Auo. In the myth of Serakokoi we saw her as the opponent of Mimiandeisei, the Aighei. It is among the Auo that the *ghasaiwin* find the souls of the children that have been robbed by the ancestors. The Auo usually build their house in a banyan tree or in a large cactus.

They are mythical beings who behave like pseudo-women. In their jealous infatuation they oust the legal spouse or carry the husband away to their dwelling in the trees. People who haven broken the sexual rules and who have had their soul stolen by the water-monsters, eventually change into Auo. In order to simulate pregnancy they swallow cooking-pots. They cannot speak with a normal articulation; according to my informants they speak with a curious nasal sound. And they are so stupid, that in their fright they drop the precious objects like drums, ancient plates, triton-shells, etc., when people startle them by knocking against the tree in which they live (123). And when they pursue the robbers, they use as their most dangerous projectiles reeds and orchids. When they dwell among men, they cannot even do the work of a young girl, spoiling the mush and making the freshly beaten sago useless by simply urinating in it. When it finally comes to a duel with the ousted spouse, they willingly accept a toy-sword made of the rib of a sago-leaf and so meet with a horrible end (119).

In spite of this stupidity and their gullibility these half-finished women are not so innocent. In this way Aimeri is also called an Auo, born by fire from a snake, whilst the Auo are his companions on his endless raiding parties. They possess the power to transform the ousted wife into a pig (108b). There even exists a certain similarity between the Auo and the *sema*. Like the Auo, the *sema* are also in the possession of riches. They help the initiandus to the chicken which produces riches (164). The two *sema* Aka Weroroi (dry atap) and Fema Weroroi (dry wrapping-leaf), or Notaperei and Notarerei, who have robbed the wife of Kuriserai (i.e. Uri) in order to devour her, fervently wish to cohabit with Kuriserai. Curiously enough, both *sema* are also called Aighei and Insosi, names used to indicate the primeval woman.

It is not clear which place the Auo exactly occupy in mythology. As unfinished women they definitely belong on the side of the night. Perhaps one may deduce from their relation with Aimeri and their resistance against the alliance between man and wife, and against the reunion of Serakokoi and Mimiandeisei, that they take a position midway between man and woman, or like Aimeri, between the initiation-demon and the initiandus. In that case they are the personification of the mythical discord, powers of darkness whose victim the dead person temporarily becomes, but powerless and therefore stupid and gullible, because Mimiandeisei still has to conquer them eventually, in other words, because, due to the ritual the heavenly bird is able to assume its radiant form again every time and because the darkness of the night is vanquished every day.[228]

THE PRIMARY INCEST

Acquiring a wife is closely connected, also in the religion of the Waropen, with initiation, because only when this critical period has been terminated the young man may go and look for a spouse. By the mere fact of its coincidence with the period of puberty, the initiation is closely connected with the growing consciousness of sex. The establishment of sexual relations is fraught with grave dangers, not only because the female sex is dangerous from a religious point of view due to the mystery of blood, which necessitates the protection of procreation and birth by means of various taboos, but also because relations have to be established with a group, which considers the barter-transactions connected with marriage as a struggle. Enough rivalry is connected with marriage to make this contractual tie a dangerous relationship.

In a mythical sense marriage is dangerous because in the last instance it is the alliance of the two component parts of the macrocosm, the alliance between two clans who are originally two brothers, the marriage of the children of brother and sister. According to the logic of Waropen

---

[228] The Auo show a striking resemblence to the Lise of Toradja lore. In his *Bare'e verhalen* (Bare'e tales), II, p. 28, Adriani writes: "In general, her character is that of the jealous woman who tries to undermine the married happiness of young couples, or of women who are pregnant for the first time. She tries to insinuate herself among the men by stripping the face of their wives and gluing it on to their own. In order to seem pregnant, they swallow the pounding block for rice. They always speak with repetitions of their own words and in an importunate and whining tone of voice. All manner of stupid things are attributed to them e.g. that they wear the necks of broken pots by way of leg-rings, that they have centipedes and millipedes in their hair in stead of lice, that their posterior is so round that they have to sit on the floor in a basket, that their children are born from their anal opening, etc. Whenever a woman with a disagreeable character has to appear, one has recourse to a Lise".

A. C. Kruyt, referring to the curiously ambiguous character of the Lise-figure, considers her to be a moon-goddess who fulfills a function in the cultivation of the rice ("De rijstgodin op Midden-Celebes, en de Maangodin" (The rice-goddess on Central Celebes and the Moon-goddess), in *Mensch en Maatschappij* (Men and society), XI, p. 109 ff.). She also reminds us of the *hiwai-abére* of the Kiwai; see G. Landtman, *The Kiwai Papuans of British New Guinea*, p. 312. Perhaps the Auo-figure is more widely distributed in the mythology of this area. I have not found any connection with the figure of the trickster.

mythology, incest is unavoidable at the beginning, if one is to start from one pair of human beings.

Incest is mentioned many times (23, 30). In the beginning of humanity we find a brother and a sister (14), or a mother and her child (29). Even if one does not place a couple consisting of brother and sister at the very first beginning, or a parent and a child, but a bird and a snake (135), or a man and a dog (29), the difficulty is not solved. Incest remains unavoidable. In another myth incest is not mentioned expressly. It is Seraiani ("it is the lord") who when drunk lays his legs on his sister's lap, whereupon he goes and commits suicide for shame. His mother revives him by boiling him up again, but he is no longer capable of living and grows into an ironwood-tree, from which it is not the crow but the cockatoo which will be able to liberate him. The sago-palm and the banana arise from his bones (85). The "grand-mother" who in the village of women boils her woman's belt for the helpful youngest Aighei, causes the successive appearance of a bow, a neck-support and a man. This man is nobody else but Seraiani who states that he has made love with his sister (201).

Incest is therefore dangerous, but it is necessary, for the sago-palm and the banana result from it. Also, the men who flee before the cannibalistic giant coalesce with a tree, from which the women liberate them (84). For the initiandi are confined in a forest, from which they are liberated by their marriage. Several myths on Japen Island concerning primeval origins begin with human beings who have appeared from trees.

In several tales humanity is originated by the death of the mother, who has to be killed by her own son. The son has to divide her body into pieces, from which the different clans appear (95). This is un-doubtedly another circumlocution for the primary incest, starting from the curious play upon words in the meaning of the word *muna*, which may mean both "to fight", "to kill", and "to make love". Waropen love-making with the accompanying blows and erotic scratching with the vicious "comb" of the sword-fish (cf. p. 88) closely approaches an actual fight.

Sexual relations are also implied in the curious request of the mother to nail her red apron (i.e. her festive and bridal apron) to the stair-pole (53). It is the very brother of the divine trickster who has to kill his mother (52). Incest even takes place in the sky, the moon being the elder brother of the sun who cohabits with his mother. During a raid the moon is killed and out of grief for this occurrence the sun

also commits suicide, after having tried in vain to overtake the incestuous couple. Since that time the moon has obtained its place in
the night-sky, whilst the sun is allowed to shine during the day-time(55).
Perhaps this incest is conceived as being continuously repeated, because
the same myth tells us that a great drought is brought about when sun
and moon converse.

With this incestuous relationship I also wish to connect those myths
which deal with the jealousy and the malevolent activities of the father,
the mother or the sister against the partner of the other party. The man
kills all his wives due to his desire for his own daughter (108a). In text
52 it is Aighei Wanai (*ghana*, kangaroo) who lets herself be killed by
her son Kuru Paisai, whilst in 105 and 106 it is Wanaibini (kangaroo-
woman) lusting for her own son, who kills all his wives.

Quite probably we may also connect this in a wider sense with the
motif of the jealous elder sister who is envious of the younger sister's
husband, whom the latter has usually acquired by performing a task
which the elder sister had refused to do (82, 113, 114, 201). The
stories about the jealous woman in a polygamous marriage who kills
her rival because the latter is younger (107), or because she has
remained childless whilst the latter did have a child (116), are perhaps
more readily explained by the actual social conditions, just like those
where the woman runs away from her husband when he takes a
concubine beside her (117), or when a man kills all his wives except
the youngest (55), as well as those tales which mention the latent discord
between the wife and her husband's sisters who live in the same *ruma*
(202). However, we find on the other hand that the neglected wife is
also the man's sister (110). Moreover, in this case the man, who is
married to the sister who was maltreated by his wife, is called Serarumbai,
whilst the brothers of the maltreated sister in text 202 bear the names of
Nurumbai and Serarumbai. These myths are therefore also quite definitely
focussed on the sacred world. It is not always easy — nor is it always
possible and permissible — to interpret the myths from one point of
view.

So, the incestuous marriage is reprehensible and dangerous, but nevertheless necessary. Whatever difficulties the candidate for marriage may
have to overcome, the marriage still has to be concluded. In this way
the myth often opposes two brothers, one of whom has to conclude the
marriage, whilst the other tries in vain to prevent it. In spite of the
action of the sun, the moon has committed incest with the mother (55).
The myth opposes two brothers who contend with each other for the

woman. But this two-sidedness of the pair of brothers has again to be taken in the mythical sense, as also becomes apparent from the arbitrary way in which one and the same name is once given to the one brother and then again to the other. In the myth every pair of brothers, or of man and wife, or of parent and child, may represent a higher unity, which may finally be dissolved in the unity of heaven and earth which is included in the Supreme Being.

It is usually the younger brother who concludes the marriage and it is the elder who in vain creates all kinds of obstacles between the plan and its execution. This accentuates once more the importance of deceit in the sacred world. It is an attempt to weight one side of the scales and still keep the balance undisturbed. The elder brother attempts e.g. to keep the younger away from the woman by leaving him on an island, but all in vain. The mother loads food, water and arms in a bamboo, placing the ancestral image of Koghei on top, and personally goes to the island where she produces human society, dying by her son's hand at her own request (50, 52, 53).

Other stratagems employed by the elder brother to keep the younger away from the woman are e.g. by keeping him a prisoner through letting his hand be caught in a tree, or by having him flung away by a tree which had been drawn down so as to act as a spring when released (51, 58, 61). But his attempts miscarry again and again. The woman Umai (star) possesses miraculous powers and every time when the elder brother returns home in the conviction of being at last the sole possessor of the woman, his younger brother proves to have rejoined her already (57). The younger flees into the earthen magic prahu (61). However, according to others the struggle for the woman is mortal. The two brothers murder each other and they are changed into rocks, which the Papuan father shows to his son when they go on their first voyage (57).

For the rest, it is due to the elder brother's own stupidity that he has not acquired the woman, because he is the first to find her when she floats on the sea in a gourd. But when he has smelt the gourd in which she dwells, he dislikes the smell and throws it away from his place in the forepart of the prahu, whereupon the younger brother in the stern fishes her out of the water and takes her unto himself (57—61). The image of the brothers in the stem and the stern of the prahu is again the representation of the two rival clans, of the "head" and the "tail" of the triton-shell. See p. 53.

THE HIDDEN FATHER

Highly popular are those myths which relate cases of supernatural conception. Generally they mention a male being who usually lives somewhere in the mountains, often on the heavenly mountain, or on the upper waters of a river, and who now lets some object or other drift down with the current or with the wind, the object being taken up or touched by a female being. This female being becomes pregnant and gives birth to a child. The object which causes this miraculous conception is a banana-leaf (95), or a leaf of the pandanus-palm (96), which is blown against the breast of the woman. It may likewise be a stem or a leaf of magic herb (90, 95), or also an areca-nut (91), the fruit of the bitangur (96), or an ancient bowl (97), which the man throws against the woman's breast. In another story it is his semen which he floats down the river in a bamboo, whereupon it is found by a woman who drinks of it (99, 100), or his comb which floats along and which scratches the woman (98). The male being dwells on Mount Ghamusupedei in the Kai area, often considered as the dwelling-place of the ancestors (91), or on Mount Inggorosai in the Woisimi region, which is also believed to be the haunt of the divine trickster Uri (100).

The pregnancy caused by this supernatural conception and the birth of the child are followed by an investigation into its paternity. But, unlike normal circumstances, the woman is unable to indicate anybody as the father of the child. Therefore all men present in the village are made to appear at a *saira* to dance there, so that the child itself may indicate the person who is his father. In the end it recognises him in an old lame man who is suffering from the itch. On this discovery all the girl's relatives flee in alarm, leaving the woman and her child to the old man. According to some, this is the cause of the distribution of mankind over the earth (96), which in other cases is explained as a flight for the monster Roponggai.

The woman is far from pleased with the child's choice and she heaps reproaches and abuse on her highly unattractive husband. The child, however, fully trusts its old father and this trust is not disappointed. The unattractive old man possesses miraculous powers. Bananas, fish or areca-nuts appear at his command (94). From the drawing of a boat loosely scratched in the sand, there springs forth a prahu when the tide comes in, and when the child is tired of boating, the old man puts down islands for him, like so many playgrounds (93). These are the romantic islets which still lie in the blue waves of Geelvink Bay as hearts of green coconut-palm leaves enclosed in a double ring of clear yellow sand and white foam.

Finally, the old man decides to reveal himself to his wife in his true form. He goes and purifies himself, or, like the Waropen put it, he goes and "bakes himself again" in the fire. He collects fire-wood from the motoa-tree and first flings his dog into the fire (94, 96, 98, 100), or he also adds his two lories (96), or the sea-eagle (89), or the bird of paradise (99), which rise again from the fire, alive. When finally the old man subjects himself to this purification, he is changed by the fire into a handsome young man; his diseased skin drops away from him, to be changed into all kinds of precious things: bracelets of different kinds (89), or blue cotton, whilst his bones are transformed into bush-knives and axes (90). In this way he returns to his wife, whilst the sea-eagle is perched on his bow. In this attractive young man she is unable to recognise her husband, but the child recognises its father. At last the woman also discovers his identity, and enamoured she lies down in her husband's lap, whilst he kindly remonstrates with her about her earlier revulsion.

I believe that at the base of this myth we again find the initiation ritual. The myth starts out from the unity of the initiandus and the initiator. The terrifying old man, who has to lean on a lance when dancing the *saira* because of his lameness, and who after his purification is praised for his vigorous dancing (94), is the initiation-demon or the initiator. The initiandus, the child, recognises the father in his hidden form, because essentially they are one. In the Numfoor - Biak version — which in view of the names alone must have greatly influenced the Waropen tales — the child is called Konoor, a name which is likewise applied to the future reincarnation of the old man, when he will eventually appear as the Messiah (cf. p. 320).

In the interpretation suggested here it is of course a difficulty that it is not the father, but the mother's brother who is marked out for the role of the initiator. This difficulty cannot be solved by the remark that the term *imai*, father, is occasionally also used to indicate the mother's brother.

I would prefer to find the solution of this difficulty in a change which was about to become consolidated in Waropen society. Of course, it is still the mother's brother who acts as the officiant in the initiation ritual, but he is steadily diminishing in importance in comparison with the father who organises the feast. The principle of social status is beginning to gain the ascendence over the clan-organisation.

In this way the father becomes the figure to which the child attaches the authority which formerly the mother's brother possessed. The Waropen father does not make himself felt as a very severe authority; he is rather the beloved older person who provides the boy with an opportunity for acquiring the qualities for which people will praise him.[229]

Since the decline of the clan initiation ritual the father is therefore the first person with whom the child will identify itself. His authority is not a strict *patria potestas,* but a safe comradeship. The father introduces the child into the circles which count.

However, this does not mean to say that there are no conflicts at all in the Waropen family, although these are not conflicts due to the imposition of a strict authority. As de saw before (p. 91) the father is kept away from mother and child before and during birth. Mother and child dwell in a position of sacral isolation and one of the important aims of the prevailing initiation ritual is the abolition of this seclusion. Gradually the child passes via a series of ceremonies from its mother's hip to its father's knee. During this period of seclusion the child has the mother all to itself and only slowly it acquires a still very small amount of independence. And then the father is waiting for it to lead it again (cf. p. 155).

Still, during the period of isolation the father forms a great danger to the child (pp. 250 and 258). In the myth we see him as the cannibal (p. 310). In this case also we have to think of an historical relation with the initiation ritual, where in these regions the anthropophagous giant is a common phenomenon.

Here, however, the cannibalistic giant is immediately identified with the father, because it is the adored father personally who puts an end to the seclusion of mother and child and who takes it away from its mother's breast.

The supernatural conception is explained simply by the fact that actually the initiator is the brother of the mother of the initiandus. Via the mythical unity of initiator and initiandus there arises a direct connection with the mythical incest, the supernatural conception being an attempt to avoid this incest.

Evidently the woman herself is also brought into the process of

---

[229] The girl is less closely connected with the father than the boy, although in the myth incest between father and daughter is not unknown. To the girl the relation of rivalry with her brother is of greater importance. The sister is jealous of the woman whom the young man marries (cf. p. 311). The brother helps her in all difficulties.

initiation, for after the discovery of her relations with the old man she
is deserted by everybody. The initiation-demon is not only present at
the *saira,* but his usual haunts are the forests. No wonder, therefore,
that in some myths the child always stretches its arms in the direction
of the upper course of the river where it knows that its father dwells.
Then the woman's brothers take her and the child to the terrifying,
hidden father (90, 91, 92). It is curious — although in this world of
ambiguous and bisexual creatures not unusual — that conversely a female
being sometimes fructifies the man; in that case the male being is the
snake Seraiomi (96, 97).

Also in the Waropen versions of this myth the figure of the morning-
star often makes an appearance, a figure already known from the
Numfoor tales on this theme. Fir it is the morning-star who possesses
the magic potion by which the woman is enabled to conceive in a
supernatural way. When the morning-star descends and is sitting in
the coconut-tree stealing palm-wine, it is caught. In vain it tries to
escape and to regain its liberty by offering all its riches, even a part
of the magic medicine. The old man may only release Sapari (the
morning-star) when he has obtained the whole of the medicine.

In Waropen mythology this episode is undoubtedly connected with
the central contrast between night and day. The continuation of the
cosmic process is ensured by the change from darkness to light, and the
ritual is meant for the purpose of making this transformation possible.
It is for this reason that the marriage ceremonial extends over the
greater part of the night, to be concluded after sunrise. The morning-
star which stands on the border between night and day, possesses the
secret which makes it possible to unite the two partners. It has to
relinquish this secret before the day has definitely arrived, as otherwise
the world-order itself would be endangered. In this way the women's
medicine at this critical moment falls into the hands of the initiation-
demon who has not yet thrown off his hidden and terrifying form.

Now he goes and subjects himself to the purification by fire, a
purification in which also the bird, the lory, the bird of paradise or
the sea-eagle, participate. It is only the reason for which also the dog
is allowed to participate in this purification which is not clear, but
neither is the position of the dog itself in the other myths.[230]

---

[230] Perhaps the dog was also allowed to stay with the initiandus in the forest, but
in any case the dog is the friend of man, accompanying Kuriserai to the
sema to kill them (133—136), whilst according to others he is the first who
dares to eat human food or to drink from the palm-wine (152). On Japen

The ugly skin of the hidden father changes into all kinds of riches, for the successful ritual ensures increase of treasure to the organisers of the feast. Perhaps there also exists a connection between this process of purification and mummification, for during the latter the women who look after the corpse take off the skin, whereupon the dead person assumes a curious white colour. Possibly the mythical significance of this process was also a rejuvenation of the dead. However, in this case we have to rely on guesswork, because at present mummification is prohibited in the Waropen area.

The hidden father does not occupy the same central position in Waropen mythology as he does in Numfoor lore. He is also a culture hero who lays the islands in their places and who is able to make boats. And finally, like other important figures in the Waropen pantheon, he has left the land of the Papuas for the country of the Strangers. According to the Waropen he did so because he was ashamed of the contempt the people had shown to him before his revelation (93, 94, 96, 99). According to the Numfoor he retired when mankind had proved not to believe in his miraculous powers and had made the plan to go looking for food on Japen Island, being afraid that they would not obtain sufficient food from Mansren Manggundi.[231] According to others, there was a woman who would not stop keening the funeral dirge for her son, saying that he had died, whilst Mansren Manggundi insisted that death did not yet exist. Since that time man has remained subject to death. In the country of the Strangers the hidden father became the father of the Europeans, who again owe their higher culture to him (96).

In the tale of the hidden father Waropen religion has attained its most profound devotion, I believe. For once there is no mention of incomprehensible large devouring monsters or of divine fools, but of the hidden unity of the miraculous god and his child, which recognises him, in spite of the lowliness of his form. The intimate union between father and child is the normal reality of the Waropen village. Although in this religion with its matter-of-fact kindliness devotion does not attain

---

it is said that the dog is guilty of having ended the mythical period, for anciently the animals were able to speak, but when the dog started to quarrel with the pig, they lost this accomplishment. According to others the mythical period was terminated because the dog observed an ugly old man having intercourse with a beautiful young woman and spread the tale.

[231] Numfoor: "the Lord himself". The Numfoor also call him occasionally "the man with the itch".

the depths of other religions, its presence here is striking enough and it readily explains the extraordinary popularity of the myths concerning the hidden father.

INITIATION

In the foregoing sections the importance of the initiation has been repeatedly mentioned, for Waropen mythology mainly starts out from the ritual of initiation which, with its feasts, its raids and anciently its sanctuary, meant practically everything in religious life.

The first myth to be mentioned in this connection is that of the snake Siroei. There is a woman who by mistake drinks a snake-egg, whereupon she gives birth to a snake which she puts in a plate. However, when this snake starts to grow, it appears to be a devouring monster which threatens to exterminate the whole of mankind. The decision is therefore taken to bring the snake outside the village and to kill it there. Weeping, his mother states that he is to be initiated. When the decorated snake has been taken away in a prahu, it is the men from the interior who in the end are willing to kill him and not the ancestors in Waropen mythology.[232]

However, after its death the snake revives again in the form of an iguana.[233] The latter animal is caught by men who decide to eat it and for this purpose they boil it in a pot. Again and again the iguana lifts its head over the rim of the pot in order to make known its true identity, but when the people hurry near on hearing the sound, the iguana quickly retires. In this way the men finally proceed to eat the snake, whereupon they all die. According to text no. 25 it is after this event that humanity was scattered.

It is this snake which is still always present at the *saira* in the shape of a roll of sago (*ghoifio*), placed on a plate like the real Siroei between the other sago-cakes. The name given to this snake in the *ruma* Sawaki was Waisimaimuni, which is the same as Simaimuni, the other name of Siroei, or Waisimai, the name used by the people of the interior for the initiation-demon (88).

The snake Siroei, the monster which devours everybody and which

---

[232] When the snake is taken away it delays the voyage by throwing several objects over board in numerous places, which then became taboo (24). According to text no. 26, the snake is first killed once in the forest, but it revives, whereupon it is killed again inside the house and now definitely.

[233] It was not proved that the iguana occupies a special place in religion; however, the drums are covered with iguana-skin.

finally is also eaten itself, is again another aspect of the unity of the
initiator and the initiandus. Due to his central position, Siroei, like the
all-devouring Roponggai, is highly honoured among the people. At
Ambumi I was shown an ancient bowl from which Roponggai was said
to have drunk, as well as other objects used by him. The inhabitants
of the Moor Islands are said to possess Siroei's drinking-cup and other
objects he had used; the sacral value attributed to these objects is so
great that they are only to be shown on the occasion of *saira*-ceremonies.

The following description of the snake represented by the roll of
sago was given by an old *ghasaiwin* from Nubuai: "He has one set of
privy parts on the side of his back and one on the side of his face (i.e.
he is bisexual) and he wears his hair in long tangled strains (like old
women), and he has two protruding teeth and a head which can go up
and down. The people do not see him and he hunts in Fonaro Creek
(near Wonti, where this woman had been born; it was therefore a
local representative of the initiation-demon). In my sleep I can see
him and when we are rowing, I see him, whilst he pursues them (i.e.
the people). Because he belongs to me, I can make a model of him for
the initiation feasts. It was he of whom we made a model at this feast.
At the *muna* it is he whom we lament. He dwells behind the sago-forest
and his wife is called Ghaisupini and the other Ereraghoni. When we
sleep I go (in my sleep) to (the river) Ghobari and to that same Fonaro
Creek".234

Obviously, there exists a relation between the devouring snake Roponggai and the devouring snake Siroei. Siroei is as huge as the snake
Roponggai, who is likewise tended in a plate by his weeping mother
(35). Siroei's mother is Nuawirakoi, who is also the mother of another
mythical creature, called Werani or Weranarai. This Weranarai is again
a brother of Roponggai. Although, therefore, the snake Roponggai is
not completely identical with the snake Siroei — because Roponggai
is only the devourer and not particularly the initiandus — the difference
between them is not much greater than between a couple of mythical
brothers.

In several myths which deal with the theme of a devouring monster,

---

234 *Riaigha ghare uraina ruaibo we rengga, ghare uraina renado we renggagha.
Woraifarema karaba. Ienasa worukigha teraio. Woraigha kapara ma bo. Afa
nungguigha kitiri ewomo, Iadara Fonaigha ghaidogha. Renakangga rasiro.
Amboangga rasirio, iadaiki wea. Matanggu rananueka rawei wa sairagha.
Sairani iari ambebabei wea. Munani fimbo angganisi wea. Minana aroifono.
Ribinani Ghaisupini, eno ani Ereraghoni. Fimbo kenakangga rara ma Ghobari,
fimbo rara ma Fonai sigha.*

it is boys who have to start the struggle, just as it is the boy Kirisi Aimeri who defeats Roponggai. Once the snake which lives in the belly of a woman has been defeated, he changes into a number of trees. The death-throes of the monster cause the earth to split, whereupon all men are engulfed except for one couple who then in their turn become the primeval couple.

The devouring monster is not always a snake, but sometimes a pig (28), an iguana (27), a crocodile (31), and even a bird (29). The corpse of the killed iguana, like that of the dead body of the mother (52), is divided among a number of compartments of a house, from which the different families take their origin (27). Of the killed crocodile it is briefly said that his head was changed into the central clan and his tail into another clan (43). The killed pig is transformed into all manner of riches (28). In another myth all attention is concentrated exclusively on the danger of eating sacred animals; there, humanity is exterminated upon having partaken of one single sacred tortoise (32).

The mystic aspect of the oneness of initiandus and initiator is expressed in the oneness of the child and the hidden father. The other, more dangerous aspect of this oneness is expressed in the figure of Siroei, and this we find also in the tales about the flight for the anthropophagous father. In vain the mother and her brother try to protect the boy against this brute. The boy has to flee and by means of a tree he has himself flung into the other world, where he finds the heavenly woman whom he is allowed to marry (82, 83).

In the myths we often find the theme of the so-called "magical flight". Among the Waropen it is not always a pursued person who drops some object or other, in order to hold back his pursuers, but he attempts to escape by jumping from one tree to the other. When the tree in which the boy has climbed is finally cut down by the anthropophagous father or the cannibalistic giant, the boy is able to jump into another tree due to the force of his noble status (82, 84).[235]

Text no. 84 relates how the cannibalistic giant catches the boys in the forest in his scrotum, which he drops over them like a net for catching bats, whereupon he locks them in in a jar. When he is about to devour them, the boys escape, but in their flight from one tree to

---

[235] In the eastern part of Netherlands New Guinea the cannibalistic giant of the men's society pursues two children who jump from platform to platform; see Jac. Bijkerk. "De geheime mannenbond op Nieuw-Guinee" (The secret men's society in N.G.) in *Tijdschrift voor Zendingswetenschap, Mededeelingen* (Journal of Missiology, Communications), LXXV, p. 116 ff.

the other they finally become entangled in the trees. From these they are liberated again by the women whom they consequently marry. The jar is the image of the closed young men's house, like the triton-shell, and of their sacral confinement in the jungle, from which they are freed to go and marry. However, sexual intercourse is fraught with considerable danger, for when the man is ridiculed by the woman, he changes into a bat.

It is not always fear of a cannibal which causes the boy or the girl to flee from society; the direct cause is often an adhortation of the father or the mother. A child is berated by its father or mother, because it thinks that it may freely eat their areca-nuts. "Go and marry the heavenly woman; she will give you as many nuts as you like", the father remarks, annoyed. For people in love give each other areca-nuts by way of a small present. After this remark the great journey to the other world starts (57, 59, 75). In the confused myth of Werani or Weranarai, Siroei's brother, the flight from society is due to his mother's or his wife's refusal to give him mush. Then Weranarai withdraws into the bamboo forest, from where the women would like to release him in order to make love with him; in this case they are women from the interior. Werani therefore is somewhere deep in the jungle (87).[236]

Occasionally it is the sister who urges the young man to go and look for a wife, for the man runs away because he is angry with her. After his return his sister is so happy that she carries her brother around on her back, just as the bridegroom is carried to the ceremonial bedstead, or the child is carried around by its mother's brother on the occasion of the *saira*. Now the sister may take off the signs of mourning, because her brother has returned from death (65, 66). But she is jealous of the heavenly woman, whom her brother has brought with him; she ridicules her, so that she withdraws (64, 82).

However, the initiandus has to overcome numerous difficulties and to pass through many trials, before he has acquired the woman. He is not only endangered by the stratagems of the elder brother and by the

---

[236] It is said of Weranarai ('he sings inside the bamboo'?) that the women like to listen to his beautiful voice; could this be a reminiscence of the use of bamboo flutes in the initiation ritual? The bamboo flute is also known in the ritual of the tribes West of the Mamberamo river, i.a. among the people of the interior beyond the Waropen. The Waropen personally have an old-fashioned flute with two stops, which they have not borrowed from the Ambonese gurus and which, according to the informants, is still in use as a toy. At my request a model of this type was made; however, in actual practice this flute has already been superseded by the school-flute with several stops; see p. 356.

difficulties of the journey to the other world, which takes months and even years, but his intelligence and his courage are also tested time and again. Moreover, the heavenly woman is a sacred being. The man who puts his head on her lap dies and he can only be revived again by a miraculous knife (75). The miraculous powers of the heavenly woman are also demonstrated when the man takes her to his village. The people there do not want to accept her because she is an "old woman" (i.e. a mythical being) and they first have to see proof of her powers, when she creates an abundance of riches, whereupon also the other young men desire such a wife (63, 64). When the initiandus returns from heaven, he too, is so sacred that the cooking-pots explode in which the people wish to cook for him (65, 66).

Examples of the trials the initiandus has to perform in the myth are e.g. that he has to cut apart the two old women who are grown together (65, 66), that he has to defeat the "grandfather" of the heavenly woman as the latter is not willing to consent immediately to the marriage (63), that he has to remove the monsters whose stench is obnoxious to the community (64), that he has to fetch jewels from the mouth of the snake or from heaven or from the edge of the village (74).

In one of the tales it is an insatiable boy, a kind of Gargantua, who is banished by the people because of his voraciousness. In banishment he performs all kinds of miraculous deeds, eliminating monsters and liberating people from the trees which grow on them. But when he finally kills the man from heaven who devours the people, this young Hercules is crushed by the falling body. However, he is revived and thereupon he also resuscitates society. Then he stops a ship and takes the rice from it which the people still eat at present (78).

Due to his noble status the initiandus is also able to perform miracles, he (or she) says: "If I am a real noble, I shall be able to perform such and such a miracle". It is the new motif of a nascent ideal of chivalry which is beginning to make itself heard, for in these times it is especially the nobility for whom the saira-ceremonial is performed.

On the other hand, the idea of a mythical death is beginning to fade. Mythical death implies revival, a revival which in mythology is usually combined with a renewed adaptation to normal life. The initiandus behaves as if he is unable to speak or to walk, etc. This theme is treated in the myth concerning the man who catches his wife in the very act of committing adultery with a crocodile; the resulting crocodile-children he kills, but the human child he places inside an orchid. There it grows up in a miraculous fashion and when it has become adult, it learns to

fish, to eat sago, to make fire, to use a knife, to hunt, and finally to go
head-hunting and to feast. Every time the child becomes acquainted
with a new element of culture, it temporarily drops dead; then it is
found again by the fathers, who give it clothes and who make it
marry (86).

But in the end the supernatural powers also assist the boy in his
tribulations. The heavenly woman points out the dangers and saves him
from death by making an image which is killed in his stead (74). In
text no. 80 the mouse-man goes to look for a wife. The girl indicates
the dangers he will have to overcome. Often it is animals who render
their assistance: the millipede stings the heavenly woman so that she
cannot escape, or it stings her first and then it gives medicine to the boy
to cure her. It helps the boy when he is given the test of collecting a
number of beans or to pick the right woman from a number of identical
girls. It transforms its own body into a house for the boy (76).

It is the honey-suckers which are able to lead the boy to heaven when
the crow proves unable to do so (76, 197). The chicken also proves
to be a helpful miraculous bird; it is to be found at the place where
the wind takes its origin and it bestows riches on the boy (75, 77,
164). When it has come into the possession of the boy, the chicken
helps him to get a wife; twice it saves his life when this woman wishes
to kill him (77).

The boy also has to know stratagems in order to obtain the woman.
She cannot escape when she has been deprived of her dress (75, 76,
197).[237] In another tale the women first play a trick upon the man.
He thinks that he has discovered the women in the water and then he
starts to dig a whole bay in order to take them out, or he tries to drug
them bij means of fish-poison, whilst in actual fact he is digging
for the reflected image of the women who are dangling their legs from
a tree over the water. Finally he invents a ruse; he lies down as if he
were dead and when the women come to lament him, together with
his mother, he suddenly lays hold of them and makes love with them
(67, 68, 69).

The myth is far from being the rigid deposit of a tradition which is
often unjustly depicted as the immutable base of the so-called static
primitive society. To the contrary, the myth is intimately connected
with reality. This is proved by the modern initiation tale where Sièn

---

[237] This theme is widely known, both in the Ternate-Tidore region and in the
Numfoor area, cf. F. J. F. van Hasselt, "Nufoorsche fabelen en vertellingen"
(Numfoor fables and stories) in *BTLV*, LXI, p. 535.

Ara (Wandamen for Malay "Tuhan Allah", God) descends from heaven to make a sewing-machine from an *aiwo* for the poor deserted boy and to tell him, how he will marry the king's daughter and become king himself (81).

It deserves notice that in this myth God reveals himself as the Father, for in the preaching of Christianity great interest was shown in the announcement of a merciful and active Lord of Heaven, Who extends His care over all men like a true Father, greatly different from the Waropen Supreme Being, the *deus otiosus*. For a society where one special group in the clan asserts to be god, *sera*, the idea of one single God for all people is very striking. Also Sièn Ara arrives in the night in this myth, to withdraw again into heaven in the morning.

Whilst the myths go into great detail concerning the trials of the boy, they are only brief as regards the initiation of the girl. The reason why the girl withdraws from society is again, that her father grudges her the areca-nuts (57), or because of his incestuous desires (60), or because his wife gives him his own child to eat (62). The girl then has to drift along on the sea in an enclosed space, a fruit (61), or a gourd (57, 58, 59, 60). The mother of the heavenly bird is banished to the sea because her husband suspects her (102), or because she has killed his first wife (104). She is bound and flung into the water, or she is enclosed in a basket or in a *dama*, the young men's house. This enclosed space, like the stone jar in which men are confined (5, 84), is the mythical indication of the triton-shell, i.e. of the young men's house.

Mostly the tale of the drifting woman continues to tell that she is first found by the elder brother, who is, however, of the opinion that she smells and therefore throws her away. Then the younger brother discovers her inside the gourd and he marries her. Manieghasi's mother has to liberate the two old women who are grown together (104). It is also of importance that the old woman who floats on the sea teaches mankind how to have sexual intercourse (62).

It will hardly be accidental that the floating woman bears the name Umai, star, for she is associated with the night-sky. Probably consciously adapted to this is the story of the neglected crippled sister who also floats around on the sea until she is completely red and loses her loin-cloth, whereupon she becomes the mother of Siki Maki (Manarmakri is the Numfoor name of the hidden father). Siki Maki has red fingernails and he waylays women (110, 111). The red apron worn by his mother is the festive apron of the dancing girls and of the brides.

The red colour is likewise reminiscent of the red glow of the sky during the sacred transformation of night into day.

THE BEGINNING OF THE MYTHICAL PERIOD

If one were to look for a logical sequence and an intelligible connection between the events in Waropen mythology, one would be sorely disappointed. The mythical period is not to be compared with an historical era in which happenings may be grouped according to certain tendencies given in time. The Waropen has no scholarly interest in the past and he is unconcerned about the lack of unity in the conception of mythology.

In this way, at one time society is originated by the death of the mother who falls by the hands of her own son, whilst at another it comes forth from the incest between brother and sister. The divine trickster may be killed, and yet depart to the country of the Strangers. Manieghasi can be the bird of heaven and at the same time the sea-eagle which hunts for fish near the village.

The mythical world just happens to be different from the profane world. The story of the supernatural conception is far from proving that therefore sexual intercourse is considered as superfluous; this possibility was flatly denied for normal life by the informants. One of the most pious pagans was able to believe in the miraculous powers of the ancestor descended from heaven, and to laugh at the same time about the possibility that one human being, even if he were an ancestor descended from heaven, would be able to open up a whole river single-handed.

The mythical period does not start logically with the creation, but with the birth of human civilisation. Geographically and culturally the Waropen can only imagine a kind of unordered world in these primeval beginnings. Regulating is at the same time creating. Although the origin of society may have already been implied by the incestuous relation between brother and sister, every village may still quite well possess a tale about its origins of its own. Nobody is shocked by the difficulty that there is very little difference between these tales of the various villages and that every village anew considers itself as the cradle of humanity. For all villagers the world over their own village has always been the centre of creation.

The myth concerning the origin of Waren starts out from a couple which dwelt in the jungle. There they kept a pig and the pig found the way to the sea (3). According to others this first couple consisted of a brother and a sister (14). Usually the discovery of the sea is directly combined with the union of the sexes, for originally the primeval women lived in the forest and the men dwelt on the coast. Then the pig acts as a guide for the women who consequently discover the men in the floating young men's house under the care of an "old man" (1, 2, 6).[238]

The pig always plays the same role, viz. that of a guide to the women. According to my informants the pig was unwilling to stay in the house because this continued to turn round and round on its single pole. The pig represents the women's dowry and it is eaten, the "old man" having to set the example (2, 14). The dead pig changes into a variety of riches (2).[239]

Before there existed a real human civilisation worth mentioning, i.e. before the two sexes had united, mankind led but a sorry existence. At that time there did of course not yet exist the co-operation between the sexes, in which, as at present, the man has been given the sea as his field of activity, whilst the woman has found her work in the sago-forest. The man catches fish, he trades and he wages war; the woman beats sago, she collects molluscs and she rules the household. It is only the products of the labour of both together which provide a diet fit for human beings. Sago or fish by itself alone is as unusual to the Waropen as potatoes without anything else to a Dutchman or rice without condiments to a Javanese.

Now the women in the forest had only sago by way of food or merely inferior plant-food, whilst the men had to try to supplement their diet of fish by eating charcoal (1, 4, 6).[240] Only when the men and the

---

[238] As regards the floating young men's house we shall have to think probably of the prahu which among the Numfoor — and occasionally among the Waropen — is the prototype of the primeval house and of the clan-organisation (cf. p. 53). To this also the story of a floating house refers (29), whilst the tales about the floating islands (8, 9, 10, 70) will also have to be associated with this tale.

[239] The pig still remains the pet of the women who take care of it; see p. 354. In text no. 28 the devouring monster which is defeated by the boys is a pig. It did not become clear what sacral significance the pig may have among the Waropen.

[240] It is not clear why the number of these primeval women has been put at ten (2). If for once we should be allowed to think of the symbolical value of numbers, I would like to draw the reader's attention to the word *nunggu*, which means both "human being" and "twenty". In that case the ten men and the ten women together constitute the number *nunggu*, twenty. The three primeval women and the three primeval men might be connected with the circulatory clan-system (6).

women had become united, a real human existence could begin. With
enthusiasm the myths describe the caution with which these wild
primeval people start to eat the delicious food (7), which really satisfies
them for the first time (14). Again it is the grandfather or the grandmother who dares to be the first to eat the unusual food (2, 5, 14).
The first creature to drink palm-wine was the dog (152).

THE END OF THE MYTHICAL PERIOD

That the time of the miracle-working ancestors is past and gone is
clearly proved every day by the life in the village. It is from this life
that religion originates, and not from cool speculations on the marvels
of creation. Religion rose from the tears wept by men and from the
longings living in their souls. It is not the desire for creating interesting
contrasts which has made the sacred world so completely different from
the profane world; every day one may see people suffering in the
village — relatives die, disaster strikes, quarrels prevail. Even in the
quiet Waropen villages nobody is likely to find a paradise.

The mighty culture-heroes who put the world in order and who
instituted the ritual, Uri, Aimeri and the hidden father, have gone
away to the country of the Strangers. This also explains to the Papuas
why their civilisation remained so much less developed than that of the
Strangers. The culture-heroes gave those gifts to the strangers which
they withheld from the Papuas.[241] And moreover it is often the
stupidity of mankind itself which has brought about this state of affairs.
If they had not ridiculed the hidden father for his lowly appearance,
if they had paid greater attention when eating the iguana who after all
was the initiation-demon himself, maybe there still would be a paradise
on this earth.

Contact with the sacred world is soon interrupted. The heavenly
woman easily finds her dress again, with which she returns to heaven;
an unfounded suspicion of her husband causes her to fly back (72).
By numerous stupidities mankind itself disorganises the world-order. At
the request of their children the people of Waren attempt to drag
the moon from the sky, but on the advice of the "grandfather" the
moon washes them away with heavenly water. The Warenners are

---

[241] And for this very reason the Christian religion is not completely strange from
the Waropen point of view; it is rather a form of their own pagan religion,
and even better, because it is the religion of the rich Westerners.

dashed to pieces and they are transformed into bananas, tortoises, dolphins, tree-kangaroos, pigs and cassowary-birds (140). Probably we may think here of the way which, according to the myth, formerly led into heaven via a bamboo, an image also known from other Indonesian religions. In another story the Warenners are changed into termites or marsupials or pieces of rock, whilst their prahus are transformed into all kinds of trees; this happens when they have gone out to kill Serawairusi who robs the possessions of men with his long arm and carries it to Mount Ghamusupedei (137, 138, 139).[242]

Often a huge catastrophe is imagined at the beginning of time, leading to the dispersion of mankind. In this way the people scattered in all directions in order to escape the anger of the snake Roponggai, or when in their stupidity they had eaten the snake Siroei, or because they were frightened by the appearance of the hidden father. When they flee, one couple usually remains, which is then unavoidably faced with the mythical incest.

The theme of a primeval catastrophe is elaborated in an individual way in the story of the flood. The direct cause of this disaster is again a rather insignificant mistake, usually the stupid curiosity of a few small girls who make fun of "grandmother" who, when working the sago in the sago-forest, has intercourse with a crocodile. Angrily the grandmother summons Aiwui (awa, high tide), by sending the crab with a message. Only the wowa, a kind of fish which is able to fill itself completely with air or water, can take in sufficient water; with this water it swims upstream past Sanggei village and then releases the flood. According to others it was the initiation-demon Gharaisimai who brought the flood along. The water drives the people in all directions, i.a. to the island of Japen, where people then began exchanging marine products against garden-produce, for at Serui and in other localities there exists a native market. According to others again, mankind died because of the flood (16).

Other faults or mistakes may also have been the cause of the flood. So, for instance, the eating of eels which were taboo for the Waropen,

---

[242] It is not clear whether this transformation into animals etc. is simply a punishment, or that it is connected in some way with totemism. It should be noted that beside animals, also fruits and trees are mentioned; in the Pacific area totems may be both animals and plants. Professor de Josselin de Jong thinks of double totemism in connection with a double unilateral tracing of relationship.

according to Van Balen (36).[243] Or the flood is summoned by a woman who is angry because her brother has been killed in a fight for some insignificant reason. Also because the people rob an old woman (17). The flood washes away the two grandmothers because they beat to death a marsupial rat which had been raised in a jar (20). The flood arrives because in the country of the dead a woman lets the torch go out and treads on an old woman in the dark (22).

After the flood one couple of human beings remain alive and they despatch birds to see whether the earth is already dry. The barbet cannot find anything and returns without having anything to report, whilst the crow feasts on the floating corpses of the people (21). For this reason the crow is unable to liberate the man from the tree into which he has grown, or to fly with him to heaven (85, 197). After the flood, a couple remains, a man and a woman, or two brothers (the ancestors of the original village of the Waropen on the Woisimi river) who then produce present-day mankind from an incestuous relationship (18, 22, 23).

So we see how death comes to mankind because of some small mistake or some little failing. Death cannot be conceived as the necessary and regular end of life, but it can only be seen as the result of some unfortunate accident. At the same time it is closely connected with initiation. After all, it is the snake Roponggai (the initiator) or the snake Siroei (the initiandus) himself who kills men, to be killed in his turn. In a mythical sense, initiation is the introduction to incest, because after the primeval catastrophe only one couple remains.

Therefore, if this primeval catastrophe is caused by the initiation and by the beginning of sexual intercourse, it follows logically that during the mythical period when people still attained the Waropen ideal of a very long life and all therefore were "old men and women", sexual intercourse cannot have existed. Waropen mythology actually draws this conclusion. It was the woman who floated around on the sea who taught mankind the normal way of procreation (62).

The initiandus can only enter into contact with the sacred world by also crossing the border which separates the land of the ancestors from that of men. The initiation-ritual has to lead to the mythical death of the initiandus, who at the same time remains alive in actual fact. In Waropen religion this concept is actually known, but it has been less elaborated than in other religions with a similar structure. Here again we may observe the individual development of the ritual among this

---

[243] *BTLV*, LXX, p. 505.

tribe, where the young men's house was torn down and where the ritual changed more and more into the show of a potlatch-ritual for children of chiefs, a ritual which tends increasingly towards greater publicity.[244]

THE RETURN OF THE MYTHICAL PERIOD

The mystic longings in the hearts of men are never extinguished; the hope for a better future will not die. The sufferings in this world cause the Waropen to look forward to another world where his ideals of a long life in a rich country will be realised. The people believe in a Saviour who will connect the wonderful mythical past with an even more beautiful future. Who this Saviour is to be is not quite clear in Papuan mythology; in any case it will be one of the culture-heroes who had gone to the country of the Strangers. Usually one expects the hidden father, who after all does agree most with the expectations of the pious heart (93). Then the Papuas will have paid sufficiently for the mistakes which expelled the culture-heroes from the land. Then they themselves will become the first, the masters of the Strangers, who in their turn will become the servants and who will have to cede their wives to the Papuas.

In the Biak-Numfoor world the belief prevails that the Saviour will be preceded by a *konoor*. Konori is likewise the name of the child which perceives the true identity of the father; the same essential unity of father and child therefore again plays a role. In the whole of the myth of the hidden father a strong Biak influence is to be felt, as becomes already apparent from a comparison of the names: Mandara-maki, for instance, is the Waropen form of the Numfoor name Manarmakri.

The last messianic movement which some years ago held the Waropen coast in tense expectation again came from the Biak Islands via Kurudu. As a sign of the speedy arrival of the lord of the world the *konoor* gave a flag to a prahu from Kurudu. Visions were seen all along the Waropen coast and prophecies were made. In Nubuai the rumour went that Tuan Ara would send a red travel-pass (i.e. an official letter) from heaven; so Christian expectations had also been absorbed. The people started to decorate their houses looking forward to a future which was

---

[244] It will be evident that in the interpretation of the myths I have tried to follow the example of K. Th. Preuss and of W. H. Rassers, although in this descriptive ethnographical study I have not considered it necessary to refer constantly to the writings of these two scholars.

to be one great feast. Various curious rumours were spread among the men who sat looking out on their front galleries, talking and waiting, i.a. about curious hauls of fishes which might be taken in at certain indicated localities. However, the whole movement spent itself quickly, once a few arrests had been made and only an embarrassed smile remained.[245]

However, the messianic expectations have remained. They will emerge every time when the circumstances provide sufficient cause, or when they find a new solution in Christianity which has now also been accepted by practically all Waropen.

---

[245] See the interesting article by F. C. Kamma, "Levend Heidendom" (Living Paganism), in *Tijdschrift voor Zendingswetenschap, Mededeelingen* (Journal of Missiology, Communications), LXXXIII, pp. 187, 289, 387.

CHAPTER EIGHT

MATERIAL CULTURE

TECHNICAL ABILITIES. ORGANISATION OF LABOUR

In this chapter I have collected the data concerning what is called in ethnology by the not quite satisfactory term "material culture". Over against "spiritual culture" the word "material" suggests a restriction I shall not maintain in this chapter. In this connection I believe some remarks on the knowledge of nature not to be out of place. The section on children's games was inserted here, because the material was insufficient for a separate chapter.

As already remarked on p. 37 Waropen technique in general is not characterised by particularly exquisite workmanship or by great skill; nor does the Waropen develop much imagination in his work. It is always the same objects to which he devotes particular attention by applying carving, whilst the motifs are practically always the same, without any variation. Only rarely will the workman apply himself to especially difficult material, e.g. by working in hard wood or bone. He is satisfied with perishable products of soft wood or even ribs of the sago-leaf. It would appear that he only proceeds with difficulty to continuous labour, with the result that it takes a great deal of time until a product is definitely finished, at least, if it has not been put into use even before having been completed.

Still, the Waropen as such are not lacking in technical capacities for producing a good piece of work. This is proved by several prahus which have often been hollowed out with rather primitive tools. In many of the larger ones one admires the slender lines, as well as the daring with which the laborious hewing has been accomplished without measuring instruments. Equipped with often successfully executed decorative pieces on stem and stern the large Waropen prahu is an elegant vessel. Also quite some pains are often taken in the carving of the paddles. If beside these one puts the carved neck-supports and the carved tools for preparing the sago, like beaters, spoons, bowls, etc. one has assembled practically everything which in a certain sense belongs to the field of applied arts.

60. Village main-street at Ambumi

61. Rest-house at Nubuai

62. Road from Napan to Weinami

63. The houses return slowly to the water

Bamboo is also carved. Examples are provided by the many bamboo tubes in which formerly the women wore the strands of their hair, examples which prove again that in and by itself the Waropen is not lacking in technical skill. But for the rest his energy does not seek an outlet in the field of bamboo-carving either, although the material is plentiful. After all, all drinking-water is fetched in bamboo tubes, all sails are attached to bamboo masts, but his artistic sense has never been directed towards these objects, partly, of course, because this material has to suffer more, perishing more quickly as a result. Probably all objects to which the Waropen devotes special care by applying carving are objects connected with the world of religion.

The same may be said of the basketry-work made by the women. The plaiting woman is not interested in skillfully playing with her material; she repeats endlessly one and the same manipulation following the ancient model, and often quite carelessly at that. Still, she is not lacking in artistic skill, as is proved by the often daring manner in which the decorations have been applied to dancing-aprons, using simple strips of white cotton on a red ground. They also obtain good results with bead-work, both in inventing patterns and in combining colours. However, a great deal of Waropen bead-work is obtained from Japen.[246] In their bead-work and in their dancing-aprons the women show a preference for rectangular motifs, whilst the men like curling and leaf-shaped patterns in their carving and painting.

One should not suppose that every Waropen makes the objects which he uses. On the contrary, there exists a very lively and continuous barter-trade which is closely connected with the system of the exchange of gifts. This barter in and by itself already presupposes a division of labour, as explained by Malinowski.[247] But differently from occidental society, we only find a slight division of labour in the Waropen organisation. If he wishes to do so, every individual, both *sera* and *waribo*, may perform any kind of labour, only observing the division of the work between the sexes: a man will not be willing to stitch atap roof-covering and a woman will not carve a prahu. But for the rest there is no work which is completely let to specialists. Every man will have made a net in his time, or tried his hand at carving a prahu. But not all men possess the same skill, as soon becomes apparent when one has the opportunity to see a larger fleet of prahus; then one will always see a few clumsy or twisted or badly hewn boats amongst them. Several

---

[246] For other objects obtained from elsewhere, see p. 232.
[247] See B. Malinowski, *Argonauts of the Western Pacific*, p. 156 ff.

men who have made more progress in this field spend a great deal of
their working-time in building prahus or in carving. Due to their special
skill they also enjoy a certain fame and it seems that there even are
certain *ruma* whose members devote themselves to some special craft.
At Nubuai the inhabitants of the house of Tanatirewo were busy all
the time carving prahus.

In the Waropen area there exists not only what Malinowski calls
"organised labour" and a certain division of labour between the sexes
as well as a certain specialisation, but there also prevails a rather developed organisation of work in "communal labour".[248] Like most other
Papuas, the Waropen like to work in groups, all participants co-operating for some definite object, without any difference or specialisation. In
this way the men build houses, and the women stitch the atap roof-
covering or work the sago. Usually the group is composed of relatives.
When the *seraruma* is built, the *sera* may exercise a certain claim on the
labour of the members of his clan. When building a house, it is usually
one man who takes the initiative, i.e. the man who has the direction
over the house (*ionéa riruma*). In that case the man is obliged to give
the participants some tobacco and food. An important form of communal
labour of this type is the raid, in which sometimes not only relatives
or members of one single clan and inhabitants of one village take part,
but where even several villages may join.

In other cases also the Waropen like working in groups, even without
a common aim. Fishermen like to lay their boats together for fishing.
When erecting the sero (fish-trap) men and women co-operate. At
Napan the whole village once came to participate in catching fish from
a trap, to the serious indignation of the *sera* who had placed this trap
together with his relatives. Women who have to collect fire-wood or
molluscs usually start out in groups.[249]

HOUSES

The word *ruma* possesses the two-fold meaning of "traditional group
of relatives" and "dwelling-space". In actual practice these two meanings
do not completely coincide, for members of the same *ruma*, group of
relatives, do not always live in the same *ruma*, dwelling-space. In this

---

[248] See B. Malinowski, *Argonauts of the Western Pacific*, p. 159.

[249] For such groups a special term is used, viz. *aira*, like e.g. *ghaidoroaira*, a
  group of women looking for molluscs in the tidal forest; *aroaira*, a group of
  people working in the sago-forest; *pareriaira*, a group of people who arrive
  to ransom a slave.

way the three brothers Ghoadei, Kokoboi and Dedui from the Nuwuri clan (Nubuai), each with their children and relatives, are living in a separate dwelling, without wishing to deny that they belong to the same family-branch. This disintegration may have been influenced by the introduction of the "small family-house" by the Administration.

Although the house has a theoretical plan, this is not systematically followed in actual practice. Small additions are made to the house at different points to suit the needs of the moment. A man may, for instance, build a stand to dry his nets or to store his prahu. Then there occurs a case of sickness, or a quarrel. And rightaway this addition is covered with a few pieces of atap roof-covering, a few days later walls have been improvised and a new room has been added to the house! To the eye the houses present a confused group of larger and smaller rooms in various stages of decay, among which from time to time some semblance of order is created by a zealous Administrative Assistent.

Originally the Waropen villages all stood in the tidal forest, over the water, even the villages in the Napan area which now present such a well-ordered spectacle on the mainland. The movement towards the water persists and so Ambumi village, for instance, is imperceptibly sliding from the land to the water. As regards the type of housing the preference remains for the *ruma bawa,* long-house, which is liked better than the modern "small family-house" which is called *toko* (Malay: store, shop). The latter with its higher roof and walls is considered cold and empty as compared with the old-fashioned houses whose tortoise-shaped roofs brooded low over the living quarters. The old-fashioned, streamlined roofs are better suited to resist the often violent winds which blow the atap roof-covering apart. Without constant control not only the *ruma* move always closer to the water, but they also become constantly longer. At Nubuai some people even started to build old-fashioned tortoise-shaped, pointed roofs again.

Between the inhabited houses stand the ruins of relinquished dwellings, here still with crooked roofs and dilapidated walls, there with only the stumps of rotten poles. These are houses that have been left by their inhabitants, either because of the great trouble caused by vermin, or because of some tragic history connected with such a house. There was e.g. the house where a widower was said to have killed himself and his children because of "shame", as his offer for a new marriage had been refused. On this spot one is afraid of the *dare* of the dead.

The theoretical plan of the house includes a central portion (*wundo*) with living-quarters (*arado*) in a long row on both sides. At the sides the floor of the rooms extends some distance outside, where it forms a kind of gangway, *ruaibo* (also meaning "beach") or *nau*. The front part of the house (*rumarengga*) has been built out in the shape of a front gallery, usually covered, whilst at the back (*rumafono*) one finds a back gallery, mostly smaller. The living-quarters are completely separate from the central portion along which they have been constructed. The floor of the rooms is usually a few feet higher, extending slightly over the central portion, so that there also a narrow gangway is formed, a not very comfortable, but practical place for sitting or standing from which the children may look down quietly on the ceremonies and dances performed in front of them in the *wundo*. The central portion of the house and the living-quarters therefore have not the floor and the supporting poles in common, but only the roof. People therefore often start by simply building a few rooms. If an extension is wanted, one builds a parallel row of rooms opposite to the first. Eventually the narrow passage between the two rows is closed and the roof is continued from the one side to the other. Then the central portion has been completed, with living-quarters on both sides.

If one wishes to build a house, one starts by collecting a number of rhizophore trunks and peeling these; they will eventually come to support the floor. The floor-supports (*ri*) have to be so long that, once they have been placed in the ground, they extend beyond the highest level of the tide, hence they should have a length of at least three metres (10 ft.). These *ri* are rammed into the soft mud without difficulty, because a number of men dance on them maintaining the same rhythm (*kikida ruma*). Across the *ri*, which have to be forked at the top, shorter poles are placed at right angles to the main axis of the house (*radaru*), and these support longer poles (*rora ba*), placed longitudinally, which are finally covered with floor-slats made of nipah-palm (*reamu*).

The floor-slats may either be tied singly to the *rora ba* or in panels by lashing a heavy piece of rattan across the floor to the *rora ba*. Often, however, so much trouble is not taken, as it is moreover convenient to have the loose-lying slats available for all kinds of other purposes. Therefore most floors, especially those in the *wundo*, show so many gaps that a stranger is strongly advised against venturing inside an unknown house in the dark. Even the inhabitants, who have acquired a marvellous agility in balancing across shaky poles and rickety frames through long years of training, occasionally make a misstep and hurt themselves seriously in this way. It is not uncommon that small children have to be fished from the water, because they have slipped through the floor-slats. Sometimes these occurrences may be comical, but they may have serious results.

Against the *ri* the longer poles are tied on which later on the roof will have to rest (*ekaini*). The largest of these are the *masa* in the central portion of the house; they will have to carry the ridge-pole (*oriruai*). Across the *ekaini*, i.e. along the longitudinal axis of the house, the rafters (*sorudo*) are tied, and perpendicularly to these the battens (*raisa*). These are covered with sheets of atap roof-covering (*aka*), starting at the bottom, the atap consisting of dried leaves of the sago-palm which have been stitched to bamboo laths of about 1½ metres (5 ft.), usually by women and rarely by men (*sira ruma,* thatching the house). The sheets of atap are tied to the battens, whereupon they are pressed down on the outside by bamboo laths, in order to prevent their lifting when a strong wind blows. The roof-ridge is finished by means of a kind of shield, consisting of two sheets of atap of which the loose ends of the leaves are plaited together. This atap shield is bent around the ridge-pole and fastened with long wooden pins (*akaunabo*) passing below the pole.[250]

Formerly the roof was shaped liked a tortoise, and narrowing to a point like a prahu, with which this type was also compared by the Waropen. At the tip of the old-fashioned roof (*rumarenggatatami*) there was a board (*rumarenggafama*) fashioned like the Biak *kob* which is illustrated in W. K. H. Feuilletau de Bruyn, *Schouten- en Padaido-Islands,* p. 33. At Weinami the point of the roof was surrounded by a large wooden ring, called *inisaufama,* moon-board. At Ambumi the roof is crowned by a small, pagoda-shaped roof (*rumawaita,* lit. the child of the house); there the interior of the house is called *rumadahia,* lit. the body of the house. In the frontpart of the house a triangular fringe of atap-leaves (*kawofesa; komupesa* N.) was suspended from the roof.[251]

The form of the roof of the modern house is different. The front is an inclined plane, shaped like a inverted V, beyond which the two sides of the roof jut out. It was said that the old-fashioned *rumarengga* was carved, but the people were unable or unwilling to indicate the motifs. Presumably it consisted of some painting on the *rumarenggafama,* which assumed ever brighter colours in the imagination concerning

---

[250] At Napan the ridge has to be closed by the family of the mother's brother, for which they receive a present.

[251] Could this triangular fringe be meant to indicate the sharp teeth in the open mouth of a mythical monster? Cf. p. 162, note 124.

"the good old days". The gentle slope of the tip of the old-fashioned house was crowned with a number of decorations shaped like fins, called *fafa* at Kai and *sakabungguro,* shark's fin at Napan.

The walls of the rooms were made of the ribs of sago-leaves or of gaba-gaba (*roibo*), laced together with wooden pins (*fafario,* to wall in a room). In the beams between which the gaba-gaba is fastened a slot has been cut by way of a runway. These beams (*saghaisoa*) run longitudinally through the house up to the front gallery, where they occasionally assume the form of a crocodile's head. In recent times cutting the groove is often omitted, the gaba-gaba being simply nailed to the poles under a bamboo frame.

In the walls an opening is left, usually leading into the *wundo* or the front or back gallery, which is used by way of a doorway (*rei*). A door-board (*famarei* or *reifama*) closes this opening which in our view is rather more like a window. A hole is made through the jamb through which the wooden doorknob protrudes into the room. By inserting a wooden pin through the knob on the inside, the door is "locked". This lock is of course not very strong, but that would not have much sense, because the wall or the roof may be opened without great difficulty, if necessary. By way of a window a square opening is cut in the roof-covering; this may be closed with a separate flap.

All wooden parts are joined by simply tying these together. The joint made by inserting the notch of one part in a hole in the other is known and practised, but no pin is inserted at the end, except by a few persons who have learnt carpentering outside the village. Although in the village some carpentering tools may be found, people mostly work with the axe (*mano*) and the bush-knife (*naiwiro*) which most Papuas use with great skill. At Nubuai there was one man who possessed saw, chisel, plane and hammer, as well as an instrument in which ran a string coiled in blacking, which was run out to mark longer lines on planks etc. This man had worked as a craftsman outside the village and occasionally still used his tools for making a chair or a table, and later for furnishing the school. The villagers mostly continued to make shift with axe and bush-knife. The stone axe (*bono*) of the normal Papuan type [252] is no longer used at present, except perhaps by some very old-fashioned men for cutting down sago-palms. The bush-knife, like all other native iron-work, comes from Biak. Nowadays many of the knives

---

[252] See A. N. J. Thomassen à Thuessink van der Hoop in *Geschiedenis van Neder-landsch Indië* (History of the Netherlands Indies), edited by F. W. Stapel, I, p. 50.

used are of European manufacture. Often the handle is taken off and replaced by a new wooden or bone handle at an angle to the back of the blade and not in line with it as in a European knife.

The Waropen house is usually situated in such a way that at low tide it is still possible to moor a prahu at the front. The rooms at the back are also less practical because it is not so easy to bring in things hidden from the curious eyes of the neighbours. In front of the house there usually lies a large floating boom, fixed between poles in such a way that it can go up and down with the tides without leaving its place. It is at this floating boom that the visitor steps out of his prahu to climb the notched tree trunk which serves as a stairway (*epamani*). There he arrives on the front gallery, occasionally on a boarded platform (*rarado*) where nets are dried or prahus are chiselled out.

If the visitor is welcome, his question: "*So mangga resi nduna*", "May I perhaps come up?" is answered by: "*Aesi ndu*", "Come up". He is given a place in the front part of the house, where a small fire is usually smouldering. For a prominent visitor a mat is spread, on which he may squat or sit with outstretched legs. Sometimes there is a table or a chair of the European type for the non-Papuan visitor, in honour of the guest, but to his great discomfort. Most visitors simply look for a place on the floor, sometimes on a piece of wood or on a bamboo. One may also sit on the stand (*pata*) in the front part of the house, where usually a few young men lie sleeping. The women have to join the other women in the back of the house where they have their place, which does not mean to say, of course, that a woman is never seen on the front gallery. If the visitor only finds women in the house, good breeding demands that he postpones his visit.

Beyond the front gallery one arrives in the *wundo,* the central portion of the house, where it is mostly cool, because the funnel-shaped construction gives rise to a certain amount of draught; moreover, the atap roof-covering provides an excellent protection against the heat. In many houses the large travelling prahu finds a place in the *wundo.* There one moreover finds a garret (*rofu, daidéa* or *ghopo*) where all kinds of gear, like fish-spears, bamboo masts, outrigger-beams for the prahus, sails, etc., may be stored. Against the walls of the *wundo* one may observe all sorts of fishing-tackle, lines, floats, material for making rope, nets, etc. On the first central pole or on the joists one likes to hang hunting-trophies: skulls and jaws of tortoises, pigs and crocodiles or tails of large fishes, captured during a lucky hunt. Somewhere in the *wundo* the ceremonial seat (*kambo*) is also often placed.

Once the visitor has carefully found his way through the *wundo* with its often carelessly laid floor, he stands in front of the openings which give access to the oppressive and smoky rooms. During the day-time the room is the normal abode for the women who watch over the fire and the small children. When the high tide floods the dried space underneath the house, the pig has also to be dragged inside the house in spite of its shrieking protests, to find a place near its enemy, the dog, who has to try to eke out his often miserable existence by scavenging some food, whiningly escaping kicks and blows. The cat, which may be found here and there, faces less difficulties in view of the extensive fish diet of the Waropen, and also because pussy can seek safety in sudden flight in case of necessity. In some houses the people even keep a chicken or a cock, exclusively for amusement, it would seem, because in this muddy country these animals do not thrive so that nothing is to be expected from breeding them. In the Kai area birds in captivity usually die quite soon for lack of suitable food; an occasional pigeon or marsupial manages to stay alive.

The hearth (*awu*) is the centre of the cheerful family-life. At Kai the hearth-plate is suspended by four rattan loops (*awusara*) from the floor-beams. When these loops are severed (see p. 95), the plate drops down with everything which happens to be on it, together with the three hard-baked lumps of clay shaped like sugar-loaves which serve as hearth-stones (*wainara*), on which the cooking-pots may be placed over the fire. At Napan the fire-plate is placed on the floor-beams like a kind of tray; it is not suspended above the floor.

At the four corners of the hearth-plate poles stand on the floor, attached to the roof above. These are the hearth-struts (*arawo*), to which one or two shelves are attached, supported by transverse bars (*arawobo*). These shelves are used for storing cooking-gear and fuel. The lower ends of the struts are often carved in the shape of an ancestral image; it is said that the two in front are male and those at the back female, but according to the informants they do not bear any particular name. Near the fire a small shelf is suspended from the roof, where sago-cakes, fish, etc. may be stored out of reach of rats, dogs and cats.[253]

---

[253] F. S. A. de Clercq and J. D. E. Schmeltz, *Ethnographische beschrijving van de West- en Noordkust van Nederlandsch Nieuw-Guinea* (Ethnographical description of the West coast and North coast of Netherlands New Guinea), plate XVII, figure 13.

Against the wall beside the fire-place (*ataratuni*, where one hangs clothes) stand a number of bamboo water-tubes (*arai*), about half a metre long (1½ ft.), in which the women and girls have to fetch fresh water. In the corner one finds paddles, fish-traps, tools for preparing sago, etc. In a more distant and secluded corner there are placed, partly on a stand, the highly valued *faisi*, cases of pandanus-leaf, mostly imported from Japen Island, in which the possessions of the inhabitants are stored, such as bracelets, beads, pieces of cloth, etc. In these modern times the small clothes-chests (*burua*) made by the Chinese are very popular, as well as tin trunks; the former are sometimes used as ceremonial seats.

Beside some pieces of ancient porcelain, most modern houses also possess imported pottery, like bowls and plates, a few enamelled dishes and iron cooking-pots, etc. Furthermore, most people possess a small mirror and sometimes a little lamp, but it is rarely possible to use these lamps, in the first place because the necessary oil is not available, and secondly, because these lamps do not stand the rough and inexpert treatment for a long time. Mostly the people are content with the glow of the hearth-fire, whilst sometimes torches of bamboo and resin are used (*nara*).

In a screened-off corner of the room is the *feretei*, the women's privy, where the woman also waits for her confinement. The men usually relieve themselves in a screened-off corner of the side or the back gallery. Children do so simply through the slats of the floor, so that the cleanliness of these floors mostly leaves quite something to be desired, all the more so because it is not contrary to good manners either to aim between the slats when ejecting the juice of one's quid. Floors and woodwork are, moreover, highly convenient for wiping one's hands when these have been dirtied, for instance if one has been unfortunate in blowing one's nose. All kinds of sick persons likewise spread their mat beside the fire, providing another reason why a stay in these dark and smoky rooms is not very attractive for the stranger.

Theory demands that there should be a separate room for every married couple and their children, but the actual situation is probably that only aged and prominent persons attain this ideal, because there are always more families in a house than there are rooms available. Not all inhabitants of the room have a fixed place where they sleep. Like most other Papuas the Waropen sleep soundly and they are satisfied with an inconceivably small place, the outrigger beams of the travelling prahu if necessary. The sleepers often lie in a confused mass through and over each other, so that the myth can tell us of a brother who laid his leg across his sister's lap by mistake.

Pillows are unusual. The woman lays her head on a bamboo or on a piece of wood; it is only the husband who uses the well-known neck-support, *runa,* carved in the shape of an iguana or a crocodile, or often with carving in the shape of a two-headed creature which supports the horizontal piece of the *runa.* This double figure, which according to the informants should be male-and-female, is often topped by another, in which they claim to recognise a double-headed snake.[254] The wife is not allowed to touch this object; the myth relates that the primeval women, the Aighei, played with the head-supports of the men. Should we assume that this object occupied a place in the mythical complex of initiation and sexual intercourse? In one of the myths we learn that formerly men conceived through the ear which during sleep may rest on the *runa.*

Material objects often possess a mythical significance. The house too has a sacral plan, as it is differentiated in a more sacred front part, where the men dwell, and a back part, where the women usually stay just as in the classification where up and down, sea and land, men and women, people of the interior and inhabitants of the coast, are placed in opposition to each other. It is in the front part of the *wundo* that the rites are performed; there the dance winds around the first *masa,* the central pillar which represents the *inggoro.*[255] In nearly all houses the front and the back have a name of their own; so the house Pedei is called Pedei as regards the front and Amosi at the back. The garret also plays a role in the myth; here the skulls of the departed are suspended and here leads the stairway on the occasion of the marriage-feast. On this garret the grandfather or grandmother dwells in the myth — those who were with the first men and the first women —, whilst on Biak the initiandus sometimes sits isolated in the garret.

When the visitor leaves the house, he says by way of taking leave: "*Rawo fino*", "I had better go", to which an *Io* muttered in agreement forms the reply.

Finally I should add that in the vicinity of the village simple lean-tos

---

[254] See the illustrations in F. S. A. de Clercq and J. D. E Schmeltz, *Ethnographische beschrijving van de West- en Noordkust van Nederlandsch Nieuw-Guinea* (Ethnographical description of the West coast and North coast of Netherlands New Guinea), plates XVIII and XXI.

[255] Among the Numfoor this pole had to be erected by an old man; this was said to give steadiness to the house.

(*rawaro*) are found here and there which serve fishermen or sago-workers when they are compelled to dwell outside the village.

PRAHUS

The Waropen passes a considerable part of his life in the prahu. He can hardly go a step outside his house, because even at low tide the mud is ankle or knee-deep; only the Napan villages are situated on dry land. This life spent in rowing has evidently had little influence on the development of the leg-muscles, because the Waropen usually possess slender and well-formed limbs, whilst on the contrary some mountain tribes near Manokwari are conspicuous for the slight development of their leg-muscles. If necessary these water-folk can manage equally well without a prahu, because everybody is able to swim. Children often swim through the village with or without a piece of wood to support them, but adults only do so reluctantly.

It is only rarely that a person leaves the village in a prahu all by himself and therefore the prahu exerts a great practical influence on the formation of the groups. Very often the crew of a prahu consists of a man and his brother or another member of the *ruma* when fish have to be speared or when palm-wine has to be tapped, or of a woman and her child when she wants to go visiting in the village, or of a man and his wife when they go to set the fish-trap or when they have some business in the forest. A prahu is most easily managed by two persons, one in the back who also takes care of the steering, and one in front who only has to row. An important personage who wants to be rowed somewhere, takes his seat in the front. The man usually lets his child sit in the front, and the husband his wife. If there is work to be done outside the village, like collecting molluscs at low tide, or beating sago, the women usually set out in groups. A group of men in a prahu is mostly going out to fish with the net. Because every individual is far from possessing a private vessel or from having one at hand, the people are constantly obliged to appeal to each other's kindness. On the other hand, this also forms a source of continual bickerings, e.g. when the woman does not arrive soon enough to fetch her husband from somewhere with the prahu. It would seem that there is a chronic shortage of prahus.

The large travelling prahus are able to stand up against rather high seas. The prahu adapts itself to the movement of the waves, and because it is tied together, it is so flexible that the chances of its breaking up are small. If the boat gets filled, it retains its buoyancy, provided people do not wait to long in jettisoning the cargo. The Biak prahu, in which i.a. the Biak smith with his family and a Papuan smithy appears everywhere, is probably more sea-worthy than the Waropen prahu, but on the other hand the Waropen boats need not make any long crossings.

There are three types of prahus, viz.:

a. *gha*, the normal dug-out with outriggers, used especially by men;

b. *soado*, the same, but flatter and without outriggers, used in the tidal forest;

c. *sandu*, the bark of the sago-palm which remains when the marrow has been taken out, especially used by women and in the tidal forest.

The *gha* is the normal type of outrigger-prahu, and is also used for long voyages. The smallest type, intended for traffic in the village and

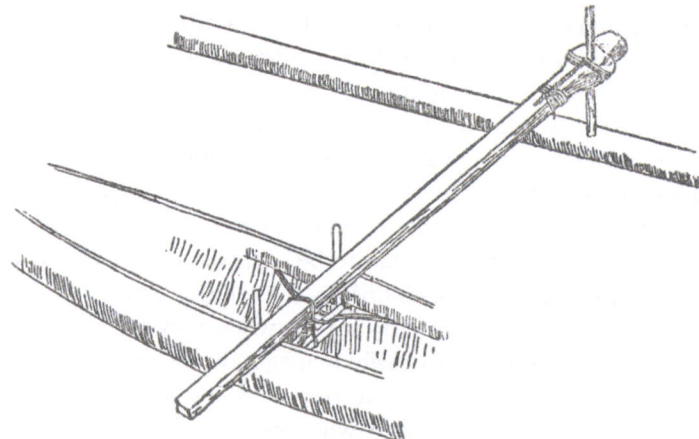

Fastening of the outrigger.

in the close vicinity, usually measures three metres (10 ft.) and has only one outrigger. The larger type, the *somandu*, lit. "two-outrigger" is used for making more extensive voyages. The making of these prahus is more or less the work of specialists; although there does not exist any rule prohibiting a man to build a prahu, in actual practice it is often the same people who undertake this work.

The tree has to be selected with some care; an augury is taken to establish that the tree selected is favourable and then a sacrifice is offered to the ancestors. The names of the trees suited for making prahus are also used as terms for "prahu", like *maighéana, rewonawo, na, sigha, marano, ainuko, sifara. Kiwona kirimaighéano,* they were rowing in their prahu made of *maighéana,* a reddish kind of wood, much appreciated. The first rough working takes place in the forest; finishing is then done in the house (*kipamgha ro,* lit. "to hew well"), sometimes on a special framework, sometimes at low tide on the ground. The tool used is the *ghamura,* the chisel-hammer, an adze-shaped tool, which is used to give the correct line to the prahu, piece by piece, without applying measuring instruments and relying only on the eye. The sketches inserted here indicate how the prahu is constructed.

Besides the oar, *nama,* which is often slightly carved, the equipment of the prahu often contains an *oa,* pole, also used to moor the prahu, a *fama* or squatting-board, and a *riwa* or *gharana,* bailer, made of a folded leaf. The oar is made of iron-wood; for adults it is about one and a half metres (5 ft.) long. Rowing is done in a sitting position, not standing. By way of an anchor, needed e.g. when fishing in deep

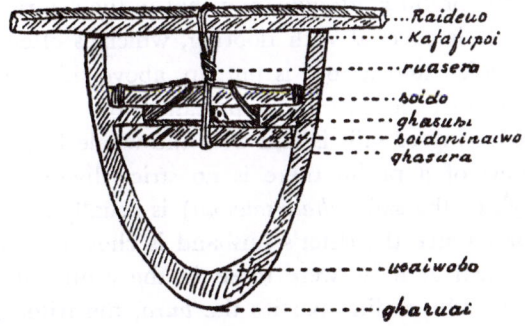

....*Raideuo*
.....·*Kafafupoi*
-----·······*ruasera*
--------*soido*
-----·····*ghasubi*
- - - ---*soidoninaiwo*
-----·····*ghasura*

----·---*uaaiwobo*
- - - - ------*gharuai*

Cross-section of a prahu with the fastening of the outrigger.

water, a heavy stone tied to a rattan rope is taken along. On the outriggers the fishing-tackle and bow and arrows are placed, and often a small fire, when people have to set out at night.

Of course, most care is bestowed on the large travelling prahu. To this belongs in the first place a sail (*rararo*) and a tripod mast (*raraiwo*), consisting of three bamboo stakes, the front one being called *mano,* male, and the two back ones *bino,* female. The sail is made of pandanus-leaf and it has been sewn by men; it is rectangular, consisting of four horizontal widths (*mui*), rolled around a bamboo pole (*raraikawono*). It is hoisted on a rattan rope (*fara*) to be unfurled downwards, its position being ruled by another rope, the *manawi* or sheet. With these sails it is impossible to sail close to the wind, but for the rest they are large enough and they are handled with relative ease, also with rising squalls.

The sails, especially in the Napan region, often carry a separate triangular piece of sail with tufts of frayed coconut-leaf, which can have no influence on the sailing. At Napan these *iaraiwaita*, lit. "child of the sail", (at Kai *rarai boma*, lit. " small sail") differ in shape with every *ruma,* every house possessing a model of its own. That of Warami is triangular, that of Waratanoi rectangular with a triangular incision at the bottom. According to informants at Kai these small sails formerly bore a picture of an ancestor; on the sails proper, animal designs used often to be drawn, like rays, "ikan sembilang", etc.

The space between the poles (*raideuo*) of the outriggers (*somano*) has to be arranged in such a way that the rower can dip his paddle freely into the water. Inside the prahu the space between two poles is called *mararo,* that in the stem (*ghare*) of the prahu *gharedo,* that in the stern (*ghafera*) of the prahu *ghaferado,* and that in the centre *edo.* For longer voyages this *edo* is covered with flooring, which is often extended onto the outrigger-poles, and a roof is put up, above this. This cabin serves as storage-space or as a place for important travellers to stay in. During the night a covering of sails is laid over the whole length of the prahu.

For the crew of a prahu there is no strict division of labour. The man in charge of the sail (*ghawisawano*) is usually one of the younger people; by preference the sister's husband is chosen for this post, as he would be laughed at if he were to sit in the centre of the prahu, my informants said. On raiding parties the *buro,* the triton-shell has to lie in the stem. The man responsible for the prahu, i.e. the hirer or the chief fellow-owner, often sits directly behind the cabin; he is also the commander (*nunggu iona ghagha*). His younger brother often sits directly in front of the cabin, in charge of the division of the food stored in the *edo* which the leader of the voyage has to provide for the crew of the prahu. The steersman sits in the stern (*nunggu iapura ghagha*); for this function one prefers to select people who possess a good *ghamaiwo,* wind-medicine, and who are a good judge of "favourable" constellations at sea, or weaker, and hence often older men.

The larger prahus have been rubbed on the outside with blacking (*asano,* carbonized ribs of sago-leaves, soot) and painted with *onda,* motifs, in *ura,* red paint, made with red earth and white *fafanarei,* ground shell, applied to a viscous layer of *kauia,* rhizophore-juice (*kipaiano ghagha,* they paint the prahu). On stem and stern boards are placed in lieu of figure-heads, often surmounted by beautifully carved pieces. For decorating the stem, especially of raiding prahus, one likes to use strings of *korombowi*, white porcelain-shells, their tinkling sound being especially admired.[256] The gunwales of the large travelling prahu are planked up with the ribs of the leaves of the sago-palm, strung together by means of flexible pins, clamped down in holes in the *susi*, the inboard chocks for fastening the poles of the outrigger. The extra freeboard needed when transporting a large amount of freight is obtained by adding atap-leaves.

Great care is spent on the maintenance of these large prahus. After each voyage they are dismantled and pulled inside the houses as a pro-

Roof and cabin of the travelling prahu.

tection against the weather; therefore they have to be rigged anew for every longer voyage (*fandi ghagha*), which takes at least half a day as a rule. Cracks are sown up, so to say, by means of rattan, pulled through holes burnt in the wall of the prahu and tightened extra by small wooden wedges which are driven between the rattan and the wall of the prahu. Broken pieces have to be carefully replaced by new ones. The seams are caulked with fibres and closed with resin or the viscous scrapings of *kaifosua,* the fruit of the most frequent tree in the tidal

---

[256] This shell is frequently used for decorative purposes and fastened to the belt, especially by women.

forest, scraped on the very coarse bark (*kairuru*) of this tree. Recently, easy-going owners of prahus have started to use painted pieces of tin for closing holes.

As soon as a new large prahu is completed, she is fully rigged and decorated, and rowed fully manned to the houses of the wife's family in the own and the other villages. When this prahu with her gaily singing crew arrives at the front of the house, the stem is pelted with all sorts of harmless projectiles like pieces of coconut-bark and ribs of sago-leaves. Thereupon the owner of the prahu is given some simple plates, which sometimes are thrown into the water from where the crew have to fetch them amongst general hilarity.

Travellers know by experience that the rowers only will really exert their strength when a song has been started. When at the hottest time of the day the prahu drifts along with flapping sails across the sparkling water, the rowers sit down, panting for breath. But when the sea is boisterous and when the rattan ties creak under the pressure of the combers, the rowers loudly sing their song against the wind. And when at nightfall the goal of the journey has been reached and they enter close to the village, the harmonious rowing-song resounds; with short beats the paddles send up the spray (*kikawaro*, lit. "they knock"), whilst many curious glances are directed towards the boat which is putting in.

The rowing-songs (*rano*) are sung as catches. First one man takes the lead in singing the song, then the men aft take over the first verse, and when they have sung part, the crew forward join in again with the beginning of the song. In this way both parties have short rests.[257] The informants compared this style of singing with a pursuit, both parties trying as it were to rouse and to overtake each other. The idea is probably that the mythical language in which these songs are composed, in this way drives the prahu forward.

Concerning the other types of prahu (*soado, sandu*) there remains little to be said. The *soado* is flatter and wider; its board cannot be raised and it is difficult to attach outriggers to it, so that it is not very suitable for long voyages. It is used especially for traffic in the tidal forest. The *sandu* is a highly primitive type of craft, but eminently suited for its purpose. It is light and it is very easy to push it across the mud. Moreover, the whole craft is in actual fact a by-product of the

---

[257] Samples of these songs were recorded on a phonograph, kindly placed at our disposal by the Berlin Museum of Ethnography, see note 203.

64. Hairdress

65. Girls boating in a *sandu*-prahu

66. Fishing by means of the sero-fishtrap

67. The trap at the end of the sero

extraction of sago, because after the removal of the marrow, the bark automatically remains. In front and at the back one merely needs to leave a piece of marrow in place, closing up the holes with some mud if necessary, and the boat is ready! It is mainly used by women.

KNOWLEDGE OF NATURE

Although the Waropen are nearly always faced with living nature, their knowledge of nature is not very developed. And on those occasions when they have pondered on some matters, their speculations rather went in a mythical direction than taking what we would call a "scientific" turn.

In this way they imagine that the various mythical beings have found a place in the night-sky as constellations, as becomes apparent from the well-known tale in which the Morning Star appears. For that matter, the sky of the night seems to have aroused more attention than the sky of the day-time, the moon more than the sun. In a mythical sense the moon is conceived as a man with two women, probably also the women whom Aimeri wished to marry. According to their mythology, there existed formerly a connection between heaven and earth, consisting of a rattan, but once upon a time, when men climbed up along this rattan in order to try to drag the moon somewhat closer, so as to have the light of day also at night, this rattan was severed (Texts 140). I did not meet with any views concerning the origin of the phases of the moon, although there is a myth which relates that the moon is the elder brother of the sun. The moon committed incest with its mother, whilst the sun pursued them in order to prevent this. By way of punishment the moon became the weaker light of the night, whilst the sun, the younger brother, became the light of day (Texts 55). In the "face" of the moon one recognises the tufts of the hair of the woman Aighei Ghininggini. According to some people, the sun disappears in the evening in a hole in the West, travelling back underground to the East. During a certain time of the year the sun rises above the horizon in the sea and then the season is unfavourable. When the new moon is observed in the West, a general shout is raised, by rapidly covering and uncovering the mouth with the hand, thus making a tremolo-sound; this custom is also known in other parts of Southern New Guinea.

During an earthquake (*dighasi*) the people make as much noise as they can, shouting, beating pieces of wood together or blowing on triton-shells. In that case the two women Inggoibini and Insosi, who hold the

world-tree (*anaeuo*) are angry because the children have been playing
(Texts 153). *Anaeuo* may be "trunk", "bottom part", "origin" of the
earth. Is earth perhaps imagined as a *dama*, the single pole of which is
upheld by these two mythical women, somewhere underground? [258]

The change of the monsoon is not so striking in Northern New Guinea
as elsewhere; moreover, the Waropen hardly practise any agriculture,
so that from a practical point of view the change of the seasons is of
less importance to them. It is known, that during the period from
April to November the prevailing winds are easterly and that during this
time the sea is steadier than in the other half of the year. When the East
wind (*raghama*) started to blow for the first time, I once heard a curious
buzzing sound in the forest behind the village, a sound which might have
been caused by a bullroarer. Thereupon a great noise and shouting and
trumpeting arose in the village, but the explanation of the sound I was
unable to find. The informants pretended that it was caused by the wind
in the trees and by the *ghéa*, a kind of bird, which flies in groups through
the village in the evening.[259] When the *ghéa* flies and when in the
morning [260] the old men see the "head" of the constellation Sawai (the
hammer-headed shark) appear above the horizon, the time has come
for trading voyages in a westerly direction, from Kai to the Moor Islands,
whilst the time is also propitious for laying out bean-gardens.

Other constellations are Ombaibai, the wife of Sawai, and his younger
brother Samamai, Wawiniki (Pleiades?), Kawoata (Aldebaran?), Gho-
wiratara (in the constellations Eagle and Dolphin?), Monafu (Orion?).
The identification of the Waropen constellations is extremely difficult
because the indications are made so hesitatingly; the starting-point for
an identification is not so much a definite constellation, but rather a
name to which the people seem to attach a number of stars.

The stars are not needed when travelling at sea, because traffic nearly

---

[258] One difficulty in obtaining information concerning cosmological ideas lay
in the fact that the people knew from the school-children that the earth had
to be a sphere; older people did not want to make themselves ridiculous in
the eyes of the children with their mythological views. Information on the
constellations was hard to obtain, partly because the old men so often had
been drinking palm-wine in the evening.

[259] It is not clear which bird is exactly meant. These birds, of the size of starlings
and with lively moving tails, could always be found in little groups in the
grass. In the beginning of the East monsoon they fly to the West — according
to the Numfoor to Ternate, in order to pay tribute. The Numfoor believe
that these birds transmit all kinds of diseases. — *Ghéa* is also the word
for "year".

[260] I was once shown the wooden model of a hammer-headed shark used as a toy.

always takes place in sight of the land, whilst the sailor can also take
his bearings from the movement of wind and waves. A division of time
into weeks or months seems to be also unknown; only recently the
Christians have come to pay attention to the division of the week in
view of the celebration of sunday. Another fixed point is the *festa
potiranggi* (Malay: festa pohon terang, the feast of the shining tree),
i.e. Christmas. Longer periods are occasionally computed in *ghafa,*
moons. The changing of the tides provides an important criterion for
the division of the day; it is known that there exists a relation between
the regularity of the tides and the phases of the moon.

Counting-stick for establishing the days of the week.

Knowledge of nature, in so far as this is not mythically determined,
is mainly directed towards practical usefulness. Of numerous trees the
names are known, as well as the use which may be made of the tree,
its leaves or its fruit. To the contrary no specific names are known for
trees which are wholly without some direct use. Again, the fishes which
constitute a major part of the Waropen diet, are known in a great
many varieties.[261] Even the children know where the different kinds
are caught, whether they are tasty, whether their meat is solid, etc.
The people can often observe from a distance, by the water, which kind
of fish to expect. They know several characteristics of fishes and they are
aware of the fact that the "bawal hitam" swims on its side, that the
"ikan kerong" produces noises, and of how the sucking-fish attaches
itself or how reef-fishes catch their prey.

The interest and the knowledge stop short, however, as soon as types of
fish are concerned which are not of direct use to the Waropen. In this
way all the striking and interesting fishes that live near the coral-reefs
and not on the muddy coast, are taken together under the name of
"reef-fish". Although it was known that fishes spawn, even the old men,
in spite of their lively interest in the problem, were unable to decide
whether fishes copulate.

---

[261] Names of numerous fishes were noted with the help of the illustrations in
H. C. Delsman and J. D. F. Hardenberg, *De Indische zeevisschen en zee-
visscherij* (Seafishes and seafishing in the Indies); the descriptions given
there concerning species served as a further means of control, as far as possible.

Equally meagre was the Waropen's knowledge of insects with which
he has little to do. The development from egg to larva and then to pupa
and insect was known, but all types of butterflies were indicated by
the term "bat-bug" (*ghaiakumbo*). One man showed me a wonderfully
coloured beetle which he carefully tried to keep alive in a lime-box.
It was not so much that he considered this animal to be an extraordinary and strikingly beautiful insect, but rather that he believed
this to be a being which had perhaps temporarily assumed the shape
of an insect and which later might mean "good luck" to him.

FISHING

Fishing is mainly man's work. There are several ways of fishing:
1. with line and hook,
2. with nets,
3. with fish-spear, bow and arrow, or harpoon,
4. with fish-poison,
5. with sero-fishtraps.

1. The line (*tasi*) bears a hook (*kare*), consisting of a small strip
of bark of the sago-palm with the thorn which grows on it. Formerly
hooks were also made of bone, but at present steel hooks have become
general. The float (*napono*) is a piece of wood or rib of a sago-palm
leaf, unless the object is to fish for *taiwu*, snake-fish, because in that
case the float is roughly carved in the shape of a fish or a crocodile.
The line is often wound on a reel, decorated with an ancestral figure.
The bait (*anakana*) consists of *inggoiwu*, a small mollusc or crab, or a
shrimp. Larger fishes are hauled in very carefully, to be taken into the
prahu by means of a landing-net (*tara*), where they are immediately
killed by a blow with the sharp edge of the paddle or with a cudgel. The
line is also used for fishing large lobsters (*aifatasi*), but then the hook is
replaced by a small net (*aifarowu*) in which the bait is hidden. The
lobster fastens itself to the bait and may be easily pulled up. Lobsters
are roasted; they seem to occur in abundance. The term for this kind of
fishing is *kiwoi tasi* (from *kiwoiwa tasi*), lit. "to row with the line",
"to fish".

2. The drift-net (*faiano*) is provided with floats (*ramu*) above and
weighted below with white porcelain-shells (*korombowi*), which are first
ground down to prevent them filling with water and so rendering it
impossible to take in the net because of their weight. Nets are made by
the men by means of a simple flat stick (*roano*). They are handled with

care, because the string used does not wear very well and takes a long time to prepare and to work. Large nets often possess a name of their own.262 At Nubuai the Kai clan alleged that formerly they had the monopoly of manufacturing drift-nets and fishing with them, and that others were only allowed to use these nets with their permission. After use the net has to be washed, controlled and dried on a jetty close to the house, to which a bamboo drying-stand (*fainisa*) has been tied. The term is *woi faiano*, to fish with the net.

The casting-net (*diara*) was used in the Kai district by only one man who had formerly worked on Amboyna as a carpenter. Although the catch obtained with this type of net seems to be quite good, it is little used because of the difficulty in obtaining the lead used for weighting it.

3. The harpoon (*damano*) possesses an iron tip, fitted in a loose haft attached to a string which is coiled in a conical basket (*damana-*

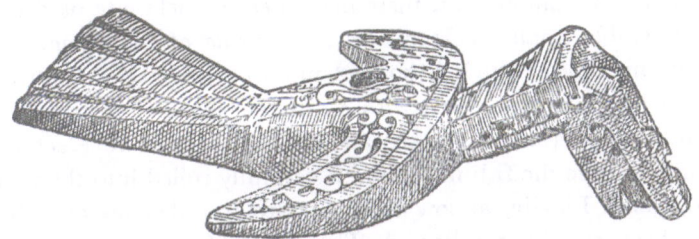

Ghafa, wooden float for the line of the harpoon.

karawano) in the stem of the prahu and run out as soon as the prey, usually a tortoise, sometimes a dugong, has been hit.263 In order to tire the prey as soon as possible, floaters have been attached to the line in several places (*ghafa*, lit. "moon") 264, sometimes representing a bird (cf. p. 160, note 120). The term is *anggiwa*, to harpoon.

In recent times a new type of harpoon has come into use. It consists of an iron rod, about one metre (3 ft. 3 in.) in length, with a barbed

262 See also C. G. Seligmann, *The Melanesians of British New Guinea*, p. 227, note 1.
263 De Clercq and Schmeltz, *Ethnographische beschrijving van de West- en Noordkust van Nederlandsch Nieuw-Guinea* (Ethnographical description of the West coast and North coast of Netherlands New Guinea), plate XXVII. This work contains excellent illustrations of a great many objects mentioned here.
264 De Clercq and Schmeltz, *op. cit.*, plate XXVI, fig. 10.

tip and sliding through a bamboo cylinder of about 30 cm. (1 ft.). A strip of the inner tube of an automobile tyre has been attached to the bamboo cylinder, enabling one to shoot the iron rod like from a catapult. The fisherman, wearing diver's goggles, dives after his prey and tries to harpoon it under water. This instrument was as yet unnamed; it was said to be of Japanese origin.

A very common method is also spearing fish (*soroma*), using a fish-spear with two or three tips (*raghéano*) or a fishing-arrow (*fasano*) with four, eight or more points.[265] One may also use the bow (*farako*). When fishing with the harpoon, the spear, the arrow or the casting net, at least two men have to co-operate, viz. one for shooting and the other to handle the prahu.

4. The sero-fishtrap (*ea*) consists of a mat, about one metre (3 ft. 3 in.) high and several metres long, made of laths of the bark of the ribs of sago-palm leaves, tied together with thin rope (*earona*). When the tide is running out, these mats, tied to sticks, are used to close the innumerable small creeks inside and outside of the village. In the centre of this fence, which is shaped more or less like a fish-trap, an opening is left, which is closed with a separate piece of matting, shaped like a horse-shoe (*osa*). It takes about half an hour to erect a sero. From time to time the fish in the *osa* is carefully rolled into the mat and taken on land. Finally, at low tide the sero is rolled up and the fish that has been caught is collected. The result of several hours work is often very meagre, providing only small fish at that; these are rolled into leaves and roasted. At Napan also a kind of hand-trap [266] is used for scouring the whole of a stretch of water by literally covering it step by step. This instrument often serves to catch very small fish which swim in shoals.

When an exceptionally low ebb is expected, a number of men some-times come together in order to block a wide creek with their sero. On such an occasion at Weinami practically all villagers turned out, evidently to the displeasure of the men whose seros had been used. It was said that this was frequently a cause for quarrels, although I was unable to prove the existence of fishing-rights. Everybody is allowed to place his sero where he wishes, and because the number of small creeks in the immediate neighbourhood is practically legion, disputes concerning their placing are very rare. Moreover, due to such quarrels one is likely to acquire the name of being an uneducated fellow, a man

---

[265] De Clercq and Schmeltz, *op. cit.*, plate XXVII, figs. 1, 13, 14, 3.
[266] De Clercq and Schmeltz, *op. cit.*, plate XXVI, fig. 4.

from the interior; at most one may indicate by means of a broken branch or by some other sign that one intends to set up one's sero at a certain place (*we sema ura ghaido*). Everybody is free to fish wherever he likes, also in the vicinity of the village. When the men of Nubuai e.g. caught a dugong close to the village of Risei, their catch was not in any way contested by the people of the latter village.

5. The fishing-poison (*saimua*) used at Napan is the kernel of the seeds of the Cerbera odollam (*womimbo*). The poison is chopped into small pieces, mixed with pounded crabs (*rafu*) and thrown into the creeks when the water is running out. The stupefied fish then appear downstream floating on the water (*sowa*, i.e. *so wa mamura ado*, to cast for, viz. in order that the fish may be stupefied).

The women, who also help in setting up the sero, have to contribute to the daily fare by collecting molluscs when at ebb-tide the tidal forest runs dry. Tying a basket to their waist with a piece of string, they turn up the mud with a rough scoop (*rekuwo*), flinging up the animals with a wooden hook (*ema*). The women also catch fish by means of a hand-trap made of thorny rattan (*patu*), with which they catch i.a. a larger type of mud-jumper (*roma*).

The combination of fish and sago provides a complete meal according to Waropen taste; sago without fish is to a Waropen what potatoes without vegetables are to a Dutchman. If there is no fish, people often do not touch the sago. This is the reason that according to the myth which describes the pre-civilised condition of humanity, the women in the jungle only ate sago and the men on the coast only fish. Only when the men and the women had come together and civilised life had begun, sago and fish together also became the normal diet (Texts 1, 4).

Although the amount of fish available does not seem to be insufficient, it would be incorrect to assume that the Waropen always enjoy plenty of fish. Only shrimps, lobsters and molluscs are to be had daily in satisfactory quantities, but for the rest the technical equipment is such, that the people have considerable trouble in collecting their daily food. It is a general misconception that the Waropen, in spite of the low level of the organisation of his work and in spite of his primitive equipment, need not exert himself in order to obtain the food he needs. The many quarrels about the sago-property and the jealousy about the catch of a large fish are already proof to the contrary. Although we offered payment in readily accepted goods, the people were unable to provide us daily with fresh fish during our stay in the village. And if the catch has been exceptionally plentiful, they do not have any other

means of preserving the fish but stringing them and smoking them —
a process which promotes their durability but spoils their taste.

Also when fishing an appeal to the sacral world is only rarely considered necessary, although the people admit generally that the ancestors
decide on the catch. Occasionally one appeals to them for a good
catch and then one sacrifices some tobacco or palm-wine. Often a

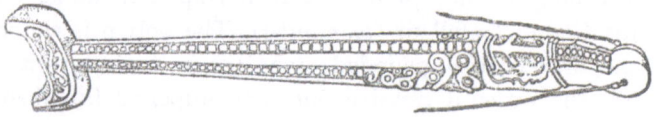

Aiwo, amulet for fishing.

curiously shaped *aiwo* is tied to the net.[267] If for some reason or other
the ancestors are angry, e.g. because someone in the village has committed incest, or because the fisherman has performed an "inauspicious"
act, the catch will be poor. With the return of the mythical primeval
times a wondrously rich haul of fish is also usually foretold.

SAGO PRODUCTION

Besides fetching fuel and drinking water, and collecting molluscs, the
tasks of the women include the monotonous work of extracting sago. As
soon as one observes that the tree is about to flower, it is cut down by the
men, not only by the women's husbands, but particularly by the women's
brothers who often come from far distant places in order to help their
sisters who have married elsewhere in working the sago. Next, the top
is cut out and the palm is split lengthwise. Formerly this was done with
the stone axe, as an occasional very old-fashioned man will perhaps
still do at present.[268]

Then the women's work starts. First they have to loosen the marrow
by beating with an adze-shaped beater (*magha*), the surface of which
has been rendered slightly concave and provided with a cutting edge.
The beaters are made of hard wood; the handle is usually carved, mostly
with an ancestral figure, just as the sago-spoons and bowls are often
decorated with carving. The women chop standing (or sitting, when
it is desired to chop the marrow more finely), with a mat (*enggana*;
rowuboio N.) across their feet in order to prevent the loosened pulp

---

[267] De Clercq and Schmeltz, *op. cit.*, plate XXXVIII, fig. 19.

[268] Cf. the excellent description by K. van der Veer in the chapter on agriculture
and horticulture in W. C. Klein, *Nieuw Guinea*, p. 469 ff.

from squirting away. Sometimes the palm is cut into pieces, to be carried to the village to be processed there.

Then the marrow is kneaded and washed in a trough made of the lower part of the rib of the sago-leaf, in which a piece of the fibrous sago-leaf sheath has been inserted to serve as a sieve. Through this sieve the water containing the flour runs into a collecting trough, usually a small prahu topped by a small basket which serves as a second strainer to remove the larger impurities. The kneaded pulp is thrown on the ground and there it spreads a penetrating sour smell which renders the sago-marsh with its mud and its mosquitoes an all the more disagreeable place to stay. Moreover one has to be careful of the bamboo caltrops (*tusi*) which have been hidden in the mud in order to keep off undesirable visitors.

Thereupon the sago is transported home in large carrying-bags (*fisara*).[269] There it is pounded together into cylindrical parcels in wrapping-leaves of the nipah or the sago-palm, for this purpose a stamper is used (*fiaisari*), likewise usually carved. When they have to be transported over long distances, these parcels — called "tumang" in Malay — are often reinforced with bamboo laths. There is no fixed standard, but the usual measurements are 60 by 30 cm (2 ft. by 1 ft.).[270] These tumangs of sago are the chief trading-product; they are exchanged with the inhabitants of Japen Island against gardening products, store-goods, pandanus-leaf cases, etc. The money-value of one tumang at Nubuai lay between about 20 and 40 cents per tumang. At Kai all sago was called *fi*, although occasionally the term *ofu* was used for raw sago, a word which at Napan was the usual term for this product.

I do not understand what Detiger means by the statement that "in the Waropen area where sago-palms grow abundantly, claims on the grounds do not exist and every stranger is free to beat sago there without more ado".[271] For the sago-trees are the property of a family-branch. At Kai they were perhaps also owned by the *ruma,* as is still the case at Napan. For that matter, quarrels about the ownership of sago-palms belong to the order of the day in the village. When one considers that one palm produces about 100 kilos of wet sago-flour, and

---

[269] The women usually put the sling of the carrying-bag across their forehead, whilst the men often carry the bag hanging from the shoulder. The carrying-pole is unknown among the Waropen.

[270] According to Van der Veer one sago-tree provides about 100 kilograms of wet sago-flour, so that one tumang of sago comes at about 10 kilos, estimating one palm at 10 to 12 tumang, according to the statements by the Waropen.

[271] *Adatrechtbundels* (Bundles of Adat Law) XXXIX, p. 433.

that moreover the palm can only be harvested after fifteen to twenty years, or with exceptional care after ten to twelve years, one can easily understand that every village must have quite an extensive sago-area at its disposal to provide for its own needs alone. Detiger's statement should be understood as referring only to the sago-forest situated outside the area of the village. One should also consider that the sago-gardens cannot be left to nature.

"A sago-forest which is left completely to nature consists of an impenetrable and dense vegetation of sago-trees of greatly different age. Young palms, closely pressed together, surround the older trees, some of which are in flower or already bear fruit and which have died off. None of these trees is able to make a proper reserve which would be important for the production of sago. The productive value of such a grove is therefore very slight. The sago-area only becomes valuable for obtaining sago when it is exploited. In order to harvest a tree, an open space has to be made for cutting down the tree; if this gets entangled in the surrounding growth when falling, much more work is needed in order to get the trunk flat on the ground than if space had been cleared beforehand. By cutting away a great number of young shoots, the remaining ones obtain the space they need in order to reach a good productivity. *The longer the sago-forest is under exploitation, the more superfluous young trees are cleared away, the better producers the remaining trees will become*".272

The work of the women is not restricted to the production of the sago, for they also have to prepare the sago as the daily food. For this purpose the sago is first cleaned in a small basket-shaped strainer (*kauo*) 273 in order to get rid of possible impurities, and again when the mush is boiling by means of a sieve-like skimmer (*tafi*).274 The sago may be baked as a cake (*sofi*) in earthenware moulds (*forna*).275 The cakes are smoothed in the mould, they are loosened with a slice and lifted from the mould by means of wooden pincers. Sometimes cakes are baked in flat earthenware dishes (*faimufi*). All the earthenware is imported, mostly from Japen Islands, as are the cooking-pots, large (*urano*) and small (*wesi*). Less dry, but also less durable than the cakes baked in moulds are the *femafi*, cakes wrapped in various kinds of leaves; these may be made in all kinds of shapes as is also done with the ceremonial

---

272 K. van der Veer, *op. cit.*, p. 470.
273 Similar strainers are illustrated in De Clercq and Schmeltz, *op. cit.*, plate XVI, figs. 1, 3, 20 and 22.
274 De Clercq and Schmeltz, *op. cit.*, plate XVI, fig. 11.
275 De Clercq and Schmeltz, *op. cit.*, plate XIX, figs. 6 and 14.

sago-cakes. By using different leaves the taste of the *femafi* may also be slightly varied. A piece of raw marrow is sometimes roasted and eaten without more ado.

Bone spoon.

The commonest way of preparing sago is to boil it as mush (*wiwiro*); for this purpose one only needs to pour boiling water onto the crushed sago which has first been diluted with some water. This method is so

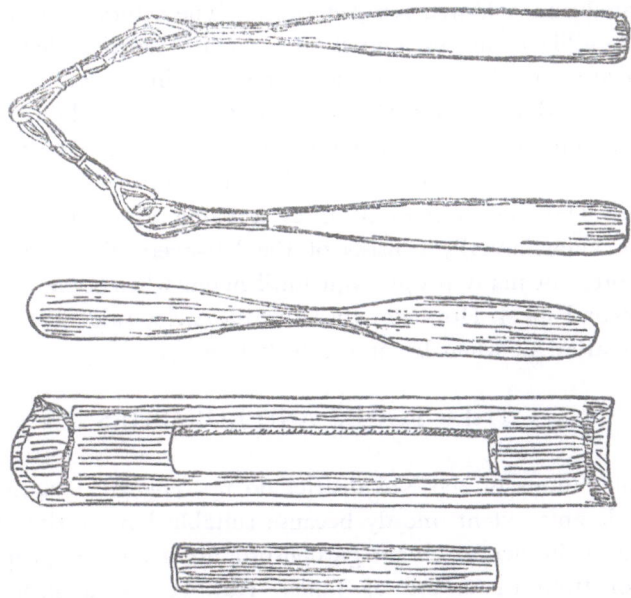

Tools for modelling sago-cakes,
and for disengaging and extracting the cakes from the mould.

simple that even a young girl can do it, so that it is doubly stupid of the Auo-women that they are even incapable of performing this task. The mush is made in a wooden bowl (*ua*) and stirred with spoons and skimmers (*wiwirari*), usually decorated with carving, like the sago-beaters.[276]

---

[276] De Clercq and Schmeltz, *op. cit.*, plate XVII.

In these modern times nearly every housewife possesses a few enamelled dishes, some iron cooking-pots and some glazed pottery, like plates, dishes and bowls. But wooden kitchen utensils and scoops made of coconut (*rakaiwi*) are also still in constant use. As far as possible the mush is eaten together with green vegetables and of course with some fish. For this food little wooden bowls are used, sometimes carved from the same piece of wood as the larger bowl, for instance in the shape of a turtle with a detachable head. The taste of the mush is improved by adding lemon or pomelo-juice, whilst the housewife renders the mush extra palatable by adding beans (*kawaruifi*), coconut (*nighaifi*), laboo (*biraiwafi*), banana (*uifi*), lard (*foinimanafi*), turtle-grease (*enifi*), etc.

The mush is eaten with two-pronged sticks (*kake*), sometimes slightly decorated and coloured red (*uwikake*). These forks are occasionally decorated — like combs sometimes are — with a wooden chain, the links of which are carved from one piece of wood, in the same way as for the pincers used to lift the sago-cakes out of the mould (cf. p. 349). The family often comes together for the meal, the husband usually eating first; when there are male guests, the husband eats separately with his guests. The most important meal is eaten at night, shortly after sunset. Breakfast usually consists of the left-overs of the meal of the night before, but many people wait until noon, when some persons take a light second meal. However, there are no strict rules, because everybody eats after all when he or she feels hungry.

OTHER FOODS AND LUXURIES

The inhabitants of the Kai district do not engage in agriculture to any considerable extent, mostly because suitable land is situated so far distant; it is simpler to obtain garden-produce from the people of the interior or from the island of Japen. However, some people possess small gardens for beans (*kawarui*), laid out in the well-known manner by cutting down the jungle and burning it as far as possible and then planting the beans in plant-holes made with a pointed stick. The beans are planted at the beginning of the dry season; they are harvested in a slightly festive manner, the men going out there in a decorated prahu. These gardens are not the property of the family-branch, but they belong to the man who has laid them out together with his brothers or his friends, or by himself. Rights to the garden originate with the exploitation; they lapse again when the land is abandoned.

Several people possess trees in the dry hinterland, trees which are their personal property, like e.g. a manggo-tree (*aifa*). There are furthermore numerous trees which are nobody's property, but whose fruit is collected and eaten; sometimes the fruit of rhizophores is eaten. Other trees with edible fruit mentioned were *desi* (the fruit of which is said to smell like excrement), *munititi, tefembo, katagheni, mema* (Malay: pala hutan, jungle-nutmeg), *kakumbo*, etc.

At Napan the people are more active in agriculture, there being sufficient land available on Nusariwe Island. In the gardens we find well-known plants and trees like areca (*rifu*), banana (*ui*), manggo (*aifa*), coconut (*nighaiwo*), sugar-cane (*kowu*), corn (*kaisitero*), kasbi (*timuri*), sweet potato (*farenggeno*), ubi (*uwi*), pineapple (*anderamano*), etc. In the Napan area a small quantity of masoi-bark (*manggasa*), obtained from the people of the interior, is taken from the forest, for export purposes.[277] The inhabitants of Kai also collect damar-resin (*kesi*) from the damar-plots in the jungle. In the Kai district small gardens are made on stands near the houses; here the people sometimes cultivate red peppers, or *nini*, a plant which produces the red pigment for dyeing the belts of the women, and occasionally flowers.

The tobacco (*sabaku*) which plays a very important role in the social life of the Waropen is the well-known Java tobacco in cakes, the so-called "tabaku lempeng". In former days the tobacco was probably obtained from the tribes in the interior, but at present, in any case, no trace of this trade was to be found any more. Tobacco is used in ritual, e.g. in the ceremonial tobacco-prahu (*sabakugha*), or in the model of the Morning Star. Tobacco is often sacrificed to the ancestors. It is especially the men who chew tobacco, together with areca, sirih, lime and sometimes gambier. To preserve the lime the men possess a lime-box made from a gourd.[278] On festive occasions a decorated gourd is sometimes used for this purpose. The tobacco is kept in bamboo tubes (*tapa*), or in braided bags, or, in recent times, in a tin box. The women chew less tobacco, but prefer to smoke their tobacco as cigarettes rolled in dried nipah-leaves (*tomarana*).

The tidal forest is not very hospitable, but it produces one thing in great abundance and that is palm-wine (*esa*), drawn from the nipah-palm (*somaro*). Unfermented, it is drunk by women and children, whilst nearly all adults drink the fermented product, fermentation having been caused by adding a piece of the aerial root of the mangrove.

---

[277] De Clercq and Schmeltz, *op. cit.*, p. 107.
[278] De Clercq and Schmeltz, *op. cit.*, p. 73 ff.

Nipah-palms occupy large areas; for tapping purposes a cut is made below the inflorescence, from which the sap runs into a bamboo tube which may be collected towards the evening of the same day. For a consideration another person is also willing to perform this service, so that in the evening one often sees a prahu loaded with cylinders full of this bitter drink going along the houses. In the house the palm-wine is decanted into gourds (*bararo*), from which it is sucked through a bamboo ending in a small strainer (*toki*).

Drunkenness is very frequent, but, although it is not believed to be recommendable, it is not considered as being worse than ridiculous. Every night the drink is taken in great quantities and the addiction to palm-wine is certainly one of the worst popular evils known in the Waropen region. Due to drunkenness, fights and ill-treatment are constantly provoked, and the people will admit that to them palm-wine is more than a tonic against all kinds of diseases and against weakness, which they like to adduce in extenuation.

Like tobacco, palm-wine is often sacrificed to the ancestors on various occasions, e.g. when fishing, when hunting for bats, when going on a raid, etc. In the ritual, however, palm-wine has no important role, and it was never proved that the dancing parties especially stimulated excessive drinking, in spite of the considerable consumption. The organiser of the feast is not obliged to provide his guests with palm-wine. In the mythology palm-wine is only rarely mentioned; it is only said that the nipah-palm came rolling down from the mountains (from heaven). The dog was the first to taste the palm-wine (Texts 152).

HUNTING AND PIG-BREEDING

Hunting is not very productive. All kinds of birds are shot at and sometimes even hit with small arrows (*kowai*). A wild boar may be shot by chance, but in general the bag is only an occasional titbit added to the daily fare. Boars are sometimes caught in snares (*dide*), set in the sago-plantations. The construction (*wana*) which serves as a trap for rats, mice and other small animals is rather more complicated than practical. The assumption that the Waropen eat the crocodiles, snakes, rats, mice, etc. which they catch was indignantly refuted; if anybody, it would be the people from the interior who did this.

The bat-hunt seems to provide better game; in any case, for this purpose special wide-meshed nets were made, with a length of as much as ten metres (33 ft.) and a width of four metres (13 ft.). They were

used to block a creek where the bats liked to fly. As soon as a bat
approaches the net, this is dropped and it was said that in the right

Boar-trap.

season, about the middle of April, a night's bag might consist of as much
as twenty bats. A triangular piece of wood with an animal-head,

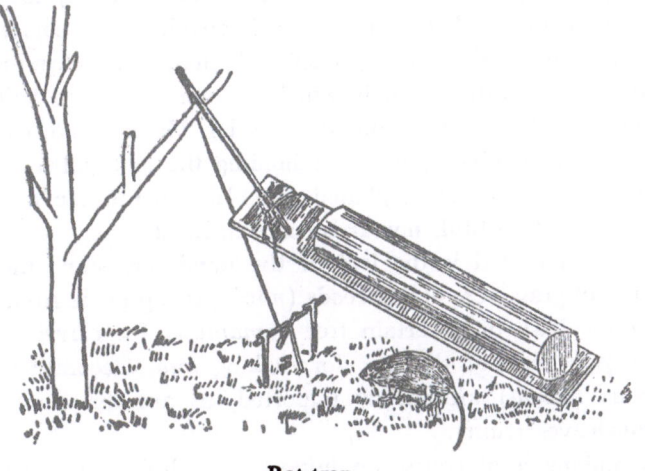

Rat-trap.

presumably a bat's, is tied to the net by way of *aiwo*. The bat also has
its special use, because one of its bones is used as a needle (*ni*).

The domestic pig is man's companion, because at high tide the only place to keep it is inside the house. The owners are very fond of their pets and they used to weep bitter tears when they finally had to surrender them to their fate; it is especially the women who have fed the animal daily who are deeply moved and they demand a small present for the shedding of their tears. In many cases the people even did not want to eat their own pig, surrendering it to another *ruma,* whose members will eventually hand their own pig to the first party. It was said that pigs are killed by drowning; they are consumed practically in their entirety, but for small children the eating of pork is believed to be dangerous. As the diet of the Waropen is rather deficient in fat, they are very fond of the lard, the "sweet" (*mana*) of the pig, which they prefer to eat together with sago.

BASKET-WORK AND ROPE-MAKING

The women of the Kai district produce a great amount of basket-work which is exported in fairly large quantities to the Napan area and to the island of Japen. Using a simple cross-stitch they especially manufacture *baka,* baskets of all sizes, *rowu,* carrying-bags, and *kauo,* sago-strainers. Beside the cross-stitch, the herring-bone stitch is also used, e.g. when making the delicate arm-rings (*aisa*) worn by men, and the twilled stitch, in which case the warp is enveloped in coloured leaves. The bandoleers for the carrying-bags and small men's carrying-bags are sometimes made with a stitch which is reminiscent of knitting; this technique may have come from the interior. No tools are used for this work, except a bamboo-splinter for holding the material together when making the very delicately plaited *aisa.* In general, the final product is neither very beautiful, nor artistically finished.

The material used is the leaf of the pandanus-palm; the fibers of a tall kind of grass, resembling reeds (*noni*); string made from pandanus or from the bark of a certain tree (*amamusi*) for carrying-bags, and the blackish fibre of the *sase* for making *aisa.* Decorative mats and rain-hoods (*saiwu*) are mostly imported, as are the chests made of pandanus-leaves (*faisi*).

Rope-making is of course an important technique for these people. The material used is the aerial root of the pandanus-palm (*rimu*); the interior is torn into ribbons and then reeled on a notched bamboo (*seka*) to form a silvery fibre. The gleaming white fibres (*ritina*) are then rolled on a plank of slightly more than two feet in length; this

68. Beating sago

69. Carrying-bags filled with sago are ready

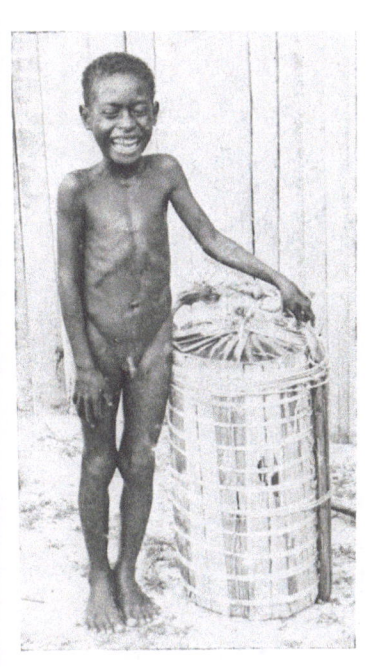

70. A large "tumang" of sago

71. Beating in a standing and in a sitting position

72.   Nubuai village at low tide

73.   The hanging gardens of Nubuai

may be leaned against the poles which support the house by means of a deep notch which fits around the pole (*somambo*).

The thinner types of rope are called *rona,* e.g. *earona,* string for putting together a sero-fishtrap; *faiandona,* yarn used for making nets. By twining this thin string one obtains the heavier kinds, like *sera,* rope for the harpoon, *sesa,* anchor-rope, etc. The heavy and stiff rope of the boar-trap (*dide*) is made of the intertwined bark of the ribs of sago-palm leaves.

The bark of the *rewi* (Malay: pohon genemon) with its edible fruit and leaves produces a strong and long-wearing type of string, which is considered especially suitable for fishing-lines and nets. The string for the women's belts (which are nothing but thin cord, wound many times around the hips) is rolled by the women themselves on their thigh. The material is produced by a certain tree (*fasi*); after having been rolled the string is wrapped spiral fashion in strips of pandanus-leaf dyed yellow or reddish brown.

For tying things (*terafai*), the expensive rope which has been made with great trouble is preferably not used. This is not necessary either, because rattan (*gharaisina*) is available in all kinds and thicknesses. For stitching the leaves of the atap roof-covering hard strips of the *noni*-grass mentioned above are used; the sheets are fastened to the roof by means of long strips of bamboo.

MUSIC AND GAMES

The most important instrument is the drum (*siwa*), mainly imported from the Haarlem and Moor Islands, where the population possesses noticeable skill in wood-work. It has been said that the Waropen also possess this skill, but proofs of this statement were never observed. The drums are shaped more or less like a goblet [279], except the "half drum" (*siwabuino*), often used by women, which actually resembles half the original model. At the dance the drums have to harmonise on two levels; for this purpose the drum-skins have to be constantly adjusted over a low fire, whilst small pieces of resin have to be stuck on to them. Evidently, the Waropen are rather demanding as regards the sound of the drum, the degree of tension of the skin and the manner of beating. To many drums they give a name of their own and even at a distance they recognise them by their sound.

---

[279] De Clercq and Schmeltz, *op. cit.,* plate XXXIII.

The drum is covered with the skin of the iguana (*moiwa*), an animal found here in large quantities. The skin is scraped clean and stretched on a frame of thin slats, to be dried in the sun. When a drum is to be re-covered, the rim of the drum is smeared with the sticky scrapings of the fruit of the mangrove, whereupon the moistened and supple iguana-skin is drawn across and fastened with a rattan. Once the skin has been thoroughly dried in the sun, it is tightly glued to the drum, so that the rattan may be removed.

Old-fashioned flute.

Another instrument used is a kind of whelk (*buro*), the triton-shell, with a round hole cut at the side of its conical tip,[280] and several gongs (*mauno*). The jew's harp (*tungge*) is also known; it may have come from the interior. The Waropen were also acquainted with the flute, which possessed only two holes, thus differing from the modern bamboo flutes imported for the schools. As far as human memory went, this flute had not been borrowed from other tribes. It was possible to play a simple melody on this instrument.

The Waropen are very little acquainted with play in the sense as defined by Huizinga "a free activity, consciously undertaken as 'un-intentional' and standing outside normal life",[281] also because things which seem "play" to us are more concentrated on the sacred realm on the one hand, and more on utility on the other. Although the many figures in the game of cat's cradle are a game, they still are a game which has its place in the funerary ritual. The hammer-headed shark is a mythical animal and yet people will show a wooden shark and reply to the question why this was carved by saying: "*kimasa wéa*", "it is only for play". In the same way one may see a *sambonama*, a rowing jumping-jack, used at one moment by children when they go boating in the village, and at another fixed by adults to the ceremonial tobacco-prahu, in the latter case also for fun. In the sphere described by Huizinga play and ritual are quite close to each other for the Waropen. Toy prahus are used during the *saira* and also to sail them in the water. Nobody sees anything strange in the fact that a grown man makes a toy airplane and then rows with it through the village, or that an

---

[280] De Clercq and Schmeltz, *op. cit.*, plate **XXXIX**, fig. 3.
[281] J. Huizinga, *Homo ludens*, p. 20.

adult plays with a little mill, consisting of a propellor which may be made
to turn by pulling a string which runs through a round nut.[282]

The last-mentioned object is the *mbumbu*; this word is also applied
to the peg-top, slightly flattened and pear-shaped and started by means
of a piece of string, whereupon the other party has to hit it with his
own top. It is also the name of the humming-top, a round fruit pierced
by a stick which produces a slight humming noise when turning.

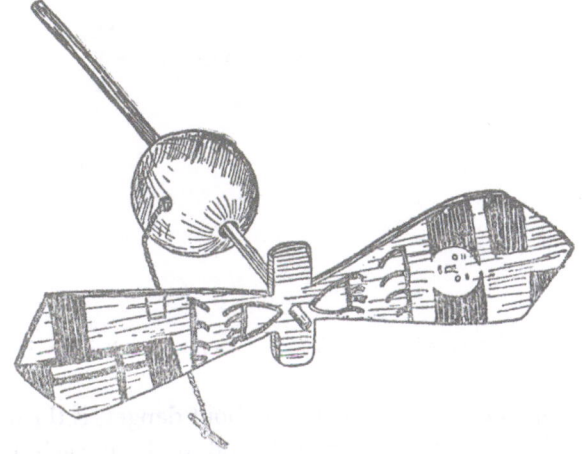

Mbumbu, top.

A game with which young people amuse themselves is shooting small
burning arrows (*wia*) during dark nights. A kind of pop-gun (*panda,*
also meaning "rifle") is likewise known; the pellets are shot out with
a popping sound. In view of the name it probably is an object which
has been adopted at a later period. In all these games it is never quite
clear whether in actual fact the game does not possess some mythical
meaning. This question arises even for simple children's games, e.g.
when at high tide small children see their mud-castle being carried away
by the slowly rising water and sing: "*Awa, awa arifinio*", "Flood, flood,
here is sago for you".

On the other hand the Waropen games are concentrated more on
direct usefulness. When the child is given a small prahu and paddle
of its own, or when it begins to assist its mother in working the sago
or its father in fishing, this may not yet be normal, real work and the
parents may still smile lightly about the childish performance, but

[282] De Clercq and Schmeltz, *op. cit.*, plate **XXXVI**, fig. 11.

nevertheless childhood is not considered as a period during which the
child is allowed to play and learn freely without being directly concerned
with the productive labour of the older people. Quite soon the girl
begins to make basket-work, not "samplers", but basket-work which is
seriously meant. And although the boy plays more than the girl, he is still
expected to do his share in the struggle for the daily food up to the
measure of his capacities.

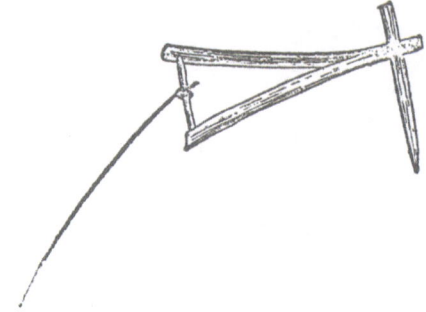

Toy for catching crayfish and similar small animals.

A highly popular round game, not without danger, is throwing pointed
sticks, often in two parties. Party A and party B each have a goal,
e.g. a banana-trunk, A throwing the sticks towards B's goal and B
throwing back the sticks which missed their aim. Every time that party B
throws a stick into the goal of party A, party B is allowed to withdraw
a stick of party A which had penetrated B's goal and cast it towards A's
goal. When all the sticks have landed into the goal of one party, the
game is finished, or, as the Numfoor put it, "the house has burnt
down".283 The throwing is also done at moving objects, e.g. at sticks
or coconut-shells which are dragged behind a prahu, for instance, this
representing a slave-hunt. When dry land is available, arrows are shot
at a rolling wooden disk (*burumaki*).

Another well-known game is "making a net for each other" (*wefaiam-
bararuko*), where the players march in a row underneath the raised arms
of two other players, to be caught one by one. Then the "prisoners" may
chose between sun and moon, whereupon they take their place behind
one of the two players who caught them. When in the end all players

---

283 The Numfoor play this game when the trading prahus have gone to
    Wandamen Bay. In the Wandamen Bay area it is played, like the ball-game, at
    the time when the chempedak-fruit is harvested. The game is also known
    in the mandated territory. See R. Neuhauss, *Deutsch Neu Guinea*, I. p. 372.

74. "Counting-out" for the flag-game

75. Playing football in the mud

76. Drilling fire

77. Napan; after the service

have been caught and have taken their place at the side of the sun or the moon, there follows a tug-of-war.[284]

Again another game is "to jump and strike each other" (*kikodi kipandaruko*). A double row of rather large squares is drawn on the ground as for our game of hop-scotch, with a base at the end where the player may turn round. The two first squares are guarded by one of the players and the others have to try and jump across these without being touched. The child which has been touched now also has to help in catching the others and it is given two other squares to guard. Still another game is "to strike towards each other", *fambararuko,* a kind of tag.

The school has likewise introduced and popularised new games, like e.g. the "flag-game" (*padi*). Two children sit face to face, each with a number of small sticks. Each child stakes one stick by placing it parallel to the stick of its opponent; the child which wins the preliminary finger-game is allowed to start by flicking its own stick with its finger. When the stick lands across the opponent's stick, the first child wins; otherwise the opponent is given a chance.

Modern young people also play a kind of card-game, the stakes being tobacco, areca-nuts or other trifles. The game seemed to consist merely of drawing cards, the court-cards being winners.

An enormous future exists, no doubt, for football which is highly popular among the Papuas, but in the Kai district the nature of the country is very unsuitable. Playing with a ball made of rattan was also an indigenous game from early times onward.

Once I observed a man very patiently rolling a ball by rolling a pellet of a viscous substance, exuded by a leaf unknown to me, across similar other leaves, the pellet growing in the process, because the substance from the other leaves adhered to it. In this way an elastic ball came into being, which might have served as a toy, but I never saw a child play with it.

Riddles (*ghadiabawa*) are not wholly unknown in Waropen literature, although, like proverbs and sayings, they do not occur very often. In any case, posing riddles is not a game which is much played. Some examples are:

1. *Ghadiabawa. Agheabara. Nunggu natio iungha wasi rongha titiwa.* Riddle. Solve it. One man drinks a well completely empty. *Nighai katawiwea.* A sprouting coconut.

---

[284] This game is also generally known in the mandated territory, cf. Neuhauss, *op. cit.,* I, p. 376.

2. *Ghadiabawa. Nunggu natio wo.* Riddle. One man rowing. *Manigha ioaina abogha ruiwi.* A bird floating along in a tree.

3. *Ghadiabawa. Nunggu natio ra ghaidarewa. Raika paikéa. Paikéieka iosaia ruo, fodaroairagha kiwuio.*

Riddle. Somebody walks all alone through the mangrove forest at low tide. And when walking he is trapped and held and remains standing in this way until high tide. And then the women, who are looking for molluscs, catch him.

Manigha. Raika kabo so aifaghaika aifa iafami. A bird. It went and then there was perhaps a lobster or something similar, and then the lobster bites it (so that it cannot fly away).

MAKING FIRE. ARMLETS. EARRINGS

The genuine Waropen method of making fire is considered to be drilling fire by means of a bamboo, provided at the lower end with a hardwood pin and twirled in a soft kind of wood until the wood starts to smoulder. The fire-drill or fire-saw is unknown.

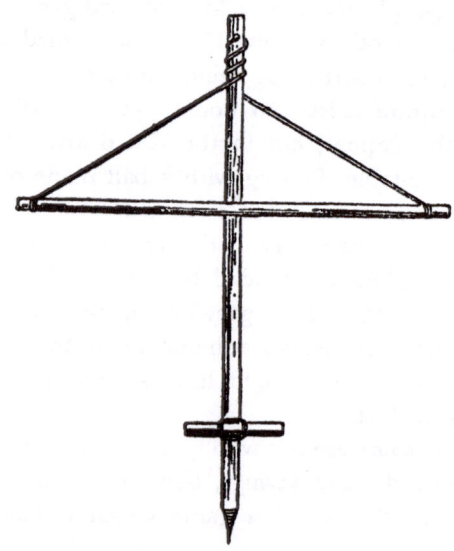

Drill for perforating shell-bands.

A second method, which is considered to be that of the people of the interior is placing one's foot on a board, putting a strong piece of rattan underneath the bord and holding the ends of the rattan in the hands. Between board and rattan a piece of dry bark is inserted and this is made to smoulder by forcefully pulling the rattan to and fro.

The third method is supposed to have come from Biak. It consists of striking a spark into a piece of dry tinder by striking a porcelain shard strongly along a bamboo.

In normal life, however, it is much simpler to take a glowing ember than to make fire in such a complicated fashion. And the Papuas are real wizards when it comes to handling glowing embers.

A kind of drill is known (*sorira*), not, however, for drilling fire, but for perforating the shell-armlets (*saparo*).[285] These highly valued ornaments are usually bought from other tribes, but the people are able to make these themselves by grinding away the top and the rim of a certain conical shell until the armlet remains. The glass-earrings (*dimbo*), also greatly appreciated, are made of ordinary bottle-glass, which is made flexible in the fire and then bent in the desired shape with great trouble.[286]

---

[285] De Clercq and Schmeltz, *op. cit.*, p. 225. R. W. Williamson, *The Mafulu*, p. 71.

[286] However, according to De Clercq a small stone or clay mould is used, in which the liquid glass is made to cool; see De Clercq and Schmeltz, *op. cit.*, plate V, figs. 13—17, 19, 22, 47.

FIGURES USED IN THE GAME OF CAT'S CRADLE.

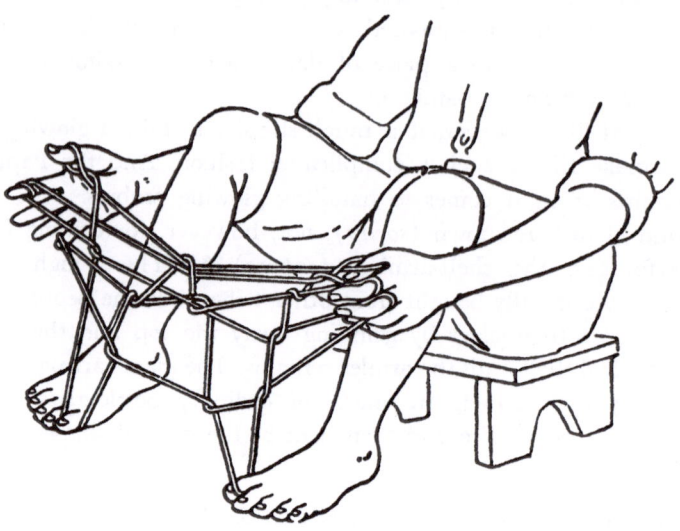

Ghaia, the bat.

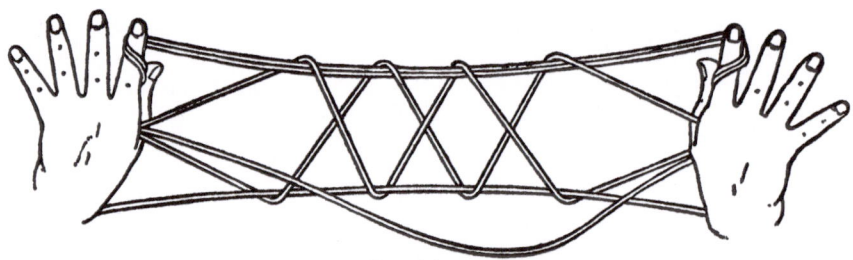

Siwerere (Napan), star.

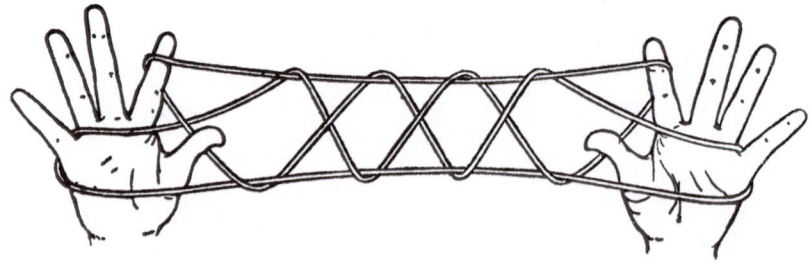

Aimura, beads.

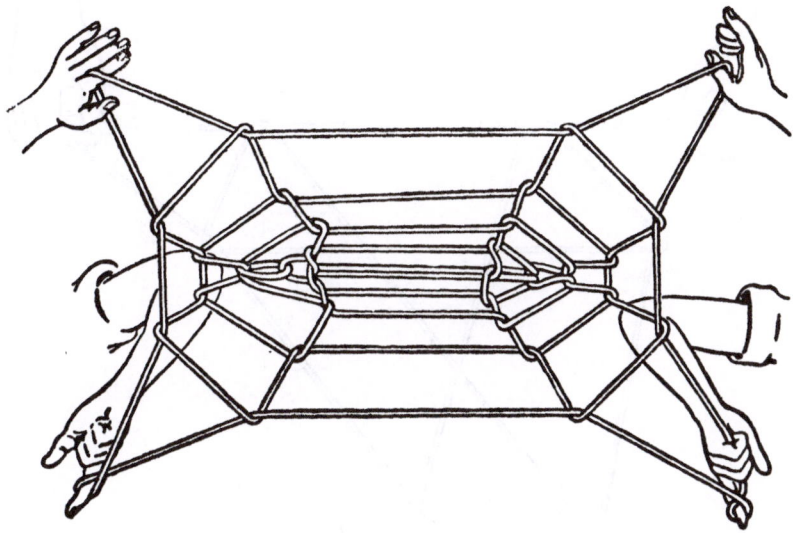

Eni, the turtle.

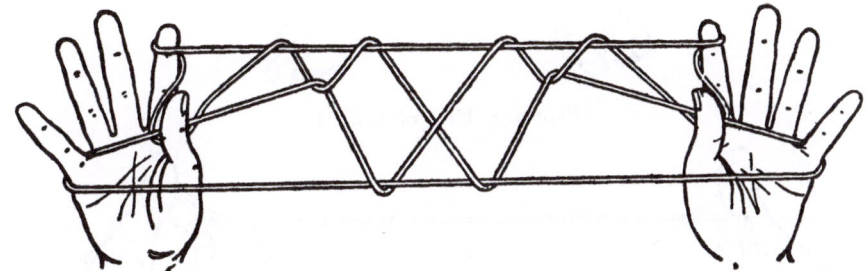

Nanami.

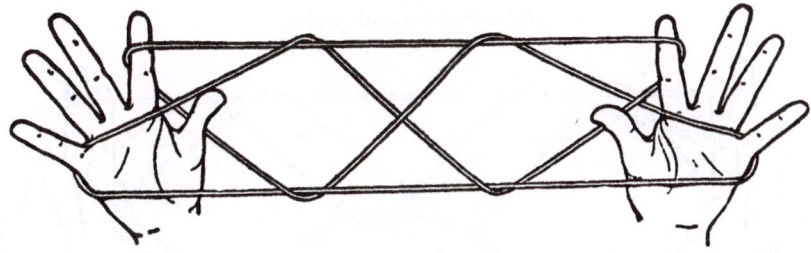

Egharo, the mat.

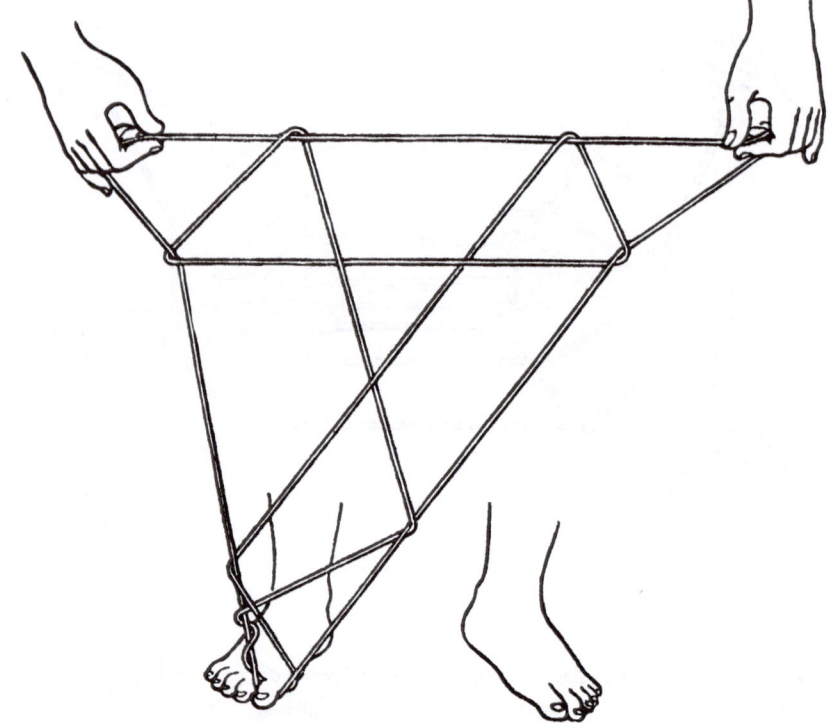

Fafara, a kind of fish(?).

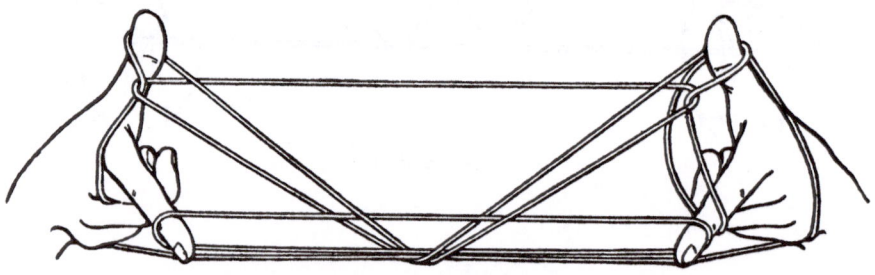

Gharata, a small prahu.

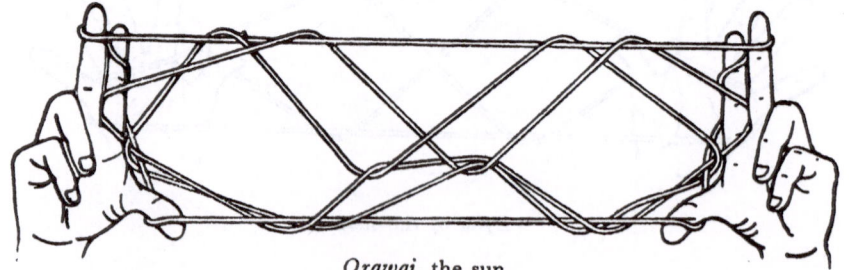

Orawai, the sun.

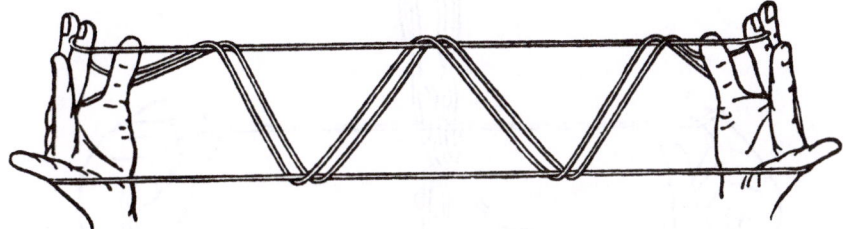

Wanggaruaibo, the radius (bone).

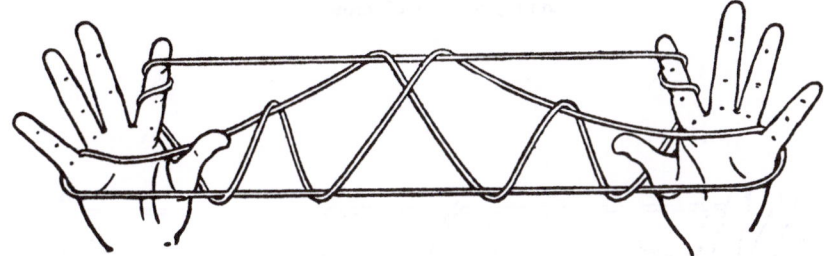

Ghomino, the slave.

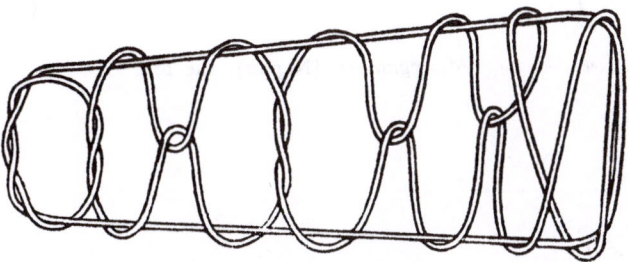

Fosuawai, pig's testicles.

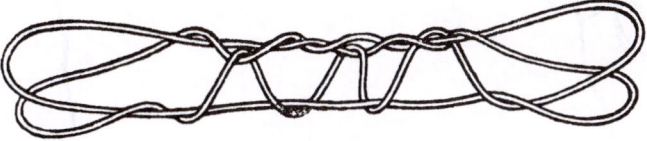

Fo, the pig.

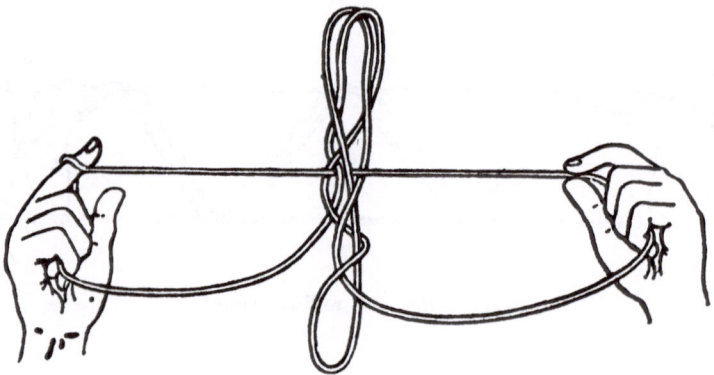

Kuro, a kind of crow.

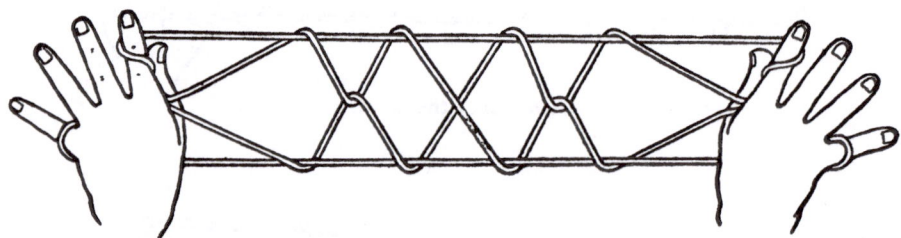

Manggamusa (Napan), the axe.

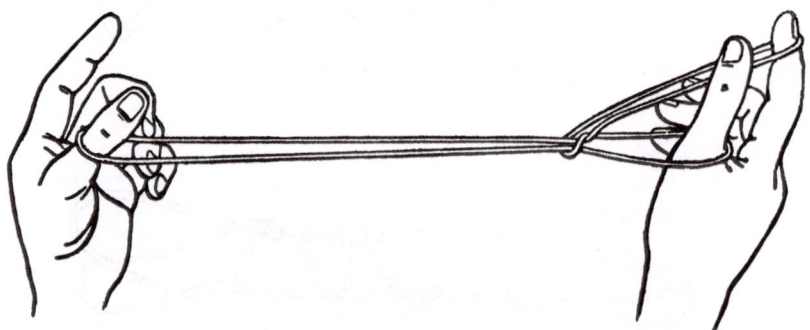

Faisano, the fishing-spear.

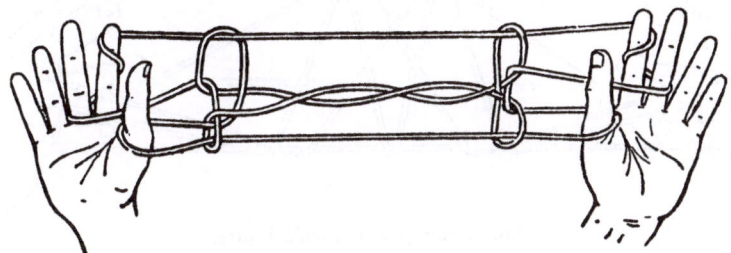

Rai, the leech.

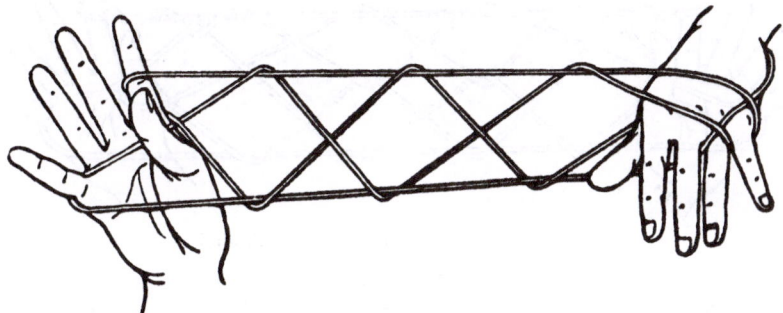

Nama, the paddle.

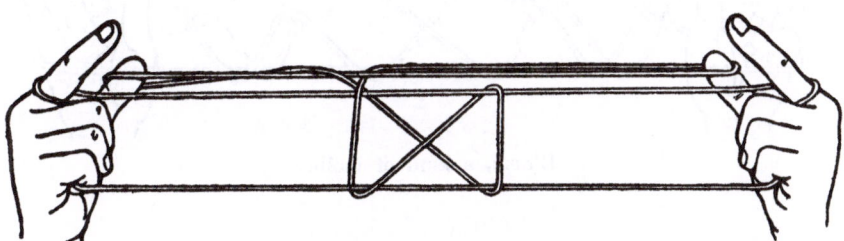

Deruri, the chin.

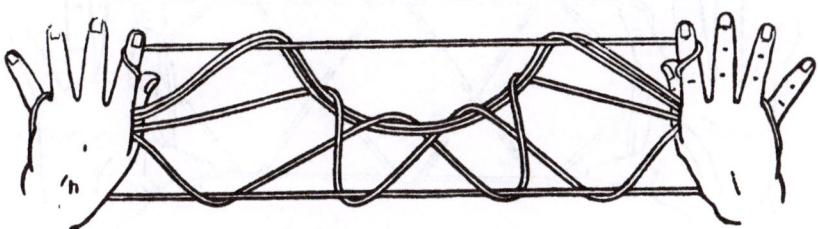

Inisaua (Napan), the moon.

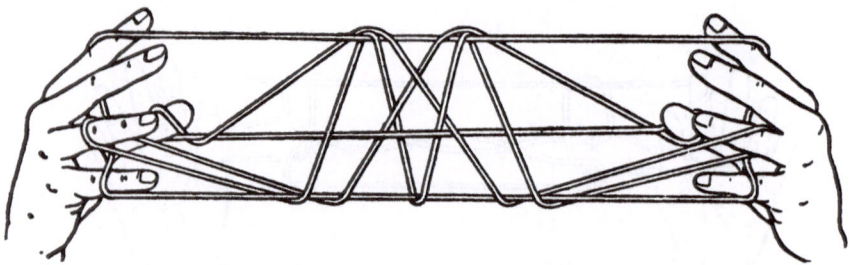

Dama, the young men's house.

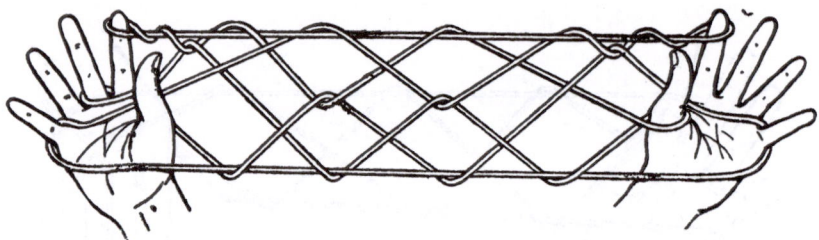

Sarana, the bird of paradise, or Aighei.

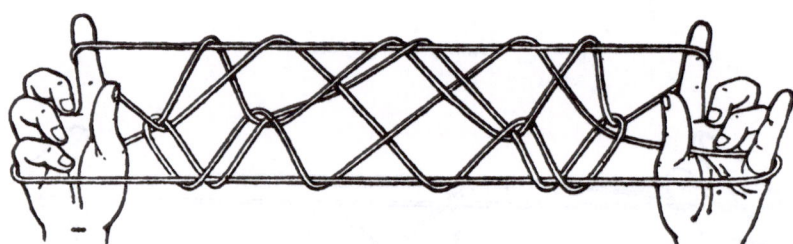

Wama, a kind of mollusc.

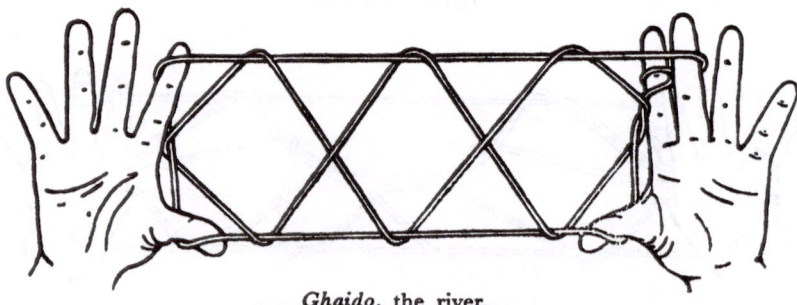

Ghaido, the river.

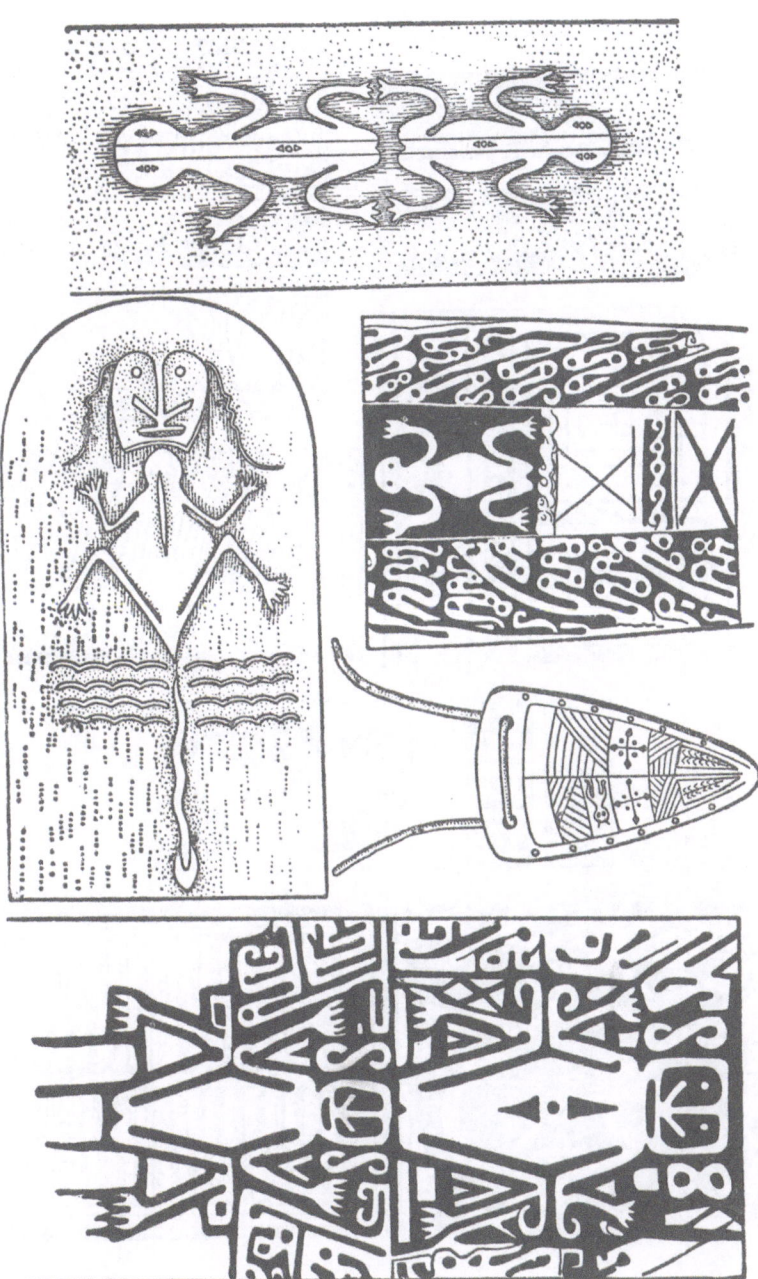

Carving on **sago-beaters** and triangular pubic coverings for young girls.

Carving on bamboo.

GLOSSARY *

Adat, custom (in the widest sense).
adat law, customary law.
ande, non-Papua, stranger (Ambonese, European, etc.).
aiwo, medicine, amulet.
arado, room.
atap (Malay), roofing thatch.
barang (Malay), goods, see p. 102.
bawa, great, prominent.
binano, 1. water-spirit; 2. person guarding a corpse.
bino, woman, wife.
  bino bawa, prominent old woman.
buro, triton-shell.
  buriworai, head of the triton-shell.
  buriferai, tail of the triton-shell.
buriniei, nose ornament.
burua, clothes-chest.
da, 1. clan; 2. group of head-hunters, raiding party.
dama, young men's house.
damasura, type of comb, worn at initiation feasts.
dare, blood shed by relatives, murdered relative.
dimara, honorific title of nobleman.
dimbo, ear-pendant.
dora, heaven, rain, rainwater.
egharo, mat, decorative mat.
ekaini, lateral pole under the roof.
embo, wedding staircase.
er (Numfoor), the Numfoor tribe is divided into four *er*.
faisi, plaited box, small chest.
Faksi (Numfoor), people of the interior.
fi, sago, bag ('tumang') of sago, sago-cake.
firuma, friend, girl-friend, cousin.
gaba-gaba (Malay), stem of the leaf of the sago-palm.
gha, prahu.
ghafa, moon.
ghasaiwini, medicine-woman.

ghéa, 1. certain bird; 2. year.
Ghoa, person from the interior.
ghomino, slave.
guru (Mol. Malay), religious teacher and schoolmaster.
heso, good fortune during the raid.
  hesoaiwo, auspicious medicine for the raid.
hoana, ancient clans among the inhabitants of Tobelo.
hongi, raid.
ikan sembilang (Malay), kind of fish, Plotosus.
ina, wedding scaffolding.
inggoro, ancestors.
kambo, ceremonial seat.
kamuki, (trading) friend.
kasbi (Malay), cassava, Manihot Utillissima.
kedondong (Malay), kind of tree, Enc. N.I.: Spondias.
keret (Numfoor), group of exogamous relatives.
keri, counting-stick.
korano, honorific title of a nobleman, village chief.
korombowi, white porcelain-shell.
kowai, light arrow for shooting birds.
kowuembo, see *embo*.
laboo (Malay), edible gourd.
maiori, honorific title of nobleman, village chief.
manduko, sea-eagle.
mano, man, husband.
  mano bawa, prominent old man.
mansren (Numfoor), nobleman, free man.
masa, central supporting pillar of the house.
masoi (Malay: mesui), massoia aromatica, (New Guinea tree).
mi, mythical, sacral, myth, taboo.
mosaba, woman of the upper stratum.
munaba, funerary feast.

---

* See further the *Woordenlijst van het Waropensch* (Waropen Vocabulary) in Verhandelingen van het Koninklijk Bataviaasch Genootschap van Kunsten en Wetenschappen (Transactions of the Royal Batavia Society of Arts and Sciences), LXXVII, 2nd part (1942).

nibung (Malay), kind of palm, On-cosperma Filamentosa.

nu, village, community.

nurawa (Napan), ancestor, ancestral image.

nyora (Mol. Malay), wife of the guru.

onda, drawing, tattooing design, letter.

owa, to dance; myth, story.

patu, hand fish-trap, headgear for initiandi.

ponisi, celluloid band.

rak (Numfoor), raiding expedition, head-hunt.

rano, men's song.

rari, blue cotton, woman's apron.

ratara, women's song.

resa, leafed branch, triumphal branch.

rewanggu, plate, old porcelain.

roséa, soul.

rowu, carrying-bag.

ruma, house, group of exogamous relatives, family-branch.

 rumarengga, front part of the house.

 rum s(e)ram (Numfoor), young men's house.

runa, neck-support.

sabaku, tobacco, cigarette.

saghara, to sing, to celebrate an initiation ceremony.

saira, men's feast, initiation feast.

samfar (Numfoor), see *saparo*.

sanadi, honorific title of a nobleman.

sandua, bark of the sago-palm, used as a prahu.

Sapari, Morning-star.

saparo, armlet or leg-ring, made of shell.

sarako, silver armlet.

sawai, hammer-headed shark, constellation of this name.

sema, evil, cannibalistic person, 'suangi', devil.

sera, nobleman, chief, god.

 serabawa, grandee, prominent nobleman.

 seraruma, house of the clan-chief.

sero (Malay), large marine fish-trap.

siwa, drum.

suangi (Malay), see *sema*.

sura, comb.

suruka (Numfoor, Wandamen), dwelling-place of the dead.

toddy, Indian word for palm-wine, here substituted for sagoweer in Dutch text.

tumang (Malay), bag of sago, sago wrapped in leaves.

waribo, young man, companion, clan-member.

wiama, young woman, girl.

wundo, central portion of the house.

ubi (Malay), kind of yam, sweet potato.

LIST OF PHOTOGRAPHS

LIST OF DRAWINGS

ABBREVIATIONS

TBG   *Tijdschrift voor Indische Taal-, Land- en Volkenkunde* (uitgegeven door het (Kon.) Bataviaasch Genootschap van Kunsten en Wetenschappen).

BTLV *Bijdragen tot de Taal-, Land- en Volkenkunde van Nederlandsch-Indië* (uitgegeven door het Koninklijk Instituut voor de Taal-, Land- en Volkenkunde van Nederlandsch-Indië; known after 1949 as the *Bijdragen tot de Taal-, Land- en Volkenkunde*).

INDEX